CHEMICAL PROCESS PRINCIPLES

Part I

MATERIAL AND ENERGY BALANCES

CHEMICAL PROCESS PRINCIPLES

Part One

MATERIAL AND ENERGY BALANCES

BY

OLAF A. HOUGEN

AND

KENNETH M. WATSON

PROFESSORS OF CHEMICAL ENGINEERING
UNIVERSITY OF WISCONSIN

NEW YORK

JOHN WILEY & SONS, Inc.

CHAPMAN AND HALL, LIMITED

LONDON

PRINTED IN THE UNITED STATES OF AMERICA

PREFACE

" In the following pages certain industrially important principles of chemistry and physics have been selected for detailed study. The significance of each principle is intensively developed and its applicability and limitations scrutinized." Thus reads the preface to the first edition of *Industrial Chemical Calculations*, the precursor of this book. The present book continues to give intensive quantitative training in the practical applications of the principles of physical chemistry to the solution of complicated industrial problems and in methods of predicting missing physicochemical data from generalized principles. In addition, through recent developments in thermodynamics and kinetics, these principles have been integrated into procedures for process design and analysis with the objective of arriving at optimum economic results from a minimum of pilot-plant or test data. The title *Chemical Process Principles* has been selected to emphasize the importance of this approach to process design and operation.

The design of a chemical process involves three types of problems, which although closely interrelated depend on quite different technical principles. The first group of problems is encountered in the preparation of the material and energy balances of the process and the establishment of the duties to be performed by the various items of equipment. The second type of problem is the determination of the process specifications of the equipment necessary to perform these duties. Under the third classification are the problems of equipment and materials selection, mechanical design, and the integration of the various units into a coordinated plot plan.

These three types may be designated as process, unit-operation, and plant-design problems, respectively. In the design of a plant these problems cannot be segregated and each treated individually without consideration of the others. However, in spite of this interdependence in application the three types may advantageously be segregated for study and development because of the different principles involved. Process problems are primarily chemical and physicochemical in nature; unit-operation problems are for the most part physical; the plant-design problems are to a large extent mechanical.

In this book only process problems of a chemical and physicochemical nature are treated, and it has been attempted to avoid overlapping into the fields of unit operations and plant design. The first part deals primarily with the applications of general physical chemistry, thermophysics, thermochemistry, and the first law of thermodynamics. Generalized procedures for estimating vapor pressures, critical constants, and heats of vaporization have been elaborated. New methods are presented for dealing with equilibrium problems in extraction, adsorption, dissolution, and crystallization. The construction and use of enthalpy-concentration charts have been extended to complex systems. The treatment of material balances has been elaborated to include the effects of recycling, by-passing, changes of inventory, and accumulation of inerts.

v

In the second part the fundamental principles of thermodynamics are presented with particular attention to generalized methods. The applications of these principles to problems in the compression and expansion of fluids, power generation, and refrigeration are discussed. However, it is not attempted to treat the mechanical or equipment problems of such operations.

Considerable attention is devoted to the thermodynamics of solutions with particular emphasis on generalized methods for dealing with deviations from ideal behavior. These principles are applied to the calculation of equilibrium compositions in both physical and chemical processes.

Because of the general absence of complete data for the solution of process problems a chapter is devoted to the new methods of estimating thermodynamic properties by statistical calculations. This treatment is restricted to simple methods of practical value.

All these principles are combined in the solution of the ultimate problem of the kinetics of industrial reactions. Quantitative treatment of these problems is difficult, and designs generally have been based on extensive pilot-plant operations carried out by a trial-and-error procedure on successively larger scales. However, recent developments of the theory of absolute reaction rates have led to a thermodynamic approach to kinetic problems which is of considerable value in clarifying the subject and reducing it to the point of practical applicability. These principles are developed and their application discussed for homogeneous, heterogeneous, and catalytic systems. Particular attention is given to the interpretation of pilot-plant data. Economic considerations are emphasized and problems are included in establishing optimum conditions of operation.

In covering so broad a range of subjects, widely varying comprehensibility is encountered. It has been attempted to arrange the material in the order of progressive difficulty. Where the book is used for college instruction in chemical engineering the material of the first part is suitable for second- and third-year undergraduate work. A portion of the second part is suitable for third- or fourth-year undergraduate work; the remainder is of graduate level.

The authors wish to acknowledge gratefully the assistance of Professor R. A. Ragatz in the revision of Chapters I and VI, and the suggestions of Professors Joseph Hirschfelder, R. J. Altpeter, K. A. Kobe, and Dr. Paul Bender.

OLAF A. HOUGEN
KENNETH M. WATSON

MADISON, WISCONSIN
August, 1943

CONTENTS

PART I

MATERIAL AND ENERGY BALANCES

$NaNO_3$ & Cl_2 from $NaCl$, NO_2

$$3NaCl + 3HNO_3 \rightarrow 3NaNO_3 + 3HCl$$
$$3HCl + 3HNO_3 \rightarrow Cl_2 + NOCl + 2H_2O$$
$$2NOCl + O_2 \rightleftharpoons Cl_2 + 2NO_2$$

High chrome alloys.

In partial liquid phase.

Ammonium Sulfate

$$2NH_3 + H_2SO_4 \rightarrow (NH_4)_2SO_4 + \circlearrowleft$$

$$NH_3 + HNO_3 \rightarrow NH_4NO_3$$
Must evaporate to crystal,

Also

$$CaCl_2 + 2NH_3 + CO_2 + H_2O \rightleftharpoons 2NH_4Cl + CaCO_3$$

$$2NH_3 + CO_2 \rightarrow \underset{NH_2}{\overset{ONH_4}{C=O}} \rightarrow \underset{NH_2}{\overset{NH_2}{C=O}} + H_2O \quad (autoclave)$$

$$CaCN_2 + 3H_2O \rightarrow CaCO_3 + 2NH_3 \qquad (steam in autoclave)$$

$$CO + 2H_2 \rightleftharpoons CH_3OH_{(g)} + \circlearrowleft$$

TABLE OF SYMBOLS

A	area
A	atomic weight
A	component A
A	total work function
a	activity
a_m	external surface per unit mass
a_p	external surface per particle
a_v	external surface per unit volume
B	component B
B	constant of Calingaert–Davis equation
B	thickness of effective film
C	component C
C	concentration per unit volume
C	degrees centigrade
C	number of components
C	over-all rate constant
C_p	heat capacity at constant pressure
C_v	heat capacity at constant volume
C_s	Sutherland constant
c	concentration of adsorbed molecules per unit mass of catalyst
c	specific heat
c	velocity of light
c_p	molal heat capacity at constant pressure
c_v	molal heat capacity at constant volume
c'	surface concentration of adsorbed molecules per unit catalyst area
D	diameter
D_{AB}	diffusivity of A and B
D_p	effective particle diameter equal to diameter of sphere having the same external surface area as particle
D'_p	effective diameter equal to diameter of sphere having same area per unit volume as particle
d	differential operator
E	energy in general
E	energy of activation, Arrhenius equation
E_A	effectiveness factor of catalysis

ix

e	base of natural logarithms
F	degrees of freedom
F	feed rate
F	force
F_e	external void fraction
F_i	internal void fraction
f	friction factor
f	fugacity
f	weight fraction
G	free energy
G	mass velocity per unit area
G	specific gravity
G	free energy per mole
$\bar{\mathrm{G}}$	partial molal free energy
ΔG	change in free energy
(g)	gaseous state
g	degeneracy
g_c	standard gravitational constant, 32.174 (ft/sec)/sec
H	enthalpy
H	Henry's constant
H	humidity
H_c	height of catalytic unit
H_d	height of mass-transfer unit
H_h	height of heat-transfer unit
H_p	height equivalent to a theoretical plate
H_p	percentage humidity
H_R	height of reactor unit
H_r	relative humidity
H_t	height of transfer unit
ΔH	change in enthalpy
ΔH_c	heat of combustion
ΔH_f	heat of formation
ΔH_r	heat of reaction
$\Delta H^{\ddagger}$	standard enthalpy of activation
H	enthalpy per mole
$\bar{\mathrm{H}}$	partial molal enthalpy
$\Delta\bar{\mathrm{H}}$	partial molal enthalpy change
h	Planck's constant
h	heat-transmission coefficient
I	inert component
I	integration constant
I	moment of inertia

J	Jacobian function
J	mechanical equivalent of heat
j_d	mass-transfer factor in fluid film
j_h	heat-transfer factor in fluid film
K	characterization factor
K	degrees Kelvin
K	distribution coefficient
K	equilibrium constant
K	vaporization equilibrium constant
K_a	equilibrium constant for adsorption
K_c	equilibrium constant, concentration units
K_g	over-all mass-transfer coefficient, pressure units
K_L	over-all mass-transfer coefficient, liquid concentration units
K_p	equilibrium constant, pressure units
K'	surface equilibrium constant
k	forward-reaction velocity constant
$\vec{k}$	thermal conductivity
k	Boltzmann constant
k_A	adsorption velocity constant
k'_A	desorption velocity constant
k_G	mass-transfer coefficient, gas film
k_L	mass-transfer coefficient, liquid film
k'	reverse-reaction velocity constant
L	mass velocity of liquid per unit area
L	total molal adsorption sites per unit mass
L_M	molal mass velocity of liquid per unit area
L'	active centers per unit area of catalyst
l	length
l_p	heat of pressure change at constant pressure
l_v	heat of expansion at constant temperature
(l)	liquid state
ln	natural logarithm
log	logarithm to base 10
M	molecular weight
M_m	mean molecular weight
m	mass
m	slope of equilibrium curve dy^*/dx
m	Thiele modulus
N	Avogadro number $6.023(10^{23})$
N	mole fraction
N_t	number of transfer units

N_G	number of transfer units, gas film
N_L	number of transfer units, liquid film
n	number of moles
P	pressure (used only in exceptional cases to distinguish pressure of pure components from partial pressures of some component in solution)
p_f	factor for unequal molal diffusion in gas film
Q	heating value of fuel
Q	partition function
q	heat *added to* a system
q_r	rate of heat flow
R	component R
R	gas constant
r	radius
r_{aA}	rate of reaction or transfer of A per unit area
r_{mA}	rate of reaction or transfer of A per unit mass
r_{vA}	rate of reaction or transfer of A per unit volume
S	component S
S	cross section
S	entropy
S	humid heat
S_p	percentage saturation
S_r	relative saturation
S_v	space velocity
$\Delta S^{\ddagger}$	entropy of activation
s	molal entropy
s	number of equidistant active sites adjacent to each other
(s)	solid state
T	absolute temperature, degrees Rankine or Kelvin
t	temperature, °F or °C
U	internal energy
U	over-all heat-transfer coefficient
u	internal energy per mole
u	velocity
V	molecular volume in Gilliland equation
V	volume
V_r	volume of reactor
v	volume per mole
w	weight
w	work *done by* system
w_e	work of expansion *done by* system

w_f	electrical work *done by* system
w_s	shaft work
$X^‡$	activated complex
x	mole fraction in liquid phase
x	mole fraction of reactant converted in feed
x	quality
y	mole fraction in vapor
y^*	mole fraction in vapor, equilibrium value
Z	elevation above datum plane
Z	height or thickness of reactor
z	compressibility factor
z	mole fraction in total system

Dimensionless Numbers

N_{Re}	Reynold's number	$\dfrac{DG}{\mu}$
N_{Pr}	Prandtl number	$\dfrac{C_p\mu}{k}$
N_{St}	Stanton number	$\dfrac{h}{C_pG}$
N_{Sc}	Schmidt number	$\dfrac{\mu}{\rho D_v}$

Subscripts

A	component A
a	air
B	component B
b	normal boiling point
C	component C
c	critical state
D	component D
D	dense arrangement
e	expansion
f	electrical and radiant
f	formation
f	fusion
G	gas or vapor
H	isenthalpic
L	liquid
L	loose arrangement

p	constant pressure
R	component R
r	reduced conditions
r	relative
S	component S
S	isentropic
s	normal boiling point
s	saturation
T	isothermal
t	temperature
t	transition
V	constant volume
v	vapor
w	water vapor

GREEK SYMBOLS

(α)	crystal form
α	coefficient of compressibility
α	proportionality factor for diffusion
α	relative volatility
α	thermal diffusivity
(β)	crystal form
β	coefficient of volumetric expansion
γ	activity coefficient
(γ)	crystal form
Δ	finite change of a property; positive value indicates an increase
δ	change in moles per mole of reactant
δ	deformation vibration
∂	partial differential operator
ϵ	energy per molecule
η	efficiency
θ	fraction of total sites covered
κ	ratio of heat capacities
Λ	heat of vaporization
λ	heat of vaporization per mole
λ	wave length
λ_f	heat of fusion per mole
μ	chemical potential
μ	viscosity
ν	frequency
ν	fugacity coefficient of gas

ν	number of ions
ν	number of molecules
ν	valence or stretching vibration
ω	expansion factor of liquid
ω	wave number
π	total pressure of mixture, used where necessary to distinguish from p
ρ	density
ρ_B	bulk density
ρ_P	particle density
ρ_C	true solid density
Σ	summation
σ	surface tension
σ	symmetry number
τ	time
ϕ	activity coefficient
ϕ	number of phases

SUPERSCRIPTS

*	ideal behavior
*	equilibrium state
°	standard state
′	pseudo state
′	reverse rate
‡	standard state of activation

CHAPTER I

STOICHIOMETRIC PRINCIPLES

The principal objective to be gained in the study of this book is the ability to reason accurately and concisely in the application of the principles of physics and chemistry to the solution of industrial problems. It is necessary that each fundamental principle be thoroughly understood, not superficially memorized. However, even though a knowledge of scientific principles is possessed, special training is required to solve the more complex industrial problems. There is a great difference between the mere possession of tools and the ability to handle them skilfully.

Direct and logical methods for the combination and application of certain principles of chemistry and physics are described in the text and indicated by the solution of illustrative problems. These illustrations should be carefully studied and each individual operation justified. However, it is not intended that these illustrations should serve as forms for the solution of other problems by mere substitution of data. Their function is to indicate the organized type of reasoning which will lead to the most direct and clear solutions. In order to test the understanding of principles and to develop the ability of organized, analytical reasoning, practice in the actual solution of typical problems is indispensable. The problems selected represent, wherever possible, reasonable conditions of actual industrial practice.

Conservation of Mass. A *system* refers to a substance or a group of substances under consideration and a *process* to the changes taking place within that system. Thus, hydrogen, oxygen, and water may constitute a system, and the combustion of hydrogen to form water, the process. A system may be a mass of material contained within a single vessel and completely isolated from the surroundings, it may include the mass of material in this vessel and its association with the surroundings, or it may include all the mass and energy included in a complex chemical process contained in many vessels and connecting lines and in association with the surroundings. In an *isolated system* the boundaries of the system are limited by a mass of material and its energy content is completely detached from all other matter and energy. Within a given isolated system the mass of the system remains constant regardless of the changes taking place within the system. This statement is known as the *law of conservation of mass* and is the basis of the so-called *material balance* of a process.

1

The state of a system is defined by numerous properties which are classified as *extensive* if they are dependent on the mass under consideration and *intensive* if they are independent of mass. For example, volume is an extensive property, whereas density and temperature are intensive properties.

In the system of hydrogen, oxygen, and water undergoing the process of combustion the total mass in the isolated system remains the same. If the reaction takes place in a vessel and hydrogen and oxygen are fed to the vessel and products are withdrawn then the incoming and outgoing streams must be included as part of the system in applying the law of conservation of mass or in establishing a material balance. The law of conservation of mass may be extended and applied to the mass of each element in a system. Thus, in the isolated system of hydrogen, oxygen, and water undergoing the process of combustion the mass of hydrogen in its molecular, atomic, and combined forms remains constant. The same is true for oxygen.

In a strict sense the conservation law should be applied to the combined energy and mass of the system and not to the mass alone. By the emission of radiant energy mass is converted into energy and also in the transmutation of the elements the mass of one element must change; however, such transformations never fall within the range of experience and detection in industrial processes so that for all practical purposes the law of conservation of mass is accepted as rigorous and valid.

Since the word *weight* is entrenched in engineering literature as synonomous with *mass*, the common practice will be followed in frequently referring to weights of material instead of using the more exact term *mass* as a measure of quantity. Weights and masses are equal only at sea level but the variation of weight on the earth's surface is negligible in ordinary engineering work.

STOICHIOMETRIC RELATIONS

Nature of Chemical Compounds. According to generally accepted theory, the chemical elements are composed of submicroscopic particles which are known as atoms. Further, it is postulated that all of the atoms of a given element have the same weight,[1] but that the atoms of different elements have characteristically different weights.

---

[1] Since the discovery of isotopes, it is commonly recognized that the individual atoms of certain elements vary in weight, and that the so-called atomic weight of an element is, in reality, the weighted average of the atomic weights of the isotopes. In nature the various isotopes of a given element are always found in the same proportions; hence in computational work it is permissible to use the weighted average atomic weight as though all atoms actually possessed this average atomic weight.

When the atoms of the elements unite to form a particular compound, it is observed that the compound, when carefully purified, has a fixed and definite composition rather than a variable and indefinite composition. For example, when various samples of carefully purified sodium chloride are analyzed, they all are found to contain 60.6 per cent chlorine and 39.4 per cent sodium. Since the sodium chloride is composed of sodium atoms, each of which has the same mass, and of chlorine atoms, each of which has the same mass (but a mass that is different from the mass of the sodium atoms), it is concluded that in the compound sodium chloride the atoms of sodium and chlorine have combined according to some fixed and definite integral ratio.

By making a careful study of the relative weights by which the chemical elements unite to form various compounds, it has been possible to compute the relative weights of the atoms. Work of this type occupied the attention of many of the early leaders in chemical research and has continued to the present day. This work has resulted in the familiar table of international atomic weights, which is still subject to periodic revision and refinement. In this table, the numbers, which are known as atomic weights, give the relative weights of the atoms of the various chemical elements, all referred to the arbitrarily assigned value of exactly 16 for the oxygen atom.

A large amount of work has been done to determine the composition of chemical compounds. As a result of this work, the composition of a great variety of chemical compounds can now be expressed by formulas which indicate the elements that comprise the compound and the relative number of the atoms of the various elements present.

It should be pointed out that the formula of the compound as ordinarily written does not necessarily indicate the exact nature of the atomic aggregates that comprise the compound. For example, the formula for water is written as H_2O, which indicates that when hydrogen and oxygen unite to form water, the union of the atoms is in the ratio of 2 atoms of hydrogen to 1 atom of oxygen. If this compound exists as steam, there are two atoms of hydrogen permanently united to one atom of oxygen, forming a simple aggregate termed a molecule. Each molecule is in a state of random motion and has no permanent association with other similar molecules to form aggregates of larger size. However, when this same substance is condensed to the liquid state, there is good evidence to indicate that the individual molecules become associated, to form aggregates of larger size, $(H_2O)_x$, x being a variable quantity. With respect to solid substances, it may be said that the formula as written merely indicates the relative number of atoms present in the compound and has no further significance. For example, the formula for cellulose is written $C_6H_{10}O_5$, but it should not be concluded that individual molecules, each of which contains only 6 atoms of carbon, 10 atoms of hydrogen and 5 atoms of oxygen exist. There is much evidence to indicate that aggregates of the nature of $(C_6H_{10}O_5)_x$ are formed, with x a large number.

It is general practice where possible to write the formula of a chemical compound to correspond to the number of atoms making up one molecule in the gaseous state. If the degree of association in the gaseous state is unknown the formula is written to correspond to the lowest possible number of integral atoms which might make up the molecule. However, where the actual size of the molecule is important care must be exercised in determining the degree of association of a compound even in the gaseous state. For example, hydrogen fluoride is commonly designated by the formula HF and at high temperatures and low pressures exists in the gaseous state in molecules each comprising one atom of fluorine and one atom of hydrogen. However, at high pressures and low temperatures even the gaseous molecules undergo association and the compound behaves in accordance with the formula $(HF)_x$, with x a function of the conditions of temperature and pressure. Fortunately behavior of this type is not common.

Mass Relations in Chemical Reactions. In stoichiometric calculations, the mass relations existing between the reactants and products of a chemical reaction are of primary interest. Such information may be deduced from a correctly written reaction equation, used in conjunction with atomic weight values selected from a table of atomic weights. As a typical example of the procedures followed, the reaction between iron and steam, resulting in the production of hydrogen and the magnetic oxide of iron, Fe_3O_4, may be considered. The first requisite is a correctly written reaction equation. The formulas of the various reactants are set down on the left side of the equation, and the formulas of the products are set down on the right side of the equation, taking care to indicate correctly the formula of each substance involved in the reaction. Next, the equation must be balanced by inserting before each formula coefficients such that for each element present the total number of atoms originally present will exactly equal the total number of atoms present after the reaction has occurred. For the reaction under consideration the following equation may be written:

$$3Fe + 4H_2O \rightarrow Fe_3O_4 + 4H_2$$

The next step is to ascertain the atomic weight of each element involved in the reaction, by consulting a table of atomic weights. From these atomic weights the respective molecular weights of the various compounds may be calculated.

Atomic Weights:

Iron....................	55.84
Hydrogen..............	1.008
Oxygen.................	16.00

Molecular Weights:

H_2O....................	$(2 \times 1.008) + 16.00 = 18.02$
Fe_3O_4.................	$(3 \times 55.84) + (4 \times 16.00) = 231.5$
H_2.....................	$(2 \times 1.008) = 2.016$

The respective relative weights of the reactants and products may be determined by multiplying the respective atomic or molecular weights by the coefficients that precede the formulas of the reaction equation. These figures may conveniently be inserted directly below the reaction equation, thus:

$$3Fe \quad + \quad 4H_2O \quad \rightarrow \quad Fe_3O_4 + \quad 4H_2$$

(3×55.84)	(4×18.02)	231.5	(4×2.016)
167.52	72.08	231.5	8.064

Thus, 167.52 parts by weight of iron react with 72.08 parts by weight of steam, to form 231.5 parts by weight of the magnetic oxide of iron and 8.064 parts by weight of hydrogen. By the use of these relative weights it is possible to work out the particular weights desired in a given problem. For example, if it is required to compute the weight of iron and of steam required to produce 100 pounds of hydrogen, and the weight of the resulting oxide of iron formed, the procedure would be as follows:

Reactants:

Weight of iron $= 100 \times (167.52/8.064)$..............	2075 lb
Weight of steam $= 100 \times (72.08/8.064)$..............	894 lb
Total..	2969 lb

Products:

Weight of iron oxide $= 100 \times (231.5/8.064)$..........	2869 lb
Weight of hydrogen...............................	100 lb
Total..	2969 lb

Volume Relations in Chemical Reactions. A correctly written reaction equation will indicate not only the relative weights involved in a chemical reaction, but also the relative volumes of those reactants and products which are in the gaseous state. The coefficients preceding the molecular formulas of the gaseous reactants and products indicate the relative volumes of the different substances. Thus, for the reaction under consideration, for every 4 volumes of steam, 4 volumes of hydrogen are produced, when both materials are reduced to the same temperature and pressure. This volumetric relation follows from Avogadro's law, which states that equal volumes of gas at the same conditions of temperature and pressure contain the same number of molecules, regardless of the nature of the gas. That being the case, and since 4 molecules of steam produce 4 molecules of hydrogen, it may be concluded that 4 volumes of steam will produce 4 volumes of hydrogen. It cannot be emphasized too strongly that this volumetric relation holds only for ideally gaseous substances, and must never be applied to liquid or to solid substances.

The Gram-Atom and the Pound-Atom. The numbers appearing in a table of atomic weights give the relative weights of the atoms of the various elements. It therefore follows that if masses of different elements are taken in such proportion that they bear the same ratio to one another as do the respective atomic weight numbers, these masses will contain the same number of atoms. For example, if 35.46 grams of chlorine, which has an atomic weight of 35.46, are taken, and if 55.84 grams of iron, which has an atomic weight of 55.84, are taken, there will be exactly the same number of chlorine atoms as of iron atoms in these respective masses of material.

The mass in grams of a given element which is equal numerically to its atomic weight is termed a *gram-atom*. Similarly, the mass in pounds of a given element that is numerically equal to its atomic weight is termed a *pound-atom*. From these definitions, the following equations may be written:

$$\text{Gram-atoms of an elementary substance} = \frac{\text{Mass in grams}}{\text{Atomic weight}}$$

$$\text{Grams of an elementary substance} = \text{Gram-atoms} \times \text{Atomic weight}$$

$$\text{Pound-atoms of an elementary substance} = \frac{\text{Mass in pounds}}{\text{Atomic weight}}$$

$$\text{Pounds of an elementary substance} = \text{Pound-atoms} \times \text{Atomic weight}$$

The actual number of atoms in one gram-atom of an elementary substance has been determined by several methods, the average result being 6.023×10^{23}. This number, known as the Avogadro number, is of considerable theoretical importance.

The Gram-Mole and the Pound-Mole. It has been pointed out that the formula of a chemical compound indicates the relative numbers and the kinds of atoms that unite to form a compound. For example, the formula $NaCl$ indicates that sodium and chlorine atoms are present in the compound in a 1 : 1 ratio.   Since the gram-atom as above defined contains a definite number of atoms, which is the same for all elementary substances, it follows that gram atoms will unite to form a compound in exactly the same ratio as do the atoms themselves, forming what may be termed a gram-mole of the compound.   For the case under consideration, it may be said that one gram-atom of sodium unites with one gram-atom of chlorine to form one gram-mole of sodium chloride.

One gram-mole represents the weight in grams of all the gram-atoms which, in the formation of the compound, combine in the same ratio as do the atoms themselves.   Similarly, one pound-mole represents the weight in pounds of all of the pound-atoms which, in the formation of the compound, combine in the same ratio as do the atoms themselves. From these definitions, the following equations may be written:

$$\text{Gram-moles of a substance} = \frac{\text{Mass in grams}}{\text{Molecular weight}}$$

$$\text{Grams of a substance} = \text{Gram-moles} \times \text{Molecular weight}$$

$$\text{Pound-moles of a substance} = \frac{\text{Mass in pounds}}{\text{Molecular weight}}$$

$$\text{Pounds of a substance} = \text{Pound-moles} \times \text{Molecular weight}$$

The value of these concepts may be demonstrated by consideration of the reaction equation for the production of hydrogen by passing steam over iron.   The reaction equation as written indicates that 3 atoms of iron unite with 4 molecules of steam, to form 1 molecule of magnetic oxide of iron and 4 molecules of hydrogen.   It may also be interpreted as saying that 3 gram-atoms of iron unite with 4 gram-moles of steam, to form 1 gram-mole of Fe_3O_4 and 4 gram-moles of H_2.   In other words, the coefficients preceding the chemical symbols represent not only the relative number of molecules (and atoms for elementary substances that are not in the gaseous state) but also the relative number of gram-moles (and of gram-atoms for elementary substances not in the gaseous state).

Relation Between Mass and Volume for Gaseous Substances.   Laboratory measurements have shown that *for all substances in the ideal gaseous state, 1.0 gram-mole of material at standard conditions (0°C, 760 mm Hg) occupies 22.4 liters.*[2]   Likewise, *if 1.0-pound-mole of the gaseous material is at standard conditions, it will occupy a volume of 359 cu ft.*

[2] The actual volume corresponding to 1 gram-mole of gas at standard conditions will show some variation from gas to gas owing to various degrees of departure from ideal behavior.   However, in ordinary work the ideal values given above may be used without serious error.

Accordingly, with respect to the reaction equation previously discussed, it may be said that 167.52 grams of iron (3 gram-atoms) will form 4 gram-moles of hydrogen, which will, when brought to standard conditions, occupy a volume of 4×22.4 liters, or 89.6 liters. Or, if English units are to be used, it may be said that 167.52 pounds of iron (3 pound-atoms) will form 4 pound-moles of hydrogen, which will occupy a volume of 4×359 cubic feet (1436 cubic feet) at standard conditions.

Illustration 1. A cylinder contains 25 lb of liquid chlorine. What volume in cubic feet will the chlorine occupy if it is released and brought to standard conditions? *Basis of Calculation:* 25 lb of chlorine.

Liquid chlorine, when vaporized, forms a gas composed of diatomic molecules, Cl_2.

Molecular weight of chlorine gas = (2×35.46) 70.92

Lb-moles of chlorine gas = $(25/70.92)$ 0.3525

Volume at standard conditions = (0.3525×359) .... 126.7 cu ft

Illustration 2. Gaseous propane, C_3H_8, is to be liquefied for storage in steel cylinders. How many grams of liquid propane will be formed by the liquefaction of 500 liters of the gas, the volume being measured at standard conditions? *Basis of Calculation:* 500 liters of propane at standard conditions.

Molecular weight of propane..................... 44.06

Gram-moles of propane = $(500/22.4)$ 22.32

Weight of propane = 22.32×44.06 .... 985 grams

The Use of Molal Units in Computations. The great desirability of the use of molal units for the expression of quantities of chemical compounds cannot be overemphasized. Since one molal unit of one compound will always react with a simple multiple number of molal units of another, calculations of weight relationships in chemical reactions are greatly simplified if the quantities of the reacting compounds and products are expressed throughout in molal units. This simplification is not important in very simple calculations, centered about a single compound or element. Such problems are readily solved by the means of the combining weight ratios, which are commonly used as the desirable means for making such calculations as may arise in quantitative analyses. However, in an industrial process successive reactions may take place with varying degrees of completion, and it may be desired to calculate the weight relationships of *all* the materials present at the various stages of the process. In such problems the use of ordinary weight units with combining weight ratios will lead to great confusion and opportunity for arithmetical error. The use of molal units, on the other hand, will give a more direct and simple solution in a form which may be easily verified.

It is urged as highly advisable that familiarity with molal units be gained through their use in *all* calculations of weight relationships in chemical compounds and reactions.

A still more important argument for the use of molal units lies in the fact that many of the physicochemical properties of materials are expressed by simple laws when these properties are on the basis of a molal unit quantity.

The molal method of computation is shown by the following illustrative problem which deals with the reaction considered earlier in this section, namely, the reaction between iron and steam to form hydrogen and the magnetic oxide of iron:

Illustration 3.   (*a*) Calculate the weight of iron and of steam required to produce 100 lb of hydrogen, and the weight of the Fe_3O_4 formed.   (*b*) What volume will the hydrogen occupy at standard conditions?

Reaction Equation:

$$3Fe + 4H_2O \rightarrow Fe_3O_4 + 4H_2$$

Basis of Calculation: 100 lb of hydrogen.

Molecular and atomic weights:

$$
\begin{array}{lr}
Fe\ldots\ldots\ldots\ldots\ldots\ldots\ldots & 55.84 \\
H_2O\ldots\ldots\ldots\ldots\ldots\ldots & 18.02 \\
Fe_3O_4\ldots\ldots\ldots\ldots\ldots & 231.5 \\
H_2\ldots\ldots\ldots\ldots\ldots\ldots\ldots & 2.016 \\
\end{array}
$$

Hydrogen produced = 100/2.016...............		49.6 lb-moles
Iron required = 49.6 × 3/4....................		37.2 lb-atoms
or	37.2 × 55.84....................	2075 lb
Steam required = 49.6 × 4/4...................		49.6 lb-moles
or	49.6 × 18.02.................	894 lb
Fe_3O_4 formed = 49.6 × 1/4...................		12.4 lb-moles
or	12.4 × 231.5...................	2870 lb
Total input = 2075 + 894......................		2969 lb
Total output = 2870 + 100.....................		2970 lb

Volume of hydrogen at standard conditions =

$$49.6 \times 359 \ldots\ldots\ldots \quad 17{,}820 \text{ cu ft}$$

In this simple problem the full value of the molal method of calculation is not apparent; as a matter of fact, the method seems somewhat more involved than the solution which was presented earlier in this section, and which was based on the simple rules of ratio and proportion. It is in the more complex problems pertaining to industrial operations that the full benefits of the molal method of calculation are realized.

Excess Reactants. In most chemical reactions carried out in industry, the quantities of reactants supplied usually are not in the exact proportions demanded by the reaction equation. It is generally desirable that some of the reacting materials be present in excess of the amounts theoretically required for combination with the others. Under such conditions the products obtained will contain some of the uncombined reactants. The quantities of the desired compounds which are formed in the reaction will be determined by the quantity of the *limiting reactant*, that is, the material which is not present in excess of that required to combine with any of the other reacting materials. The amount by which any reactant is present in excess of that required to combine with the limiting reactant is usually expressed as its *percentage excess*. The percentage excess of any reactant is defined as the percentage ratio of the excess to the amount theoretically required for combination with the limiting reactant.

The definition of the amount of reactant theoretically required is sometimes arbitrarily established to comply with particular requirements. For example, in combustion calculations, the fuel itself may contain some oxygen, and the normal procedure in such instances is to give a figure for percentage of excess oxygen supplied by the air which is based on the *net* oxygen demand, which is the total oxygen demanded for complete oxidation of the combustible components, minus the oxygen in the fuel.

Degree of Completion. Even though certain of the reacting materials may be present in excess, many industrial reactions do not proceed to the extent which would result from the complete reaction of the limiting material. Such partial completion may result from the establishment of an equilibrium in the reacting mass or from insufficient time or opportunity for completion to the theoretically possible equilibrium. The *degree of completion* of a reaction is ordinarily expressed as the percentage of the limiting reacting material which is converted or decomposed into other products. In processes in which two or more successive reactions of the same materials take place, the degree of completion of each step may be separately expressed.

In those instances where excess reactants are present and the degree of completion is 100%, the material leaving the process will contain not only the direct products of the chemical reaction but also the excess reactants. In those instances where the degree of completion is below 100%, the material leaving the process will contain some of each of the reactants as well as the direct products of the chemical reactions that took place.

BASIS OF CALCULATION

Normally, all the calculations connected with a given problem are presented with respect to some specific quantity of one of the streams of material entering or leaving the process. This quantity of material is designated as the basis of calculation, and should always be specifically stated as the initial step in presenting the solution to the problem. Very frequently the statement of the problem makes the choice of a basis of calculation quite obvious. For example, in Illustration 3, the weights of iron, steam, and magnetic oxide of iron involved in the production of 100 pounds of hydrogen are to be computed. The simplest procedure obviously is to choose 100 pounds of hydrogen as the basis of calculation, rather than to select some other basis, such as 100 pounds of iron oxide, for example, and finally convert all the weights thus computed to the basis of 100 pounds of hydrogen produced.

In some instances, considerable simplification results if 100 units of one of the streams of material that enter or leave the process is selected as the basis of computation, even though the final result desired may be with reference to some other quantity of material. If the compositions are given in weight per cent, 100 pounds or 100 grams of one of the entering or leaving streams of material may be chosen as the basis of calculations, and at the close of the solution, the values that were computed with respect to this basis can be converted to any other basis that the statement of the problem may demand. For example, if it were required to compute the weight of CaO, MgO, and CO_2 that can be obtained from the calcination of 2500 pounds of limestone containing 90% $CaCO_3$, 5% $MgCO_3$, 3% inerts, and 2% H_2O, one procedure would be to select 2500 pounds of the limestone as the basis of calculation, and if this choice is made, the final figures will represent the desired result. An alternative procedure is to select 100 pounds of limestone as the basis of calculation, and then, at the close of the computation, convert the weights computed on the desired basis of 2500 pounds of limestone. In this very simple illustration, there is little choice between the two procedures, but in complex problems, where several streams of material are involved in the process and where several of the streams are composed of several components, simplification will result if the second procedure is adopted. It should be added that if the compositions are given in mole per cent, it will prove advantageous to choose 100 pound-moles or 100 gram-moles as the basis of calculation.

In presenting the solutions to the short illustrative problems of this chapter, it may have appeared superfluous to make a definite statement as to the basis of calculation. However, since such a statement is of extreme importance in working out complex problems, it is considered desirable to follow the rule of *always* stating the basis of calculation at the beginning of the solution, even though the problem may be relatively simple.

METHODS OF EXPRESSING THE COMPOSITION OF MIXTURES AND SOLUTIONS

Various methods are possible for expressing the composition of mixtures and solutions. The different methods that are in common use may be illustrated by considering a binary system, composed of components which will be designated as A and B. The following symbols will be used in this discussion:

m = total weight of the system.

m_A and m_B = the respective weights of components A and B.

M_A and M_B = the respective molecular weights of components A and B, if they are compounds.

A_A and A_B = the respective atomic weights of components A and B, if they are elementary substances.

V = volume of the system, at a particular temperature and pressure.

V_A and V_B = the respective pure component volumes of components A and B. The pure component volume is defined as the volume occupied by a particular component if it is separated from the mixture, and brought to the same temperature and pressure as the original mixture.

Weight Per Cent. The weight percentage of each component is found by dividing its respective weight by the total weight of the system, and then multiplying by 100:

$$\text{Weight per cent of } A = \frac{m_A}{m} \times 100$$

This method of expressing compositions is very commonly employed for solid systems, and also for liquid systems. It is not used commonly for gaseous systems. Percentage figures applying to a solid or to a liquid system may be assumed to be in weight per cent, if there is no definite specification to the contrary. One advantage of expressing composition on the basis of weight per cent is that the composition values do not change if the temperature of the system is varied (assuming there is no loss of material through volatilization or crystallization, and that no chemical reactions occur). The summation of all the weight percentages for a given system of necessity totals exactly 100.

Volumetric Per Cent. The per cent by volume of each component is found by dividing its pure component volume by the total volume of the system, and then multiplying by 100.

$$\text{Volumetric per cent of } A = \left(\frac{V_A}{V}\right) \times 100$$

This method of expressing compositions is almost always used for gases at low pressures, occasionally for liquids (particularly for the ethyl alcohol:water system), but very seldom for solids.

The analysis of gases is carried out at room temperature and atmospheric pressure. Under these conditions, the behavior of the mixture and of the individual gaseous components is nearly ideal, and the sum of the pure component volumes will equal the total volume. That is, $V_A + V_B + \cdots = V$. This being the case, the percentages total exactly 100. Furthermore, since changes of temperature produce the same relative changes in the respective partial volumes as in the total volume, the volumetric composition of the gas is unaltered by changes in temperature. Compositions of gases are so commonly given on the basis of volumetric percentages that if percentage figures are given with no specification to the contrary, it may be assumed that they are per cent by volume.

With liquid solutions, it is common to observe that on mixing the pure components a shrinkage or expansion occurs. In other words, the sum of the pure component volumes does not equal the sum of the individual volumes. In such instances, the percentages will not total exactly 100. Furthermore, the expansion characteristics of the pure components usually are not the same, and are usually different from that of the mixture. This being the case, the volumetric composition of a liquid solution will change with the temperature. Accordingly, a figure for volumetric per cent as applied to a liquid solution should be accompanied by a statement as to the temperature. For the alcohol:water system, the volumetric percentages are normally given with respect to a temperature of 60°F. If the actual determination is made at a temperature other than 60°F, a suitable correction is applied.

Mole Fraction and Mole Per Cent. If the components A and B are compounds, the system is a mixture of two kinds of molecules. The total number of A molecules or moles present divided by the sum of the A and the B molecules or moles represents the mole fraction of A

in the system.   By multiplying the mole fraction by 100, the mole per cent of A in the system is obtained.   Thus,

$$\text{Mole fraction of } A = \frac{m_A/M_A}{m_A/M_A + m_B/M_B}$$

$$\text{Mole per cent of } A = \text{Mole fraction} \times 100$$

The summation of all the mole percentages for a given system totals exactly 100.   The composition of a system expressed in mole per cent will not vary with the temperature, assuming there is no loss of material from the system, and that no chemical reactions or associations occur.

Illustration 4.   An aqueous solution contains 40% Na_2CO_3 by weight.   Express the composition in mole per cent.

Basis of Calculation: 100 grams of solution.
   Molecular Weights:

$$Na_2CO_3 = 106.0 \qquad\qquad H_2O = 18.02$$

Na_2CO_3 present = 40 gm, or $40/106$ = ...   0.377 gm-moles
H_2O present = 60 gm, or $60/18.02$ = ...   3.33 gm-moles
        Total. . . . .   3.71 gm-moles

Mole per cent Na_2CO_3 = $(0.377/3.71) \times 100$ =   10.16
Mole per cent H_2O = $(3.33/3.71)$ $\times 100$ =   89.8
                                                                            ─────
                                                                            100.0

Illustration 5.   A solution of naphthalene, $C_{10}H_8$, in benzene, C_6H_6, contains 25 mole per cent of naphthalene.   Express the composition of the solution in weight per cent.

Basis of Calculation: 100 gm-moles of solution.
   Molecular Weights:

$$C_{10}H_8 = 128.1 \qquad\qquad C_6H_6 = 78.1$$

$C_{10}H_8$ present = 25 gm-moles, or 25×128.1 = ...   3200 gm
C_6H_6 present = 75 gm-moles, or 75×78.1 = ...   5860 gm
        Total. . . . .   9060 gm

Weight per cent of $C_{10}H_8$ = $(3200/9060) \times 100$ =   35.3
Weight per cent of C_6H_6 = $(5860/9060) \times 100$ =   64.7
                                                                            ─────
                                                                            100.0

In the case of ideal gases, the composition in mole per cent is exactly the same as the composition in volumetric per cent.   This deduction follows from a consideration of Avogadro's law.   It should be emphasized that this relation holds only for gases, and does not apply to liquid or to solid systems.

Illustration 6.   A natural gas has the following composition, all figures being in volumetric per cent:

$$
\begin{array}{lr}
\text{Methane, CH}_4\ldots\ldots\ldots\ldots\ldots\ldots & 83.5\% \\
\text{Ethane, C}_2\text{H}_6\ldots\ldots\ldots\ldots\ldots\ldots & 12.5\% \\
\text{Nitrogen, N}_2\ldots\ldots\ldots\ldots\ldots\ldots & \underline{4.0\%} \\
& 100.0\%
\end{array}
$$

Calculate:

(a) The composition in mole per cent.
(b) The composition in weight per cent.
(c) The average molecular weight.
(d) Density at standard conditions, as pounds per cubic foot.

Part (a) It has been pointed out that for gaseous substances, the composition in mole per cent is identical with the composition in volumetric per cent. Accordingly, the above figures give the respective mole per cents directly, with no calculation.

Part (b) Calculation of Composition in Weight Per Cent.

Basis of Calculation: 100 lb-moles of gas.

	Lb-Moles	Molecular Weight	Weight in Pounds	Weight Per Cent
CH₄	83.5	16.03	$83.5 \times 16.03 = 1339$	$(1339/1827) \times 100 = 73.3$
C₂H₆	12.5	30.05	$12.5 \times 30.05 = 376$	$(376/1827) \times 100 = 20.6$
N₂	4.0	28.02	$4.0 \times 28.02 = \underline{112}$	$(112/1827) \times 100 = \underline{6.1}$
	100.0		1827	100.0

Part (c) The molecular weight of a gas is numerically the same as the weight in pounds of one pound-mole. Therefore, the molecular weight equals 1827/100, or 18.27.

Part (d) Density at Standard Conditions, as lb per cu ft.

Volume at standard conditions = $100 \times 359 = 35{,}900$ cu ft
Density at standard conditions = $1827/35{,}900 = 0.0509$ lb per cu ft

Atomic Fraction and Atomic Per Cent.   The general significance of these terms is the same as for mole fraction and mole per cent, except that the atom is the unit under consideration rather than the molecule. Thus,

$$
\text{Atomic fraction of } A = \frac{(m_A/A_A)}{(m_A/A_A) + (m_B/A_B)}
$$

$$
\text{Atomic per cent of } A = \text{Atomic fraction} \times 100
$$

The summation of all of the atomic percentages for a given system is exactly 100. The composition, expressed in atomic per cent, will not vary with temperature, provided that no loss of material occurs. The composition of a system expressed in atomic per cent will remain the same regardless of whether or not reactions occur within the system.

Mass of Material per Unit Volume of the System. Various units are employed for mass and for volume. Masses are commonly expressed in grams or pounds and the corresponding gram-moles or pound-moles. For volume, the common units are liters, cubic feet, and U. S. gallons. Some common combinations for expression of compositions are grams per liter, gram-moles per liter, pounds per U. S. gallon, and pound-moles per U. S. gallon.

This general method of indicating compositions finds its widest application in dealing with liquid solutions, both in the laboratory and in plant work. This is primarily due to the ease with which liquid volumes may be measured.

Mass of Material per Unit Mass of Reference Substance. One component of the system may be arbitrarily chosen as a reference material, and the composition of the system indicated by stating the mass of each component associated with unit mass of this reference material. For example, in dealing with binary liquid systems, compositions may be expressed as mass of solute per fixed mass of solvent. Some of the common units employed are:

1. Pounds of solute per pound of solvent.
2. Pound-moles of solute per pound-mole of solvent.
3. Pound-moles of solute per 1000 pounds of solvent.

The concentration of a solution expressed in the latter units is termed its *molality*.

In dealing with problems involving the drying of solids, the moisture content is frequently indicated as pounds of water per pound of moisture-free material. In dealing with mixtures of condensable vapors and so-called permanent gases, the concentration of the condensable vapor may be indicated as pounds of vapor per pound of vapor-free gas, or as pound moles of vapor per pound-mole of vapor-free gas.

In all the instances cited, the figure which indicates the composition is, in reality, a dimensionless ratio; hence the metric equivalents have the same numerical value as when the above-specified English units are employed.

For processes involving gain or loss of material, calculations are simplified if the compositions are expressed in this manner. In instances of this kind, the reference component chosen is one which passes through the process unaltered in quantity. Compositions expressed in these terms are independent of temperature and pressure.

Illustration 7. A solution of sodium chloride in water contains 230 grams of NaCl per liter at 20°C. The density of the solution at this temperature is 1.148 gm/cc. Calculate the following items:

(a) The composition in weight per cent.
(b) The volumetric per cent of water.
(c) The composition in mole per cent.
(d) The composition in atomic per cent.
(e) The molality.
(f) Pounds NaCl per pound H_2O.

Basis of Calculation: 1000 cc of solution.

Total weight = 1000×1.148................... 1148 gm
NaCl = 230 gm, or 230/58.5................... 3.93 gm-moles
H_2O = 1148 − 230 = 918 gm, or 918/18.02.... 50.9 gm-moles
    Total.............................. 54.8 gm-moles

(a) Composition in Weight Per Cent:
Weight per cent NaCl = $(230/1148) \times 100$.... 20.0
Weight per cent H_2O = $(918/1148) \times 100$..... 80.0
                           100.0

(b) Volumetric Per Cent Water:
Density of pure water at 20°C = 0.998 gm/cc
Volume of pure water = 918/0.998 = 920 cc
Volumetric per cent of water = $(920/1000) \times 100 = 92.0$

(c) Composition in Mole Per Cent:
Mole per cent NaCl = $(3.93/54.8) \times 100 = $ 7.17
Mole per cent H_2O = $(50.9/54.8) \times 100 = $ 92.8
                                 100.0

(d) The Composition in Atomic Per Cent:
Gm-atoms of sodium........................ 3.93
Gm-atoms of chlorine....................... 3.93
Gm-atoms of hydrogen = 2×50.9........... 101.8
Gm-atoms of oxygen........................ 50.9
    Total................................. 160.6

Atomic per cent of sodium   = $(3.93/160.6) \times 100 = $ 2.45
Atomic per cent of chlorine   = $(3.93/160.6) \times 100 = $ 2.45
Atomic per cent of hydrogen   = $(101.8/160.6) \times 100 = $ 63.4
Atomic per cent of oxygen   = $(50.9/160.6) \times 100 = $ 31.7
                                          100.0

(e) The Molality:
Molality = lb-moles of NaCl/1000 lb H_2O
          = $3.93 \times (1000/918) = 4.28$
(f) Lb NaCl/lb H_2O = 230/918 = 0.251

DENSITY AND SPECIFIC GRAVITY

Density is defined as mass per unit volume. Density values are commonly expressed as grams per cubic centimeter or as pounds per cubic foot. The density of water at 4°C is 1.0000 gm/cc, or 62.43 lb/cu ft.

The specific gravity of a solid or liquid is the ratio of its density to the density of water at some specified reference temperature.   The temperatures corresponding to a value of specific gravity are generally symbolized by a fraction, the numerator of which is the temperature of the liquid in question, and the denominator, the temperature of the water which serves as the reference.   Thus the term sp. gr. 70/60°F indicates the specific gravity of a liquid at 70°F referred to water at 60°F, or the ratio of the density of the liquid at 70°F to that of water at 60°F.   It is apparent that if specific gravities are referred to water at 4°C (39.2°F) they will be numerically equal to densities in grams per cubic centimeter.

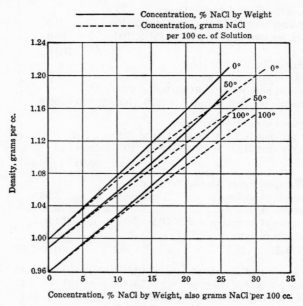

Fig. 1.   Densities of aqueous sodium chloride solutions.

The densities of solutions are functions of both concentration and temperature.   The relationships between these three properties have been determined for a majority of the common systems.   Such compilations as the International Critical Tables contain extensive tabulations giving the densities of solutions of varying concentrations at specified temperatures.   These data are most conveniently used in graphical form in which density is plotted against concentration.   Each curve on such a chart will correspond to a specified, constant temperature.   The density of a solution of any concentration at any temperature may be readily estimated by interpolation between these curves.   In Fig. 1

are plotted the densities of solutions of sodium chloride at various temperatures.

For a given system of solute and solvent the density or specific gravity at a specified temperature may serve as an index to the concentration. This method is useful only when there is a large difference between the densities of the solutions and the pure solvent. In several industries specific gravities have become the universally accepted means of indicating concentrations, and products are purchased and sold on the basis of specific gravity specifications. Sulfuric acid, for example, is marketed almost entirely on this basis. Specific gravities are also made the basis for the control of many industrial processes in which solutions are involved. To meet the needs of such industries, special means of numerically designating specific gravities have been developed. Several scales are in use in which specific gravities are expressed in terms of *degrees* which are related to specific gravities and densities by more or less complicated and arbitrarily defined functions.

Baumé Gravity Scale. Two so-called Baumé gravity scales are in common use, one for use with liquids lighter and the other for liquids heavier than water. The former is defined by the following expression:

$$\text{Degrees Baumé} = \frac{140}{G} - 130$$

where G is the specific gravity at 60/60°F. It is apparent from this definition that a liquid having the density of water at 60°F (0.99904 gram per cubic centimeter) will have a gravity of 10° Baumé. Lighter liquids will have *higher* gravities on the Baumé scale. Thus, a material having a specific gravity of 0.60 will have a gravity of 103° Baumé.

The Baumé scale for liquids heavier than water is defined as follows:

$$\text{Degrees Baumé} = 145 - \frac{145}{G}$$

Gravities expressed on this scale increase with increasing density. Thus, a specific gravity of 1.0 at 60/60°F corresponds to 0.0° Baumé, and a specific gravity of 1.80 corresponds to 64.44° Baumé. It will be noted that both the Baumé scales compare the densities of liquids at 60°F. In order to determine the Baumé gravity, the specific gravity at 60/60°F must either be directly measured or estimated from specific-gravity measurements at other conditions. The Baumé gravity of a liquid is thus independent of its temperature. Readings of Baumé hydrometers at temperatures other than 60°F must be corrected for temperature so as to give the value at 60°F.

API Scale. As a result of confusion of standards by instrument manufacturers, a special gravity scale has been adopted by the American Petroleum Institute for expression of the gravities of petroleum products. This scale is similar to the Baumé scale for liquids lighter than water as indicated by the following definition:

$$\text{Degrees API} = \frac{141.5}{G} - 131.5$$

As on the Baumé scale, a liquid having a specific gravity of 1.0 at 60/60° F has a gravity of 10°. However, a liquid having a specific gravity of 0.60 has an API gravity of 104.3 as compared with a Baumé gravity of 103.3. The gravity of a liquid in degrees API is determined by its density at 60°F and is independent of temperature. Readings of API hydrometers at temperatures other than 60°F must be corrected for temperature so as to give the value at 60°F.

API gravities are readily converted by Fig. B in the Appendix.

Twaddell Scale. The Twaddell scale is used only for liquids heavier than water. Its definition is as follows:

$$\text{Degrees Twaddell} = 200(G - 1.0)$$

This scale has the advantage of a very simple relationship to specific gravities.

Numerous other scales have been adopted for special industrial uses; for example, the Brix scale measures directly the concentration of sugar solutions.

If the stem of a hydrometer graduated in specific-gravity units is examined, it is observed that the scale divisions are not uniform. The scale becomes compressed and crowded together at the lower end. On the other hand, a Baumé or API hydrometer will have uniform scale graduations over the entire length of the stem.

For gases, water is unsatisfactory as a reference material for expressing specific-gravity values because of its high density in comparison with the density of gas. For gases, it is conventional to express specific-gravity values with reference to dry air at the same temperature and pressure as the gas.

Triangular Plots. When dealing with mixtures or solutions of three components equivalent values of any particular property of the solution can be shown related to composition by contour lines drawn upon a triangular coordinate diagram.

In Fig. 2 is shown such a diagram with contour lines of specific volumes at 25°C constructed for the system carbon tetrachloride, ethyl dibromide, and toluene. On a triangular chart covering the complete range of compositions, apex A represents pure component A, in this instance carbon tetrachloride having a specific volume of 0.630, apex B represents pure component B, in this instance ethyl dibromide having a specific volume of 0.460, and apex C represents pure component C, in this instance toluene having a specific volume of 1.159. Any point on the base line AB corresponds to a binary solution of A and B. For example point a represents 75% $C_2H_4Br_2$ and 25% CCl_4, this composition having a

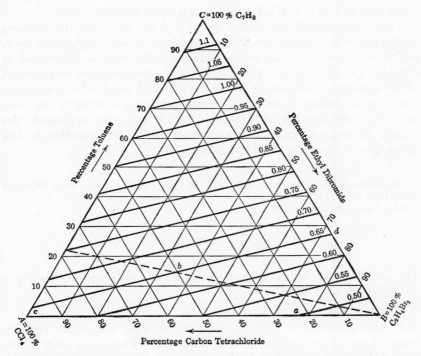

FIG. 2. Specific volumes of ternary solutions of carbon tetrachloride, ethyl dibromide and toluene at 25/4°C. (From *International Critical Tables*, III, 196.)

specific volume of 0.50. Similarly, base lines BC and CA represent all possible combinations of the binary solutions of B and C, and of C and A, respectively.

Any point within the area of the triangle represents a definite composition of a ternary mixture of A, B, and C. For example, point b corresponds to a mixture of 50% A, 35% B, and 15% C. From the scale of compositions it will be seen that point b may be considered as located on the intersection of three lines, parallel to the three base lines, respectively.

The line parallel to BC passing through b is 50% of the perpendicular distance from line BC to A, the line parallel to AC passing through b is 35% of the distance from line AC to B, and the line parallel to AB passing through b is 15% of the distance from line AB to C.

Contour lines on a triangular plot represent equivalent values of some property, in this case specific volume. For example line cd, marked 0.65, represents all possible solutions having a specific volume of 0.65. Point b lies on this line and corresponds to 50% A (CCl_4), 35% B ($C_2H_4Br_2$) and 15% C (C_7H_8), having a specific volume of 0.65.

Triangular co-ordinate charts are useful for following the changes taking place in composition and properties of ternary systems in operations of extraction, evaporation, and crystallization, as illustrated in Chapter V. For example, line Bb represents the change in composition of the solution as component B alone is removed or added to some solution initially falling on this line. Thus, moving in a straight line from an initial point within the diagram toward one apex represents the change in composition as one component only corresponding to the given apex is added to the initial solution.

Illustration 8. It is desired to calculate the final specific volume when 20 grams of $C_2H_4Br_2$ are added to 100 grams of solution corresponding to b on Fig. 2.

The resultant solution will contain

$$\frac{20 + 35}{120} = 45.8\% \ C_2H_4Br_2$$

This point lies on line bB and will be seen to correspond to a specific volume of 0.62 at 12.5% C_7H_8 and 41.7% CCl_4.

In the preceding, a triangular chart covering the entire possible range of percentage compositions is described. Frequently only a portion of the chart is required or the scale on each base line is altered to a narrow range of composition. Under these circumstances the scale used defines the significance of new apices and base lines.

Conversion of Units. The conversion of units and symbols from one system to another often presents a troublesome operation in technical calculations. Both the metric and English units are intentionally employed in this book in order to bridge the gap between scientific and industrial applications.

In nearly every handbook tables of conversion factors will be found, and these are recommended for use whenever available and adequate. A short list of the more important factors is included in the Appendix, page 438. A few simple rules will be given for guidance where calculation of conversion factors becomes necessary.

Most scientific units may be expressed in terms of simple dimensions, such as length, weight, time, temperature, and heat. In conversion the unit is first expressed in terms of its simplest dimensions combined with the known numerical or symbolic value of the unit. Thus, the viscosity of a liquid is μ grams per second-centimeter. In the English system the value will be expressed in pounds per second-foot. Each of the dimensions is replaced separately by the dimensions of the desired system together with its corresponding conversion factor. Thus, since 1 gram = 0.002204 lb and 1 cm = 0.0328 ft

$$\mu \frac{\text{grams}}{(\text{sec}) (\text{cm})} = \mu \frac{0.002204 \text{ lb}}{1 (\text{sec}) 0.0328 (\text{ft})} = 0.0670\mu \frac{\text{lb}}{(\text{sec}) (\text{ft})}$$

Similarly a pressure of 1 atmosphere =

$$\frac{14.7 \text{ lb}}{(\text{in.})^2} = \frac{14.7 (453.6) \text{ grams}}{(2.54)^2 (\text{cm})^2} = 1035 \frac{\text{grams}}{(\text{cm})^2}$$

since   1 lb = 453.6 grams and 1 in. = 2.54 cm

The gas constant R =

$$\frac{82.06 (\text{atm}) (\text{cm})^3}{(\text{gram mole}) (°\text{K})} = \frac{(82.06) (1 \text{ atm}) (0.0328 \text{ ft})^3}{(0.002205 \text{ lb-mole}) (1.8°\text{R})} = 0.729 \frac{(\text{atm}) (\text{ft})^3}{(\text{lb-mole})°\text{R}}$$

Also since 1 atmosphere = 14.7 lb per in.^{2} = 14.7 × 144 or 2120 lb per ft^2,

$$R = \frac{(0.729) (2120 \text{ lb/ft}^2) (\text{ft})^3}{(\text{lb-mole}) (°\text{R})} = 1540 \frac{(\text{ft}) (\text{lb})}{(\text{lb-mole}) (°\text{R})}$$

PROBLEMS

Tables of Common Atomic Weights and Conversion Factors will be found in the Appendix.

While the simple stoichiometric relations included in the following group of problems may easily be solved by the rules of ratio and proportion, it is nevertheless recommended that the molal method of calculation be adhered to as a preparation for the more complex problems to be encountered in succeeding chapters.

In all instances, the basis of calculation should be stated definitely at the start of the solution.

1. $BaCl_2 + Na_2SO_4 = 2NaCl + BaSO_4$.

    (a) How many grams of barium chloride will be required to react with 5.0 grams of sodium sulfate?

    (b) How many grams of barium chloride are required for the precipitation of 5.0 grams of barium sulfate?

(c) How many grams of barium chloride are equivalent to 5.0 grams of sodium chloride?

(d) How many grams of sodium sulfate are necessary for the precipitation of the barium of 5.0 grams of barium chloride?

(e) How many grams of sodium sulfate have been added to barium chloride if 5.0 grams of barium sulfate are precipitated?

(f) How many pounds of sodium sulfate are equivalent to 5.0 pounds of sodium chloride?

(g) How many pounds of barium sulfate are precipitated by 5.0 pounds of barium chloride?

(h) How many pounds of barium sulfate are precipitated by 5.0 pounds of sodium sulfate?

(i) How many pounds of barium sulfate are equivalent to 5.0 pounds of sodium chloride?

2. How many grams of chromic sulfide will be formed from 0.718 grams of chromic oxide according to the equation:

$$2Cr_2O_3 + 3CS_2 = 2Cr_2S_3 + 3CO_2$$

3. How much charcoal is required to reduce 1.5 pounds of arsenic trioxide?

$$As_2O_3 + 3C = 3CO + 2As$$

4. Oxygen is prepared according to the equation $2KClO_3 = 2KCl + 3O_2$. What is the yield of oxygen when 7.07 grams of potassium chlorate are decomposed? How many grams of potassium chlorate must be decomposed to liberate 2.0 grams of oxygen?

5. Sulfur dioxide may be produced by the reaction:

$$Cu + 2H_2SO_4 = CuSO_4 + 2H_2O + SO_2$$

(a) How much copper, and (b) how much 94% H_2SO_4 must be used to obtain 32 pounds of sulfur dioxide?

6. A limestone analyzes $CaCO_3$ 93.12%, $MgCO_3$ 5.38%, and insoluble matter 1.50%.

(a) How many pounds of calcium oxide could be obtained from 5 tons of the limestone?

(b) How many pounds of carbon dioxide are given off per pound of this limestone?

7. How much superphosphate fertilizer can be made from one ton of calcium phosphate, 93.5% pure? The reaction is

$$Ca_3(PO_4)_2 + 2H_2SO_4 = 2CaSO_4 + CaH_4(PO_4)_2$$

8. How much potassium chlorate must be taken to produce the same amount of oxygen as will be produced by 1.5 grams of mercuric oxide?

9. Regarding ammonium phosphomolybdate, $(NH_4)_3PO_4 \cdot 12MoO_3 \cdot 3H_2O$, as made up of the radicals NH_3, H_2O, P_2O_5 and MoO_3, what is the percentage composition of the molecule with respect to these radicals?

10. How many pounds of salt are required to make 1500 pounds of salt cake (Na_2SO_4)? How many pounds of Glauber's salt ($Na_2SO_4 \cdot 10H_2O$) will this amount of salt cake make?

11. In the reactions:

$$2KMnO_4 + 8H_2SO_4 + 10FeSO_4 = 5Fe_2(SO_4)_3 + K_2SO_4 + 2MnSO_4 + 8H_2O$$
$$K_2Cr_2O_7 + 7H_2SO_4 + 6FeSO_4 = 3Fe_2(SO_4)_3 + K_2SO_4 + Cr_2(SO_4)_3 + 7H_2O$$

How many grams of potassium dichromate are equivalent to 5.0 grams of potassium permanganate? How many grams of potassium permanganate are equivalent to 3.0 grams of potassium dichromate?

12. A compound whose molecular weight is 103 analyzes: C = 81.5%; H = 4.9%; N = 13.6%. What is its formula?

13. What is the weight of one liter of methane under standard conditions?

14. If 15 grams of iron react thus: $Fe + H_2SO_4 = FeSO_4 + H_2$, how many liters of hydrogen are liberated at standard conditions?

15. A solution of sodium chloride in water contains 23.0% NaCl by weight. Express the concentration of this solution in the following terms, using data from Fig. 1.

    (a) Gram-moles of NaCl per 1000 grams of water (molality).
    (b) Mole fraction of NaCl.
    (c) Gram-moles of NaCl per liter of solution at 30°C.
    (d) Pounds of NaCl per U. S. gallon of solution at 40°C.

16. In the following table are listed various aqueous solutions, with the density at 20°C and 80°C given, as well as the composition. For one or more of the solutions assigned from the table, report the composition expressed in the following ways:

    (a) Weight per cent.
    (b) Mole per cent.
    (c) Pounds of solute per pound of solvent.
    (d) Pound-moles of solute per pound of solvent.
    (e) Grams solute per 100 ml of solution at 80°C.
    (f) Grams solute per 100 ml of solution at 20°C.
    (g) Gram-moles solute per liter of solution at 20°C.
    (h) Pounds of solute per U. S. gallon of solution at 68°F (20°C).
    (i) Pound-moles of solute per U. S. gallon of solution at 68°F (20°C).
    (j) Molality.
    (k) Normality.

Solute	Composition of Solution	Density, g/ml	
		20°C	80°C
HCl	Weight % HCl = 30	1.149	1.115
H_2SO_4	mole % H_2SO_4 = 36.8	1.681	1.625
HNO_3	lb HNO_3/lb H_2O = 1.704	1.382	1.296
NH_4NO_3	lb-mole NH_4NO_3/lb H_2O = 0.01250	1.226	1.187
$ZnBr_2$	gm $ZnBr_2$/100 ml of solution at 20°C = 130.0	2.002	1.924
$CdCl_2$	gm $CdCl_2$/100 ml of solution at 80°C = 68.3	1.575	1.519
$MgCl_2$	gm-mole $MgCl_2$/1 of solution at 20°C = 3.99	1.269	1.245
$CaBr_2$	lb $CaBr_2$/U. S. gal at 68°F (20°C) = 4.03	1.381	1.343
$SrCl_2$	lb-mole $SrCl_2$/U. S. gal at 68°F (20°C) = 0.02575	1.396	1.358
LiCl	molality = 10.13	1.179	1.158
KCl	normality = 2.70	1.118	1.088

17. An aqueous solution of sodium chloride contains 28 grams of NaCl per 100 cc of solution at 20°C. Express the concentration of this solution in the following terms, using data from Fig. 1.

(a) Percentage NaCl by weight.

(b) Mole fraction of NaCl.

(c) Pound-moles of NaCl per 1000 pounds of water (molality).

(d) Pound-moles of NaCl per U. S. gallon of solution at 0°C.

18. It is desired to prepare a solution of sodium chloride in water, having a molality of 2.00. Calculate the weight of sodium chloride which should be placed in a 1000-cc volumetric flask in order that the desired concentration will be obtained by subsequently filling the flask with water, keeping the temperature of the solution at 30°C.

19. For the operation of a refrigeration plant it is desired to prepare a solution of sodium chloride containing 20% by weight of the anhydrous salt.

(a) Calculate the weight of sodium chloride which should be added to one gallon of water at 30°C in order to prepare this solution.

(b) Calculate the volume of solution formed per gallon of water used, keeping the temperature at 30°C.

20. (a) A solution has a gravity of 80° Twaddell. Calculate its specific gravity and its gravity in degrees Baumé.

(b) An oil has a specific gravity at 60/60°F of 0.651. Calculate its gravity in degrees API and degrees Baumé.

21. Make the following conversion of units:

(a) An energy of 8 ft-lb to kilogram-meters.

(b) An acceleration of 32.2 $\dfrac{\text{ft}}{(\sec)^2}$ to $\dfrac{\text{meters}}{(\sec)^2}$.

(c) A pressure of 100 mm of Hg to inches of water.

(d) Thermal conductivity of k $\dfrac{\text{Btu}}{(\text{hr})(°F)(\text{ft})}$ to $\dfrac{\text{kcal}}{(\text{hr})(°C)(\text{m})}$.

(e) The gas constant R, from $\dfrac{82.06 \ (\text{atm})(\text{cm})^3}{(\text{g-mole})(°K)}$ to $\dfrac{\text{Btu}}{(\text{lb-mole})(°R)}$.

CHAPTER II

BEHAVIOR OF IDEAL GASES

In the preceding chapter consideration was given to problems pertaining to the transformation of matter from one physical or chemical state to another. It has been pointed out that matter is essentially indestructible despite all the transformations it may undergo and that the mass of a given system remains unaltered. However, in order to complete the quantitative study of a system it is necessary to consider one other property.

The properties of a moving ball, a swinging pendulum, or a rotating flywheel are different from those of the same objects at rest. The differences lie in the motion of the bodies and in the ability of the moving objects to perform *work*, which is defined as the action of a force moving under restraint through a distance. Likewise, the properties of a red-hot metal bar are different from those of the same bar when cold. The red-hot bar produces effects on the eye and the touch very different from those of the cold bar. The essential property or physicochemical concept necessary to complete the quantitative description of matter is termed *energy*.

Energy. All matter and the properties of matter are manifestations of energy. Energy is the capacity of matter to perform work and to affect the senses. For example, the motions of bodies cited above, the warming of an object by solar radiation, and the change in composition of a storage battery when it generates electricity are all manifestations of energy. Energy is distributed throughout the universe in a variety of forms, all of which may be directly or indirectly converted into one another.

Under the classification of *potential energy* are included all forms of energy not associated with motion but resulting from the position and arrangement of matter. The energy possessed by an elevated weight, a compressed spring, a charged storage battery, a tank of gasoline, or a lump of coal are examples of potential energy. Similarly, potential energy is stored within an atom as the result of forces of attraction among its subatomic parts. Thus potential energy can be further classified as *external potential* energy, which is inherent in matter as a result of its position relative to the earth, or as *internal potential energy*, which resides within the structure of matter.

27

In contrast, energy associated with motion is referred to as *kinetic energy*. The energy represented by the flow of a river, the flight of a bullet, or the rotation of a flywheel are examples of kinetic energy. Also individual molecules possess kinetic energy by virtue of their translational, rotational, and vibrational motions. Similar to the subclassification of potential energy, kinetic energy is subclassified as *internal kinetic energy*, such as associated with molecular and atomic structure, and as *external kinetic energy*, such as associated with the external motion of visible objects.

In addition to the forms of energy associated with composition, position, or motion of matter, energy exists in the forms of electricity, magnetism, and radiation, which are associated with electronic phenomena.

The science pertaining to the transformation of one form of energy to another is termed *thermodynamics*. Early studies of the transformation of energy lead to the realization that although energy can be transformed from one form to another it can never be destroyed and that the total energy of the universe is constant. This principle of the *conservation of energy* is referred to as the *first law of thermodynamics*. Many experimental verifications have served to establish the validity of this law.

Temperature and Heat. Energy may be transferred not only from one form to another but also from one aggregation of matter to another without change of form. The transformation of energy from one form to another or the transfer of energy from one body to another always requires the influence of some *driving force*. As an example, if a hot metal bar is placed in contact with a cold one, the former will be cooled and the latter warmed. The sense of " hotness " is an indication of the internal kinetic energy of matter. The driving force which, even in the absence of electrical, magnetic, or mechanical forces, produces a transfer of energy is termed *temperature* and that form of energy which is transferred from one body to another as a result of a difference in temperature is termed *heat*.

The Kinetic Theory of Gases. As the basis of the kinetic theory it is assumed that all matter is composed of tiny particles which by their behavior determine its physical and chemical properties. A gas is believed to be composed of molecules each of which is a material body and separate from all others. These particles are free to move about in space according to Newton's laws of the motion of material bodies. It is furthermore assumed that each particle behaves as a perfectly elastic sphere. As a consequence of this assumption there is no change in total kinetic energy or momentum when two particles collide or when a particle strikes an obstructing or confining surface. On the basis of these assumptions it is possible to explain many physical phenomena by considering that *each particle of matter is endowed with a certain inherent kinetic energy of translation*. As a result of this energy the particles will be in constant motion, striking against and rebounding from one another and from obstructing surfaces.

The energy which is represented by the sum of the energies of the component particles of matter is termed the total internal energy. When heat is added to a gas, additional kinetic energy is imparted to its component particles. The average quantity of kinetic energy of *translation* which is possessed by the particles of a gas determines its temperature. At any specified temperature the particles of a gas possess definitely fixed, average kinetic energies of translation which may be varied only by a change in temperature resulting from the addition or removal of heat. Thus, an increase in temperature signifies an increase in average kinetic energy of translation which in turn is accompanied by increased speeds of translation of the particles. Conversely, when, by any means, the kinetic energies of translation of the particles of a gas are increased, the temperature is raised.

The theory outlined above accounts for the pressure which is exerted by a gas against the walls of a confining vessel. The translational motion of the particles is assumed to be entirely random, in every direction, and it may be assumed for ordinary cases that the number of particles per unit volume will be constant throughout the space. These assumptions are justified when the number of particles per unit volume is very large. Then, each element of area of confining surface will be subjected to continual bombardment by the particles adjacent to it. Each impact will be accompanied by an elastic rebound and will exert a pressure due to the change of momentum involved. In a pure, undissociated gas all particles may be considered to be of the same size and mass. On the basis of these assumptions the following expression for pressure may be derived from the principles of mechanics.[1]

$$p = \frac{2}{3} \frac{\nu}{V} \left(\frac{1}{2} m u^2 \right) \tag{1}$$

where ν = number of molecules under consideration
  V = volume in which ν molecules are contained
  m = mass of each molecule
  u = average translational velocity of the molecules

[1] This derivation may be found in simplified form in any good physics or physical chemistry text or in more rigorous form in the more advanced books dealing with kinetic theory.

From the definition of the molal units of quantity it was pointed out that one mole of a substance will contain a definite number of single molecules, the same for all substances. Then

$$\nu = nN \tag{2}$$

where

n = number of moles in volume V
N = number of molecules in a mole, a universal constant equal to 6.023×10^{23} for the gram-mole

Combining (1) and (2),

$$pV = n\tfrac{2}{3}N(\tfrac{1}{2}mu^2) = n\tfrac{2}{3}u_t \tag{3}$$

where u_t represents the total translational kinetic energy possessed by one mole of gas.

From extensive experimental investigations the ideal gas law has been empirically developed. In fact, the definition of the absolute scale of temperature is based on this relationship.

$$pv = RT \tag{4}$$

or

$$pV = nRT \tag{5}$$

where

R = a proportionality factor
T = absolute temperature
v = volume of one mole of gas
n = number of moles of gas
V = volume of n moles of gas

Rearranging (4)

$$R = \frac{pv}{T} \tag{6}$$

Assuming the validity of the Avogadro principle that equimolal quantities of all gases occupy the same volume at the same conditions of temperature and pressure, it follows from Equation (6) that the gas law factor R is a universal constant. The Avogadro principle and the ideal gas law have been experimentally shown to approach perfect validity for all gases under conditions of extreme rarefaction, that is, where the number of molecules per unit volume is very small. The constant R may be evaluated from a single measurement of the volume occupied by a known molal quantity of any gas at a known temperature and at a known reduced pressure.

Combining Equations (3) and (5):

$$\frac{2}{3} \, \mathrm{u}_t = RT \tag{7}$$

or

$$\frac{1}{2} \, mu^2 = \frac{3}{2} \frac{R}{N} \, T \tag{8}$$

Equation (7) states that the average kinetic energy of translation of a molecule in the gaseous state is directly proportional to the absolute temperature.   The absolute zero is the temperature at which the kinetic energies of all molecules become zero and molecular motion ceases. From the fact that R, the gas law constant, and N, the Avogadro number, are universal constants, it follows that Equation (7) must apply to all gases.   In other words, *the average translational kinetic energy with which a gas molecule is endowed is dependent only upon the absolute temperature and is independent of its nature and size.*   This conclusion is of far-reaching significance.   It follows that a molecule of hydrogen possesses the same average translational kinetic energy as does a molecule of bromine at the same temperature.   Since the bromine molecule has eighty times the mass of the hydrogen molecule the latter must move at a correspondingly higher velocity of translation.   If the temperature increases, the squares of the velocities of translation of both molecules will be increased in the same proportion.

From the theory outlined in the preceding paragraphs it is possible to form a definite mental picture of the mechanical nature of a gas.   The actual component parts of the gas are invisible and of an abstract and rather theoretical nature.   The kinetic theory merely presents a mechanical analogy by which the phenomena of the gaseous state are explained in terms of the familiar laws of energy and of the behavior of particles of rigid matter of tangible dimensions.   The analogy calls to mind a box in which are contained energized, elastic marbles which are in constant motion, colliding with one another and with the confining walls.   An increase in temperature merely signifies an increased velocity of motion in each marble.   A clear mental picture of such an analogy is of great value in fully understanding the properties of matter and in making use intelligently of thermodynamic relationships.

Extension of the Kinetic Theory.   Although the kinetic theory was originally developed to explain the behavior of gases, it has been extended and found to apply with good approximation wherever small particles of matter are permitted to move freely in space.   It has been shown that all such particles *may be considered as endowed with the same* *kinetic energy of translation when at the same temperature regardless of composition or size.* This principle is believed to apply not only to the molecules of all gases but also to the molecules of all liquids and of substances which are dissolved in liquids. It has been extended still further and shown to apply also to particles of solid matter of considerable size suspended in gases or liquids. Thus, at any selected temperature, a molecule of hydrogen gas, a molecule of iodine vapor, a molecule of liquid water, and a molecule of liquid mercury all are supposed to possess the same translational kinetic energy, indicated by Equation (3). Furthermore, this same energy is possessed by a molecule of sulfuric acid in solution in water and by each of the ions formed by the dissociation of such a molecule. A colloidal particle of gold, containing hundreds of atoms, or a speck of dust suspended in air, each presumably has the same translational kinetic energy as does a molecule of hydrogen gas at the same temperature. The larger particles must therefore exhibit correspondingly slower velocities of translational motion. This generalization is of the greatest importance in the explanation of such phenomena as diffusion, heat conduction, osmotic pressure, and the general behavior of colloidal systems.

The Gas Law Units and Constants. In the use of the gas law equations great care must be exercised that consistent units are employed for the expression of both the variable and constant terms. Temperature must always be expressed on an absolute scale. Two such scales are in common use. The Kelvin scale corresponds, in the size of its unit degree, to the Centigrade scale. The zero of the Centigrade scale corresponds to 273.1 degrees on the Kelvin scale. Thus:

$$x°C = (x + 273.1)°K \text{ (Kelvin)} \tag{9}$$

The Rankine scale of absolute temperature corresponds, in the size of its unit degree, to the Fahrenheit scale. The zero of the Fahrenheit scale corresponds to 460 degrees on the Rankine scale. Thus:

$$x°F = (x + 460)°R \text{ (Rankine)} \tag{10}$$

The Avogadro number N denoting the number of molecules in a mole is one of the most important of physical constants and has been carefully determined by a variety of methods. The accepted value is 6.023 $\times 10^{23}$ for the number of molecules in one gram-mole, or 2.73×10^{26} molecules per pound-mole.

Equation (7) may be solved for the gas constant R:

$$R = \frac{2}{3} \frac{u_t}{T} \tag{11}$$

From Equation (11) it is seen that R represents two-thirds of the translational kinetic energy possessed by one mole per degree of absolute temperature. The numerical value of R has been carefully determined and may be expressed in any desired energy units. Following are values corresponding to various systems of units:

Units of Pressure	Units of Volume	R
Per gram-mole (temperatures: Kelvin)		
Atmospheres	Cubic centimeters	82.06
Per pound-mole (temperatures: Rankine)		
Pounds per square inch	Cubic inches	18,510
Pounds per square inch	Cubic feet	10.71
Atmospheres	Cubic feet	0.730

APPLICATIONS OF THE IDEAL GAS LAW

When substances exist in the gaseous state two general types of problems arise in determining the relationships between weight, pressure, temperature, and volume. The first type is that in which are involved only the last three variables — pressure, temperature, and volume. For example, a specified volume of gas is initially at a specified temperature and pressure. The conditions are changed, two of the variables in the final state being specified, and it is desired to calculate the third. For such calculations it is not required to know the weight of the gas. The second, more general type of problem involves the weight of the gas. A specified weight of substance exists in the gaseous state under conditions, two of which are specified and the third is to be calculated. Or, conversely, it is desired to calculate the weight of a given quantity of gas existing at specified conditions of temperature, pressure, and volume.

Problems of the first type, in which weights are not involved, may be readily solved by means of the proportionality indicated by the gas law. Equation (5) may be applied to n moles of gas at conditions p_1, V_1, T_1 and also at conditions p_2, V_2, T_2.

$$p_1 V_1 = nRT_1$$
$$p_2 V_2 = nRT_2$$

Combining:

$$\frac{p_1 V_1}{p_2 V_2} = \frac{T_1}{T_2} \tag{12}$$

This equation may be applied directly to any quantity of gas. If the three conditions of state 1 are known, any one of those of state 2 may be calculated to correspond to specified values of the other two. Any units of pressure, volume, or absolute temperature may be used, the only requirement being that the units in both initial and final states be the same.

Equation (5) is in form to permit direct solution of problems of the second type, in which are involved both weights and volumes of gases. With weights expressed in molal units the equation may be solved for any one of the four variables if the other three are known. However, this calculation requires a value of the constant R expressed in units to correspond to those used in expressing the four variable quantities. So many units of expression are in common use for each variable quantity that a very large table of values of R would be required or else the variable quantities would have to be converted into standard units. Either method is inconvenient.

It proves much more desirable to separate such calculations into two steps. As a primary constant, the *normal molal volume* is used instead of R. The normal molal volume is the volume occupied by one mole of a gas at arbitrarily selected *standard conditions*, assuming that the ideal gas law is obeyed. The normal molal volume at any one set of standard conditions, if the validity of Equation (5) is assumed, must be a universal constant, the same for all gases. The volume, at the standard conditions, of any weight of gas is the product of the number of moles present and the normal molal volume. The general type of problem involving weights and volumes at any desired conditions may then be solved in two steps. In one, the differences between the properties of the gas at standard conditions and at those specified in the problem are determined by Equation (12). In the other step the relationship between volume at standard conditions and weight is determined by means of the normal molal volume constant.

Standard Conditions. An arbitrarily specified standard state of temperature and pressure serves two purposes. It establishes the normal molal volume constant required in the system of calculation described in the preceding section. It also furnishes convenient specifications under which quantities of gases may be compared when expressed in terms of volumes. Some such specification is necessary because of the fact that the volume of a gas depends not only on the quantity but on the temperature and pressure as well.

Several specifications of standard conditions are in more or less common use but the one most universally adopted is that of a temperature of 0°C and a pressure of one atmosphere. It is recommended that these conditions be adopted as the standard for all calculations. Under these conditions the normal molal volumes are as follows (the abbreviation S.C. is used to designate the standard conditions):

Volume of 1 gram-mole S.C.   = 22.41 liters

Volume of 1 pound-mole S.C. = 359 cubic feet

These important constants should be memorized. The conditions of the standard state may be expressed in any desired units as in the following table:

STANDARD CONDITIONS

Temperature	Pressure
0°Centigrade	1 atmosphere
273°Kelvin	760 mm of mercury
32°Fahrenheit	29.92 in. of mercury
492°Rankine	14.70 lb per sq in.

There are many substances which cannot actually exist in the gaseous state at these specified conditions. For example, at a temperature of 0°C water cannot exist in a stable gaseous form at a pressure greater than 4.6 mm of mercury. Higher pressures cause condensation. Yet, it is convenient to refer to the hypothetical volume occupied by water vapor at standard conditions. In such a case the volume at standard conditions indicates the hypothetical volume which would be occupied by the substance if it could exist in the vapor state at these conditions and if it obeyed the ideal gas law.

Gauge Pressure. All ordinary pressure gauges indicate the magnitude of pressure above or below that of the atmosphere. In order to obtain the *absolute pressure* which must be used in the gas law, the pressure of the atmosphere must be added to the *gauge pressure*. The average atmospheric pressure at sea level is 14.70 pounds per square inch or 29.92 inches of mercury.

Gas Densities and Specific Gravities. The density of a gas is ordinarily expressed as the weight in grams of one liter or the weight in pounds of one cubic foot. Unless otherwise specified the volumes are at the standard conditions of 0°C and a pressure of 1.0 atmosphere. On this basis air has a normal density of 1.293 grams per liter or of 0.0807 pound per cubic foot.

The *specific gravity* of a gas is usually defined as the ratio of its density to that of air at the same conditions of temperature and pressure.

The gas law expresses the relationship between four properties of a gas: mass, volume, pressure, and temperature. In order to calculate any one of these properties the others must be known or specified. Four different types of problems arise, classified according to the property being sought. The following illustrations show the application of the recommended method of calculation to each of these types of problems.

For establishment of correct ratios to account for the effects of pressure and temperature a simple rule may be followed which offers less opportunity for error than attempting to recall Equation (12). *The ratio of pressures or temperatures should be greater than unity when the changes in pressure or temperature are such as to cause increase in volume. The ratios should be less than unity when the changes are such as to cause decrease in volume.*

Illustration 1 (*Volume Unknown*). Calculate the volume occupied by 30 lb of chlorine at a pressure 743 mm of Hg and 70°F.

Basis: 30 lb of chlorine or 30/71 = 0.423 lb-mole.

Volume at S.C. = 0.423 × 359 = 152 cu ft.

$$V_2 = V_1 \frac{p_1}{p_2} \times \frac{T_2}{T_1}$$

70°F = 530°Rankine.

Volume at 743 mm Hg, 70°F = 152 × $\frac{760}{743}$ × $\frac{530}{492}$ = 167 cu ft.

Illustration 2 (*Weight Unknown*). Calculate the weight of 100 cu ft of water vapor, measured at a pressure of 15.5 mm of Hg and 23°C.

Basis: 100 cu ft of water vapor at 15.5 mm Hg, 23°C.

Volume at S.C. = 100 $\frac{15.5}{760}$ × $\frac{273}{296}$ = 1.88 cu ft.

Moles of H_2O = 1.88 ÷ 359 = 0.00523 lb-mole.

Weight of H_2O = 0.00523 × 18 = 0.0942 lb.

Illustration 3 (*Pressure Unknown*). It is desired to compress 10 lb of carbon dioxide to a volume of 20 cu ft. Calculate the pressure in pounds per square inch which is required at a temperature of 30°C assuming the applicability of the ideal gas law.

Basis: 10 lb of CO_2 or 10/44 = 0.228 lb-mole.

Volume at S.C. = 0.228 × 359 = 81.7 cu ft.

From Equation (12):

$$p_2 = p_1 \frac{V_1}{V_2} \times \frac{T_2}{T_1}$$

30°C = 303°K.

Pressure at 20 cu ft, 30°C = 14.7 $\frac{81.7}{20}$ × $\frac{303}{273}$ = 66.6 lb per sq in.

Illustration 4 (*Temperature Unknown*). Assuming the applicability of the ideal gas law, calculate the maximum temperature to which 10 lb of nitrogen, enclosed in a 30 cu ft chamber, may be heated without the pressure exceeding 150 lb per sq in.

Basis: 10 lb of nitrogen or 10/28 = 0.357 lb-mole.

Volume at S.C. = 0.357 × 359 = 128.1 cu ft.

$$T_2 = T_1 \frac{p_2}{p_1} \times \frac{V_2}{V_1}$$

Temperature at 30 cu ft, 150 lb/sq in. =

$$273 \frac{150}{14.7} \times \frac{30}{128.1} = 652°K \text{ or } 379°C$$

Dissociating Gases. Certain chemical compounds when in the gaseous state apparently do not even approximately follow the relationships deduced above. The tendency of hydrogen fluoride to associate into large molecules was mentioned in Chapter I. Ammonium chloride, nitrogen peroxide, and phosphorus pentachloride exhibit an abnormality opposite in effect which has been definitely proved to result from dissociation of the molecules into mixtures containing two or more other compounds. Ammonium chloride molecules in the vapor state separate into molecules of hydrogen chloride and ammonia:

$$NH_4Cl = NH_3 + HCl$$

Thus gaseous ammonium chloride is not a pure gas but a mixture of three gases, NH_4Cl, HCl, and NH_3. By decomposition, two gas particles are produced from one, and the pressure or volume of the gas increases above that which would exist had no decomposition taken place. For this reason, when one gram-mole of ammonium chloride is vaporized the volume occupied will be much greater than that indicated by Equation (5). However, when proper account is taken of the fact that in the gaseous state there is actually more than one gram-mole present, it is found that the ideal gas law applies. Conversely, from the apparent deviation from the gas law the percentage of dissociation can be calculated if the chemical reaction involved is known.

Illustration 5. When heated to 100°C and 720 mm pressure 17.2 grams of N_2O_4 gas occupy a volume of 11,450 cc. Assuming that the ideal gas law applies, calculate the percentage dissociation of N_2O_4 to NO_2.

$$\text{Gram-moles of } N_2O_4 \text{ initially present} = \frac{17.2}{92} = 0.187$$

Let x = gram-moles of N_2O_4 dissociated. Then $2x$ = gram-moles of NO_2 formed. Total gram-moles present after dissociation =

$$0.187 - x + 2x = \frac{11,450}{22,400} \times \frac{273}{373} \times \frac{720}{760} = 0.355$$

Solving, $x = 0.168$.

$$\text{Percentage dissociation} = \frac{0.168}{0.187} \times 100 = 90\%$$

GASEOUS MIXTURES

In a mixture of different gases the molecules of each component gas are distributed throughout the entire volume of the containing vessel and the molecules of each component gas contribute by their impacts to the total pressure exerted by the entire mixture. The total pressure is equal to the sum of the pressures exerted by the molecules of each component gas. These statements apply to all gases, whether or not their behavior is ideal. In a mixture of ideal gases the molecules of each component gas behave independently as though they alone were present in the container. Before considering the actual behavior of gaseous mixtures it will be necessary to define two terms commonly employed, namely, *partial pressure* and *pure-component volume*. By definition, the *partial pressure of a component gas which is present in a mixture of gases is the pressure that would be exerted by that component gas if it alone were present in the same volume and at the same temperature as the mixture*. By definition, *the pure-component volume of a component gas which is present in a mixture of gases is the volume that would be occupied by that component gas if it alone were present at the same pressure and temperature as the mixture*.

The partial pressure as defined above does not represent the actual pressure exerted by the molecules of the component gas when present in the mixture except under certain limiting conditions. Also, the pure-component volume does not represent the volume occupied by the molecules of the component gas when present in the mixtures, for obviously the molecules are distributed uniformly throughout the volume of the mixture.

The pure-component volume generally has been termed *partial volume* in the past. However, the latter term is currently used to designate the differential increase in volume when a component is added to a mixture. Pure-component volumes and partial volumes are not necessarily the same except under ideal conditions.

Laws of Dalton and Amagat. From the simple kinetic theory of the constitution of gases it would be expected that many properties of gaseous mixtures would be additive. The additive nature of partial pressures is expressed by *Dalton's law*, which states that *the total pressure exerted by a gaseous mixture is equal to the sum of the partial pressures*, that is:

$$p = p_A + p_B + p_C + \cdots \tag{13}$$

where p is the total pressure of the mixture and p_A, p_B, p_C, etc., are the partial pressures of the component gases as defined above.

Similarly, the additive nature of pure-component volumes is given by the *law of Amagat*, or Leduc's law, which states that the *total volume occupied by a gaseous mixture is equal to the sum of the pure-component volumes*, that is:

$$V = V_A + V_B + V_C + \cdots \tag{14}$$

where V is the total volume of the mixture and V_A, V_B, V_C, etc., are the pure-component volumes of the component gases as defined above. It will be shown later that each of these laws is correct where conditions are such that the mixture and each of the components obey the ideal gas law.

Where small molal volumes are encountered, such that the ideal gas law does not apply either Dalton's or Amagat's law may apply, but both laws apply simultaneously only for ideal gases. Under such conditions pressures may not be additive, because the introduction of additional molecules into a gas-filled container may appreciably affect the pressure exerted by those already there. The presence of new molecules will reduce the space available for the free motion of those originally present and will exert attractive forces on them. Similarly, if quantities of two gases at the same pressure are allowed to mix at that same pressure, the like molecules of each gas will be separated by greater distances and will be in the presence of unlike molecules, which condition may alter the order of attractive forces existing between them. As a result, the volume of the mixture may be quite different from the sum of the original volumes. These same effects are present but negligible under conditions of large molal volumes and wide separation of molecules.

Where conditions are such that the ideal gas law is applicable:

$$p_A = \frac{n_A RT}{V} \tag{15}$$

where

V = total volume of mixture

n_A = number of moles of component A in mixture

Similar equations represent the partial pressures of components B, C, etc. Combining these equations with Dalton's law, Equation (13):

$$p = (n_A + n_B + n_C + \cdots) \frac{RT}{V} \tag{16}$$

This equation relates the pressure, temperature, volume, and molal quantity of any gaseous mixture under such conditions that the mixture and each of the components follow the ideal gas law and Dalton's law. By combining Equations (15) and (16) a useful relationship between total and partial pressure is obtained.

$$p_A = \frac{n_A}{n_A + n_B + n_C \cdots} p = N_A p \tag{17}$$

The quantity $N_A = n_A/(n_A + n_B + n_C + \cdots)$ is the mole fraction of component A. Equation (17) then signifies that, *where the ideal gas law may be applied, the partial pressure of a component of a mixture is equal to the product of the total pressure and the mole fraction of that component.*

Where conditions are such that the ideal gas law is applicable

$$pV_A = n_A RT \tag{18}$$

$$pV_B = n_B RT \tag{19}$$

where V_A, V_B, etc., are the pure-component volumes as defined above. Adding these equations,

$$p(V_A + V_B + \cdots) = (n_A + n_B + \cdots)RT \tag{20}$$

Combining Equations (18) and (19) with Amagat's law (Equation 14),

$$\frac{V_A}{V} = \frac{n_A}{n_A + n_B + \cdots} \tag{21}$$

or

$$V_A = N_A V \tag{22}$$

Equation (22) signifies that, *where the ideal gas law may be applied, the pure-component volume of a component of a gaseous mixture is equal to the product of the total volume and the mole fraction of that component.*

From Equations (16) and (20) it is evident that when the ideal gas law is valid both Amagat's and Dalton's laws apply, that is, both pure-component volumes and partial pressures are additive.

Average Molecular Weight of a Gaseous Mixture. A certain group of components of a mixture of gases may in many cases pass through a process without being changed in composition or weight. For example, in a drying process, dry air merely serves as a carrier for the vapor being removed and undergoes no change in composition or in weight. It is frequently convenient to treat such a mixture as though it were a single gas and assign to it an average molecular weight which may be used for calculation of its weight and volume relationships. Such an average molecular weight has no physical significance from the standpoint of the molecular theory and is of no value if any component of the mixture takes part in a reaction or is altered in relative quantity. The average molecular weight is calculated by adopting a unit molal quantity of the mixture as the basis of calculation. The weight of this molal quantity is then calculated and will represent the average molecular weight. By this method the average molecular weight of air is found to be 29.0.

Illustration 6. Calculate the average molecular weight of a flue gas having the following composition by volume:

CO_2.....................................	13.1%
O_2......................................	7.7%
N_2......................................	79.2%
	100.0%

Basis: 1 gram-mole of the mixture.

CO_2 = 0.131 gram-mole or...... 5.76 grams
O_2 = 0.077 gram-mole or...... 2.46 grams
N_2 = 0.792 gram-mole or...... 22.18 grams

Weight of 1 gram-mole = ...30.40 grams, which is the average molecular weight.

Densities of Gaseous Mixtures. If the composition of a gas mixture is expressed in molal or weight units the density is readily determined by selecting a unit molal quantity or weight as the basis and calculating its volume at the specified conditions of temperature and pressure. This method may be applied to mixtures which do or do not follow the ideal gas law. Where the ideal gas law is applicable, a more direct method is first to obtain the volume of the basic quantity of mixture at standard conditions by multiplying the number of moles by the normal molal volume. The volume at the specified conditions is then calculated from the ideal gas law.

Illustration 7. Calculate the density in pounds per cubic foot at 29 in. of Hg and 30°C of a mixture of hydrogen and oxygen which contains 11.1% H_2 by weight.

Basis: 1 lb of mixture.

H_2 = 0.111 lb or............................ 0.0555 lb-mole
O_2 = 0.889 lb or............................ 0.0278 lb-mole
Total molal quantity......................... 0.0833 lb-mole
Volume at S.C. = 0.0833 × 359.............. 29.9 cu ft

Volume at 29 in. Hg, 30°C = $29.9 \times \dfrac{29.92 \times 303}{29.0 \times 273}$ = 34.2 cu ft

Density at 29 in. Hg, 30°C = $\dfrac{1}{34.2}$ .... 0.0292 lb per cu ft

If the composition of a mixture of gases is expressed in volume units, the ideal gas law is ordinarily applicable. In this case the volume analysis is the same as the molal analysis, and the density is readily calculated on the basis of a unit molal quantity of the mixture. The weight of the basic quantity is first calculated and then its volume at the specified conditions.

Illustration 8. Air is assumed to contain 79.0% nitrogen and 21.0% oxygen by volume. Calculate its density in grams per liter at a temperature of 70°F and a pressure of 741 mm of Hg.

Basis: 1.0 gram-mole of air.

O_2 = 0.210 gram-mole or........................... 6.72 grams
N_2 = 0.790 gram-mole or........................... 22.10 grams
Total weight..................................... 28.82 grams
Volume at S.C.................................... 22.41 liters

$$\text{Volume, 741 mm Hg, 70°F} = 22.41 \times \frac{760 \times 530}{741 \times 492} \ldots \ldots \quad 24.8 \text{ liters}$$

$$\text{Density} = \frac{28.82}{24.8} = 1.162 \text{ grams per liter (741 mm Hg, 70°F)}$$

The actual density of the atmosphere is slightly higher owing to the presence of about 1 per cent of argon which is classed as nitrogen in the above problem. The mixture of nitrogen and inert gases in the atmosphere may be termed *atmospheric nitrogen*. The average molecular weight of this mixture is 28.2.

VOLUME CHANGES WITH CHANGE IN COMPOSITION

Such operations as gas absorption, drying, and some types of evaporation involve changes in the compositions of gaseous mixtures, owing to the addition or removal of certain components. In a drying operation a stream of air takes on water vapor. In the scrubbing of coal gas, ammonia is removed from the mixture. It is of interest to calculate the relationships existing between the initial and final volumes of the mixture and the volume of the material removed or added to the mixture in such a process. The situation is ordinarily complicated by changes of temperature and pressure concurrent with the composition changes. Solution may be carried out by the methods of Chapter I if the quantities specified in the problem are first converted to weight or molal units. The quantities which are unknown may then be calculated in these same units. The last step will then be the conversion of the results from molal or weight units into volumes at the specified conditions of temperature and pressure. The relationships between molal units and volumes under any conditions are expressed by Equations (16) to (22). This method of solution may be applied with the use of either the ideal gas law or more accurate equations. The following illustration demonstrates the method for a case in which the ideal gas law is applicable. As in the problems of Chapter I, the calculations must be based on a definite quantity of a component which passes through the process unchanged.

Illustration 9. Combustion gases having the following molal composition are passed into an evaporator at a temperature of 200°C and a pressure of 743 mm of Hg.

Nitrogen.........................	79.2%
Oxygen...........................	7.2%
Carbon dioxide...................	13.6%
	100.0%

Water is evaporated, the gases leaving at a temperature of 85°C and a pressure of 740 mm of Hg with the following molal composition:

Nitrogen	48.3%
Oxygen	4.4%
Carbon dioxide	8.3%
Water	39.0%
	100.0%

(a) Calculate the volume of gases leaving the evaporator per 100 cu ft entering.

(b) Calculate the weight of water evaporated per 100 cu ft of gas entering.

Solution:

Basis: 1 gram-mole of the entering gas.

N_2	0.792 gram-mole
O_2	0.072 gram-mole
CO_2	0.136 gram-mole

Total volume (743 mm Hg, 200°C) calculated from Equations (14) and (20):

$$p = 743/760 \text{ or } 0.978 \text{ atm}$$
$$T = 473°K$$
$$R = 82.1 \text{ cc atm per } °K$$

$$V = \frac{(n_A + n_B + n_C)\, RT}{p} = \frac{(0.792 + 0.072 + 0.136)\, 82.1 \times 473}{0.978}$$

$$= \frac{1.0 \times 82.1 \times 473}{0.978} = 39,750 \text{ cc or } 1.40 \text{ cu ft}$$
$$(743 \text{ mm Hg, } 200°C)$$

This 1.0 g-mole of gas entering forms 61% by volume of the gases leaving the evaporator.

$$\text{Gases leaving} = \frac{1.0}{0.61} = 1.64 \text{ gram-moles.}$$

Water leaving = 1.64 − 1.0 = 0.64 gram-mole.

Volume of gas leaving, from Equations (14) and (20):

$$p = 740/760 = 0.973 \text{ atm}$$
$$T = 358°K$$
$$R = 82.1 \text{ cc atm per } °K$$

$$V = \frac{(0.792 + 0.072 + 0.136 + 0.64) \times 82.1 \times 358}{0.973}$$

$$= \frac{1.64 \times 82.1 \times 358}{0.973} = 49,500 \text{ cc or} \ldots \ldots \ldots \ldots \ldots 1.75 \text{ cu ft}$$

Volume of gas leaving per 100 cu ft entering,

$$\frac{1.75 \times 100}{1.40} \ldots \ldots \ldots \ldots \ldots \ldots \ldots \ldots \quad 125 \text{ cu ft (740 mm Hg, 85°C)}$$

Weight of water leaving evaporator = 0.64 × 18 = 11.5 grams or 0.0254 lb

Weight of water evaporated per 100 cu ft of gas entering,

$$\frac{0.0254 \times 100}{1.40} \dots\dots\dots\dots\dots\dots\dots\dots\dots\dots\dots\dots\dots\dots\dots \text{1.81 lb}$$

Pure-component Volume Method.   Where the ideal gas law may be applied, the above method of calculation is unnecessarily tedious. In this case the solution may be carried out without conversion to molal or weight units by application of pure-component volumes. The total volume of any ideal mixture may be obtained by adding together the pure-component volumes of its components. Similarly, the removal of a component from a mixture will decrease the total volume by its pure-component volume. Care must be taken in the use of this method that all volumes which are added together are expressed *at the same conditions of temperature and pressure.*   A process involving changes in temperature and pressure as well as composition is best considered as taking place in two steps: first, the change in composition at the initial conditions of temperature and pressure; and second, the change in volume of the resultant mixture to correspond to the final conditions of temperature and pressure. Again the entire calculation must be based on a definite quantity of a component which passes through the process without change in quantity. This procedure is indicated in the following illustration:

Illustration 10.   In the manufacture of hydrochloric acid a gas is obtained which contains 25% HCl and 75% air by volume. This gas is passed through an absorption system in which 98% of the HCl is removed. The gas enters the system at a temperature of 120°F and a pressure of 743 mm of Hg and leaves at a temperature of 80°F and a pressure of 738 mm of Hg.

(*a*) Calculate the volume of gas leaving per 100 cu ft entering the absorption apparatus.

(*b*) Calculate the percentage composition by volume of the gases leaving the absorption apparatus.

(*c*) Calculate the weight of HCl removed per 100 cu ft of gas entering the absorption apparatus.

Solution:

Basis: 100 cu ft of entering gas (743 mm Hg, 120°F) containing 75 cu ft of air which will be unchanged in quantity.

Pure-component vol. of HCl.............................   25 cu ft
Pure-component vol. of HCl absorbed....................   24.5 cu ft
Pure-component vol. of HCl remaining...................   0.50 cu ft

Vol. of gas remaining:
    75 + 0.50.......................   75.5 cu ft (743 mm, 120°F)

Vol. of gas leaving:
$$75.5 \times \frac{743}{738} \times \frac{540}{580} \dots\dots\dots\dots\dots \text{70.8 cu ft (738 mm 80°F)}$$

Composition of gases leaving:

HCl, 0.5/75.5. . . . .   0.66%

Air. 99.34%

Vol. at S.C. of HCl absorbed $= 24.5 \times \dfrac{743}{760} \times \dfrac{492}{580}$. . . . .   20.3 cu ft

HCl absorbed $= 20.3/359 = 0.0565$ lb-mole or. . . . .   2.07 lb

Partial Pressure Method.   In certain types of work, especially where condensable vapors are involved, it is convenient to express the compositions of gaseous mixtures in terms of the partial pressures of the various components.   Where data are presented in this form, problems of the type discussed above may be more conveniently solved by considering only the pressure changes resulting from the changes in composition. The addition or removal of a component of a mixture may be considered as producing only a change in the partial pressure of all of the other components.   The *actual* volume occupied by each of these components will always be exactly the same as that of the entire mixture.   The volume of the mixture may then always be determined by application of the gas law to any components which pass through the process unchanged in quantity and whose partial pressures are known at both the initial and final conditions. The use of this method is shown in the following illustration:

Illustration 11.   Calcium hypochlorite is produced by absorbing chlorine in milk of lime.   A gas produced by the Deacon chlorine process enters the absorption apparatus at a pressure of 740 mm of Hg and a temperature of 75°F.   The partial pressure of the chlorine is 59 mm of Hg, the remainder being inert gases.   The gas leaves the absorption apparatus at a temperature of 80°F and a pressure of 743 mm of Hg with a partial pressure of chlorine of 0.5 mm of Hg.

(*a*) Calculate the volume of gases leaving the apparatus per 100 cu ft entering.

(*b*) Calculate the weight of chlorine absorbed, per 100 cu ft of gas entering.

Solution:

Basis: 100 cu ft of gas entering (740 mm Hg, 75°F).

Partial pressure of inert gases entering $= 740 - 59$. . . . . . .   681 mm Hg

Partial pressure of inert gases leaving $= 743 - 0.5$. . . . . . .   742.5 mm Hg

Actual volume of inert gas entering.   100 cu ft

Actual volume of inert gases leaving $= 100 \times \dfrac{681}{742.5} \times \dfrac{540}{535} = 92.5$ cu ft.   This is also the total volume of gases leaving (743 mm Hg, 80°F).

The actual volumes of chlorine entering and leaving are also 100 and 92.5 cu ft, respectively.

Volume at S.C. of chlorine entering $= 100 \dfrac{59 \times 492}{760 \times 535}$ . . . .   7.14 cu ft

Volume at S.C. of chlorine leaving $= 92.5 \dfrac{0.5 \times 492}{760 \times 540}$ . . . . .   0.055 cu ft

Volume at S.C. of chlorine absorbed $= 7.14 - 0.055$...... **7.08 cu ft**

Chlorine absorbed $= \dfrac{7.08}{359} = 0.0197$ lb-mole or.......... **1.40 lb**

GASES IN CHEMICAL REACTIONS

In a great many chemical and metallurgical reactions gases are present, either in the reacting materials or in the products or in both. Quantities of gases are ordinarily expressed in volume units because of the fact that the common methods of measurement give results directly on this basis. The general types of reaction calculations must, therefore, include the complications introduced by the expression of gaseous quantities and compositions in volume units.

In Chapter I methods are demonstrated for the solution of reaction calculations through the use of molal units for the expression of quantities of reactants and products. Where this is the scheme of calculation, the introduction of volumetric data adds but few complications. By the use of the normal molal volume constants combined with the proportions of the ideal gas law it is easy to convert from molal to volume units, and the reverse. The methods of conversion have been explained in the preceding sections.

The same general methods of solution are followed as were described in Chapter I. All quantities of active materials, whether gaseous, solid, or liquid, are expressed in molal units and the calculation carried out on this basis. Results are thus obtained in molal units which may readily be converted to volumes at any desired conditions. The most convenient choice of a quantity of material to serve as the basis of calculation is determined by the manner of presentation of the data. In general, if the data regarding the basic material are in weight units, a unit weight is the best basis of calculation. If the data are in volume units, a unit molal quantity is ordinarily the most desirable basis.

Illustration 12. Nitric acid is produced in the Ostwald process by the oxidation of ammonia with air. In the first step of the process ammonia and air are mixed together and passed over a catalyst at a temperature of 700°C. The following reaction takes place:

$$4NH_3 + 5O_2 = 6H_2O + 4NO$$

The gases from this process are passed into towers where they are cooled and the oxidation completed according to the following theoretical reactions:

$$2NO + O_2 = 2NO_2$$

$$3NO_2 + H_2O = 2HNO_3 + NO$$

The NO liberated is in part reoxidized and forms more nitric acid in successive repetitions of the above reactions. The ammonia and air enter the process at a temperature of 20°C and a pressure of 755 mm Hg. The air is present in such proportion that the oxygen will be 20% in excess of that required for complete oxidation of the ammonia to nitric acid and water. The gases leave the catalyzer at a pressure of 743 mm of Hg and a temperature of 700°C.

(a) Calculate the volume of air to be used per 100 cu ft of ammonia entering the process.

(b) Calculate the percentage composition by volume of the gases entering the catalyzer.

(c) Calculate the percentage composition by volume of the gases leaving the catalyzer, assuming that the degree of completion of the reaction is 85% and that no other decompositions take place.

(d) Calculate the volume of gases leaving the catalyzer per 100 cu ft of ammonia entering the process.

(e) Calculate the weight of nitric acid produced per 100 cu ft of ammonia entering the process, assuming that 90% of the nitric oxide entering the tower is oxidized to nitric acid.

Basis of Calculation: 1.0 lb-mole of NH_3.

$$NH_3 + 2O_2 = HNO_3 + H_2O$$

(a) O_2 required. . . . .   2.0 lb-moles

   O_2 supplied $= 2.0 \times 1.2$. . . . .   2.4 lb-moles

   Air supplied $= \dfrac{2.4}{0.210}$ . . . .   11.42 lb-moles

Therefore:

   Vol. of air $= 11.42 \times$ (vol. of ammonia at same conditions)

or:

   Vol. of $NH_3 = 359 \times \dfrac{293 \times 760}{273 \times 755} = 388$ cu ft (20°C, 755 mm Hg)

   Vol. of air $= 11.42 \times 388 = 4440$ cu ft (20°C, 755 mm Hg)

   Vol. of air per 100 cu ft of $NH_3 = \dfrac{4440 \times 100}{388} = 1142$ cu ft

(b)   Gases entering process $= N_2, O_2, NH_3$.

   N_2 present in air $= 0.790 \times 11.42$. . . . .   9.02 lb-moles

   Total quantity of gas entering catalyzer $= 11.42 + 1 = 12.42$ lb-moles

   Composition by volume:

   $NH_3 = 1.0/12.42$. 8.0%

   O_2 $= 2.4/12.42$. 19.3%

   N_2 $= 9.02/12.42$. . . . .   <u>72.7%</u>

                                                                                          100.0%

(c)   Gases leaving catalyzer, N_2, NH_3, O_2, NO, and H_2O.

NH$_3$ oxidized in catalyzer......................... 0.85 lb-mole

NH$_3$ leaving catalyzer............................ 0.15 lb-mole

O_2 consumed in catalyzer = (5/4) × 0.85........... 1.06 lb-moles

O_2 leaving catalyzer = 2.40 − 1.06................ 1.34 lb-moles

NO formed in catalyzer......................... 0.85 lb-mole

H_2O formed in catalyzer = (6/4) × 0.85........... 1.275 lb-moles

Total quantity of gas leaving catalyzer = 9.02 + 0.15
+ 1.34 + 0.85 + 1.275........................ 12.64 lb-moles

Composition by volume:

NO   = 0.85/12.64.............................. 6.7%

H$_2$O = 1.275/12.64.......................... 10.1%

NH$_3$ = 0.15/12.64.......................... 1.2%

O$_2$ = 1.34/12.64............................ 10.6%

N$_2$ = 9.02/12.64............................ 71.4%

Basis of Calculation: 100 cu ft of NH$_3$ entering the process.

(d)   Moles of NH$_3$ = $\dfrac{1.0 \times 100}{388}$..................... 0.258 lb-mole

Moles of gas leaving catalyzer = 0.258 × 12.64..... 3.26 lb-moles

Vol. at S.C. of gas leaving catalyzer = 3.26 × 359... 1170 cu ft

Vol. of gas leaving catalyzer = 1170 × $\dfrac{760 \times 973}{743 \times 273}$... 4270 cu ft

(700°C, 743 mm Hg) per 100 cu ft of NH$_3$ entering.

(e)   NO produced in catalyzer = 0.258 × 0.85.......... 0.219 lb-mole

NO oxidized in tower = 0.219 × 0.90.............. 0.197 lb-mole

HNO$_3$ formed = 0.197 lb-mole or 0.197 × 63....... 12.4 lb

Range of Applicability of the Ideal Gas Law.   The ideal gas law is applicable only at conditions of low pressure and high temperature corresponding to large molal volumes.   At conditions resulting in small molal volumes the simple kinetic theory breaks down and volumes calculated from the ideal law tend to be too large.   In extreme cases the calculated volume may be five times too great, an error of 400 per cent.

If an error of 1 per cent is permissible the ideal gas law may be used for diatomic gases where gram-molal volumes are as low as 5 liters (80 cubic feet per pound-mole) and for gases of more complex molecular structure such as carbon dioxide, acetylene, ammonia, and the lighter hydrocarbon vapors, where gram-molal volumes exceed 20 liters (320 cubic feet per pound-mole).

The actual behavior of gases under high-pressure conditions is discussed in Chapter XII where rigorous methods of calculation are presented.

PROBLEMS

Pressures are absolute unless otherwise stated

1. It is desired to market oxygen in small cylinders having volumes of 0.5 cu ft and each containing 1.0 lb of oxygen. If the cylinders may be subjected to a maximum temperature of 120°F, calculate the pressure for which they must be designed, assuming the applicability of the ideal gas law.

2. Calculate the number of cubic feet of hydrogen sulfide, measured at a temperature of 30°C and a pressure of 29.1 in. of Hg, which may be produced from 10 lb of iron sulfide (FeS).

3. An automobile tire is inflated to a gauge pressure of 35 lb per sq in. at a temperature of 0°F. Calculate the maximum temperature to which the tire may be heated without the gauge pressure exceeding 50 lb per sq in. (Assume that the volume of the tire does not change.)

4. Calculate the densities in pounds per cubic foot at standard conditions and the specific gravities of the following gases: (a) methane, (b) hydrogen, (c) acetylene, (d) bromine.

5. The gas acetylene is produced according to the following reaction by treating calcium carbide with water:

$$CaC_2 + 2H_2O = C_2H_2 + Ca(OH)_2$$

Calculate the number of hours of service which can be derived from 1.0 lb of carbide in an acetylene lamp burning 2 cu ft of gas per hour at a temperature of 75°F and a pressure of 743 mm of Hg.

6. A natural gas has the following composition by volume:

CH_4	94.1%
N_2	3.0%
H_2	1.9%
O_2	1.0%
	100.0%

This gas is piped from the well at a temperature of 20°C and a pressure of 30 lb per sq in. It may be assumed that the ideal gas law is applicable.

    (a) Calculate the partial pressure of the oxygen.

    (b) Calculate the pure-component volume of nitrogen per 100 cu ft of gas.

    (c) Calculate the density of the mixture in pounds per cubic foot at the existing conditions.

7. A gas mixture contains 0.274 lb-mole of HCl, 0.337 lb-mole of nitrogen, and 0.089 lb-mole of oxygen. Calculate the volume occupied by this mixture and its density in pounds per cubic foot at a pressure of 40 lb per sq in. and a temperature of 30°C.

8. A chimney gas has the following composition by volume:

CO_2	10.5%
CO	1.1%
O_2	7.7%
N_2	80.7%

Using the ideal gas law, calculate:

   (a) Its composition by weight.
   (b) The volume occupied by 1.0 lb of the gas at 67°F and 29.1 in. of Hg pressure.
   (c) The density of the gas in pounds per cubic foot at the conditions of part (b).
   (d) The specific gravity of the mixture.

9. By electrolyzing a mixed brine a mixture of gases is obtained at the cathode having the following composition by weight:

$$Cl_2 \ldots\ldots\ldots\ldots\ldots\ldots\ldots\ldots\ldots\ldots\ldots\ldots\ldots\ldots \quad 67\%$$
$$Br_2 \ldots\ldots\ldots\ldots\ldots\ldots\ldots\ldots\ldots\ldots\ldots\ldots\ldots\ldots \quad 28\%$$
$$O_2 \ldots\ldots\ldots\ldots\ldots\ldots\ldots\ldots\ldots\ldots\ldots\ldots\ldots\ldots\ldots \quad 5\%$$

Using the ideal gas law, calculate:

   (a) The composition of the gas by volume.
   (b) The density of the mixture in grams per liter at 25°C and 740 mm of Hg pressure.
   (c) The specific gravity of the mixture.

10. A mixture of ammonia and air at a pressure of 745 mm of Hg and a temperature of 40°C contains 4.9% NH_3 by volume. The gas is passed at a rate of 100 cu ft per min through an absorption tower in which only ammonia is removed. The gases leave the tower at a pressure of 740 mm of Hg, a temperature of 20°C, and contain 0.13% NH_3 by volume. Using the ideal gas law, calculate:

   (a) The rate of flow of gas leaving the tower in cubic feet per minute.
   (b) The weight of ammonia absorbed in the tower per minute.

11. A volume of moist air of 1000 cu ft at a total pressure of 740 mm of Hg and a temperature of 30°C contains water vapor in such proportions that its partial pressure is 22.0 mm of Hg. Without changing the total pressure, the temperature is reduced to 15°C and some of the water vapor removed by condensation. After cooling it is found that the partial pressure of the water vapor is 12.7 mm of Hg. Using the partial pressure method, calculate:

   (a) The volume of the gas after cooling.
   (b) The weight of water removed.

12. Air is passed into a dryer for the drying of textiles at a rate of 1000 cu ft per min. The air enters the dryer at a temperature of 160°F and contains water vapor exerting a partial pressure of 8.1 mm of Hg. The temperature of the air leaving is 80°F, and the partial pressure of the water is 18 mm of Hg. The total pressure of the wet air may be taken as constant at the barometric value of 745 mm of Hg.

   (a) Calculate the volume of gas leaving the dryer per minute.
   (b) Calculate the weight of water removed per minute from the material in the dryer.

13. A producer gas has the following composition by volume:

$$CO \ldots\ldots\ldots\ldots\ldots\ldots\ldots\ldots\ldots\ldots\ldots\ldots\ldots \quad 23.0\%$$
$$CO_2 \ldots\ldots\ldots\ldots\ldots\ldots\ldots\ldots\ldots\ldots\ldots\ldots \quad 4.4\%$$
$$O_2 \ldots\ldots\ldots\ldots\ldots\ldots\ldots\ldots\ldots\ldots\ldots\ldots\ldots \quad 2.6\%$$
$$N_2 \ldots\ldots\ldots\ldots\ldots\ldots\ldots\ldots\ldots\ldots\ldots\ldots\ldots \quad 70.0\%$$

   (a) Calculate the cubic feet of gas, at 70°F and 750 mm of Hg pressure, per pound of carbon present.
   (b) Calculate the volume of air, at the conditions of part (a), required for the combustion of 100 cu ft of the gas at the same conditions if it is desired

that the total oxygen present before combustion shall be 20% in excess of that theoretically required.

(c) Calculate the percentage composition by volume of the gases leaving the burner of part (b), assuming complete combustion.

(d) Calculate the volume of the gases leaving the combustion of parts (b) and (c) at a temperature of 600°F and a pressure of 750 mm of Hg per 100 cu ft of gas burned.

14. The gas from a sulfur burner has the following composition by volume:

$$
\begin{aligned}
SO_3 &\dots\dots\dots\dots\dots\dots\dots\dots\dots\dots & 1.1\% \\
SO_2 &\dots\dots\dots\dots\dots\dots\dots\dots\dots\dots & 8.2\% \\
O_2 &\dots\dots\dots\dots\dots\dots\dots\dots\dots\dots & 10.0\% \\
N_2 &\dots\dots\dots\dots\dots\dots\dots\dots\dots\dots & 80.7\%
\end{aligned}
$$

(a) Calculate the volume of the gas at 350°F and 29.2 in. of Hg formed per pound of sulfur burned.

(b) Calculate the percentage excess oxygen supplied for the combustion above that required for complete oxidation to SO_3.

(c) From the above gas analysis calculate the percentage composition by volume of the air used in the combustion.

(d) Calculate the volume of air at 70°F and 29.2 in. of Hg supplied for the combustion per pound of sulfur burned.

15. A furnace is to be designed to burn coke at the rate of 200 lb per hour. The coke has the following composition:

$$
\begin{aligned}
Carbon &\dots\dots\dots\dots\dots\dots\dots\dots\dots\dots & 89.1\% \\
Ash &\dots\dots\dots\dots\dots\dots\dots\dots\dots\dots & 10.9\%
\end{aligned}
$$

The grate efficiency of the furnace is such that 90% of the carbon present in the coke charged is burned. Air is supplied in 30% excess of that required for the complete combustion of all the carbon charged. It may be assumed that 97% of the carbon burned is oxidized to the dioxide, the remainder forming monoxide.

(a) Calculate the composition, by volume, of the flue gases leaving the furnace.

(b) If the flue gases leave the furnace at a temperature of 550°F and a pressure of 743 mm Hg, calculate the rate of flow of gases, in cubic feet per minute, for which the stack must be designed.

16. Coke containing 87.2% carbon and 12.8% ash is burned on a grate. It is found that 6% of the carbon in the coke charged is lost with the refuse. The composition by volume of the stack gases from the furnace is as follows:

$$
\begin{aligned}
CO_2 &\dots\dots\dots\dots\dots\dots\dots\dots\dots\dots & 12.0\% \\
CO &\dots\dots\dots\dots\dots\dots\dots\dots\dots\dots & 0.2\% \\
O_2 &\dots\dots\dots\dots\dots\dots\dots\dots\dots\dots & 8.8\% \\
N_2 &\dots\dots\dots\dots\dots\dots\dots\dots\dots\dots & 79.0\%
\end{aligned}
$$

(a) Calculate the volume of gases, at 540°F and 29.3 in. of Hg pressure, formed per pound of coke charged.

(b) Calculate the per cent of excess air supplied above that required for complete oxidation of the carbon charged.

(c) Calculate the degree of completion of the oxidation, to the dioxide, of the carbon burned.

(d) Calculate the volume of air, at 70°F and 29.3 in. Hg, supplied per pound of coke charged.

17. In the fixation of nitrogen by the arc process, air is passed through a magnetically flattened electric arc. Some of the nitrogen is oxidized to NO, which on cooling oxidizes to NO_2. Of the NO_2 formed, 66% will be associated to N_2O_4 at 26°C. The gases are then passed into water-washed absorption towers where nitric acid is formed by the following reaction:

$$H_2O + 3NO_2 = NO + 2HNO_3$$

The NO liberated in this reaction will be reoxidized in part and form more nitric acid.

In the operation of such a plant it is found possible to produce gases from the arc furnace in which the nitric oxide is 2% by volume, while hot. The gases are cooled to 26°C at a pressure of 750 mm of Hg before entering the absorption apparatus.

(a) Calculate the complete analysis by volume of the hot gases leaving the furnace assuming that the air entering the furnace was of average atmospheric composition.

(b) Calculate the partial pressures of the NO_2 and N_2O_4 in the gas entering the absorption apparatus.

(c) Calculate the weight of HNO_3 formed per 1000 cu ft of gas entering the absorption system if the conversion to nitric acid of the combined nitrogen in the furnace gases is 85% complete.

Basis: 1 lb mole air.

NO = .02 × 1 = .02 lb mole

N_2 = .89 -

.89 lb mole, N_2
.21 lb mol O_2

.78 N_2 .78 N_2
.20 O_2 .1950,
.02 NO   .02 NO_2

CHAPTER III

VAPOR PRESSURES

Liquefaction and the Liquid State. Molecules in the gaseous state of aggregation exhibit two opposing tendencies. The translational kinetic energy possessed by each molecule represents a continual, random motion which tends to separate the molecules from one another and to cause them to be uniformly distributed throughout the entire available space. On the other hand, the attractive forces between the molecules tend to draw them together into a concentrated mass, not necessarily occupying the entire space which is available.

The first tendency, that of dispersion, is dependent entirely on the temperature. An increase in the temperature will increase the translational kinetic energy of each molecule and will therefore give it an increased ability to overcome the forces tending to draw it toward other molecules. The second tendency, that of aggregation, is determined by the magnitudes and nature of the attractive forces between the molecules and by their proximity to one another. These intermolecular attractive forces are believed to be of such a nature that they increase to definite maxima as the distances between molecules are diminished. This behavior is shown in Fig. 3, in which are plotted attractive forces as ordinates and distances of separation between two molecules as abscissas. The greatest attractive force between the two molecules exists when they are separated by a relatively small distance S_2. If

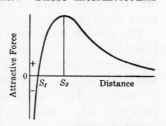

FIG. 3. Attractive force between molecules.

the distance of separation is diminished below S_2, the attractive force rapidly decreases and will reach high negative values corresponding to repulsion. At a distance of separation S_1 the attractive force becomes zero, corresponding to a position of equilibrium. If unaffected by other forces, molecules will group themselves together, separated from one another by distances equal to S_1. Any attempt made to crowd them closer together will meet with repulsive forces. In order to separate them by a distance greater than S_2 it would be necessary to overcome the maximum attractive force by heating or expansion.

When a gas is isothermally compressed and the distances of separation between the molecules are decreased, the attractive forces increase toward their maximum value. If these attractive forces become so large that the potential energy of the attraction of one molecule for another is greater than its kinetic energy of translation, the molecules will be held together to form a dense aggregation which is termed a *liquid*. The characteristic which differentiates a liquid from a gas is the fact that the liquid possesses a definite volume and does not necessarily occupy the entire available space. The individual molecules of the liquid are in motion, owing to their inherent kinetic energies, but this motion takes the form of vibrations, alternately increasing and decreasing the distances of separation, as though energized, vibrating marbles were attached to each other by short rods of rubber.

Critical Properties. Whether or not a substance can exist in the liquid state is dependent on its temperature. If the temperature is sufficiently high that the kinetic energies of translation of the molecules exceed the *maximum* potential energy of attraction between them, the liquid state of aggregation is impossible. The temperature at which the molecular kinetic energy of translation equals the maximum potential energy of attraction is termed the *critical temperature, t_c*. Above the critical temperature the liquid state is impossible for a single component and compression results only in a highly compressed gas, retaining all the properties of the gaseous state. Below the critical temperature a gas may be liquefied if sufficiently compressed.

The pressure required to liquefy a gas at its critical temperature is termed the *critical pressure, p_c*. The critical pressure and temperature fix the *critical state* at which there is no distinction between the gaseous and liquid states. The volume at the critical state is termed the *critical volume, v_c*. The density at the critical state is the *critical density, d_c*. In Table XI, page 234, are values of the critical data for the more common gases.

Reduced Conditions. At conditions equally removed from the critical state many properties of different substances are similarly related. This has given rise to the concept of reduced temperature, reduced pressure, and reduced volume. *Reduced temperature* is defined as the ratio of the existing temperature of a substance to its critical temperature, both being expressed on an absolute scale. Similarly, *reduced pressure* is the ratio of the existing pressure of a substance to its critical pressure, and the reduced volume the ratio of the existing molal volume to its critical molal volume. Thus,

$$\text{Reduced temperature} = T_r = T/T_c$$
$$\text{Reduced pressure} = p_r = p/p_c$$
$$\text{Reduced volume} = v_r = v/v_c$$

Under conditions of equal reduced pressure and equal reduced temperature, substances are said to be in corresponding states. It will be later shown that many properties of gases and liquids, for example, the compressibilities of different gases, are nearly the same at corresponding states, that is, at equal reduced conditions.

Vaporization. As pointed out above, the liquid state results when conditions are such that the potential energies of attraction between molecules exceed their kinetic energies of translation. These conditions are brought about when the temperature of a substance is lowered, decreasing the kinetic energies of translation, or when the molecules are crowded close together, increasing the energies of attraction. On the basis of this theory the surface of a liquid may be pictured as a layer of molecules, each of which is bound to the molecules below it by the attractive forces among them. One of the surface molecules may be removed only by overcoming the attractive forces holding it to the others. This is possible if the molecule is given sufficient translational kinetic energy to overcome the maximum potential energy of attraction and to enable it to move past the point of maximum attraction. Once it has passed this distance of maximum attraction, the molecule is free to move away from the surface under the effect of its translational energy and to become a gas molecule.

In the simple kinetic-theory mechanisms which have been discussed, it is frequently assumed that all molecules of a substance at a given temperature are endowed with the same kinetic energies and move at the same speeds. Actually it has been demonstrated that this is not the case and that molecular speeds and energies vary over wide ranges above and below the average values. In every liquid and gas there are always highly energized molecules moving at speeds much higher than the average. When such a molecule comes to the surface of a liquid, with its velocity directed away from the main body, it may have sufficient energy to break away completely from the forces tending to hold it to the surface. This phenomenon of the breaking away of highly energized molecules takes place from every exposed liquid surface. As a result, molecules of the liquid continually tend to assume the gaseous or vapor state. This phenomenon is termed *vaporization* or *evaporation*.

When a liquid evaporates into a space of limited dimensions the space will become filled with the vapor which is formed. As vaporization proceeds, the number of molecules in the vapor state will increase and cause an increase in the pressure exerted by the vapor. It will be recalled that the pressure exerted by a gas or vapor is due to the impacts of its component molecules against the confining surfaces. Since the

original liquid surface forms one of the walls confining the vapor, there
will be a continual series of impacts against it by the molecules in the
vapor state. The number of such impacts will be dependent on or will
determine the pressure exerted by the vapor. However, when one of
these gaseous molecules strikes the liquid surface it comes under the
influence of the attractive forces of the densely aggregated liquid molecules and will be held there, forming a part of the liquid once more.
This phenomenon, the reverse of vaporization, is known as *condensation*.
The rate of condensation is determined by the number of molecules
striking the liquid surface per unit time, which in turn is determined by
the pressure or density of the vapor. It follows that when a liquid
evaporates into a limited space, two opposing processes are in operation.
The process of vaporization tends to change the liquid to the gaseous
state. The process of condensation tends to change the gas which is
formed by vaporization back into the liquid state. The rate of condensation is increased as vaporization proceeds and the pressure of the vapor
increases. If sufficient liquid is present, the pressure of the vapor must
ultimately reach such a value that the rate of condensation will equal
the rate of vaporization. When this condition is reached, a dynamic
equilibrium is established and the pressure of the vapor will remain unchanged, since the formation of new vapor is compensated by condensation. If the pressure of the vapor is changed in either direction from this
equilibrium value it will adjust itself and return to the equilibrium conditions owing to the increase or decrease in the rate of condensation
which results from the pressure change. The pressure exerted by the
vapor at such equilibrium conditions is termed the *vapor pressure* of the
liquid. All materials exhibit definite vapor pressures of greater or less
degree at all temperatures.

The magnitude of the equilibrium vapor pressure is in no way dependent on the amounts of liquid and vapor as long as any free liquid surface is present. This results from both the rate of loss and the rate of
gain of molecules by the liquid being directly proportional to the area
exposed to the vapor. At the equilibrium conditions when both rates
are the same, a change in the area of the surface exposed will not affect
the conditions in the vapor phase.

The nature of the liquid is the most important factor determining the
magnitude of the equilibrium vapor pressure. Since all molecules are
endowed with the same kinetic energies of translation at any specified
temperature, the vapor pressure must be entirely dependent on the magnitudes of the maximum potential energies of attraction which must be
overcome in vaporization. These potential energies are determined by
the intermolecular attractive forces. Thus, if a substance has high

intermolecular attractive forces the rate of loss of molecules from its surface should be small and the corresponding equilibrium vapor pressure low. The magnitudes of the attractive forces are dependent on both the size and nature of the molecules, usually increasing with increased size and complexity. In general, among liquids of similar chemical natures, the vapor pressure at any specified temperature decreases with increasing molecular weight.

Superheat and Quality. A vapor which exists above its critical temperature is termed a gas. The distinction between a vapor and a gas is thus quite arbitrary, and the two terms are loosely interchanged. For example, carbon dioxide at room temperature is below its critical temperature and, strictly speaking, is a vapor. However, such a material is commonly referred to as a gas.

A vapor which exists under such conditions that its partial pressure is equal to its equilibrium vapor pressure is termed a *saturated vapor*, whether it exists alone or in the presence of other gases. The temperature at which a vapor is saturated is termed the *dew point* or *saturation temperature*. A vapor whose partial pressure is less than its equilibrium vapor pressure is termed a *superheated vapor*. The difference between its existing temperature and its saturation temperature is called its *degrees of superheat*.

If a saturated vapor is cooled or compressed, condensation will result, and what is termed a *wet vapor* is formed. If the vapor is in turbulent motion considerable portions of the condensed liquid will remain in mechanical suspension as small drops in the vapor and be carried with it. The *quality* of a wet vapor is the percentage which the weight of vapor forms of the total weight of vapor and entrained liquid associated with it. Thus, wet steam of 95 per cent quality is a mixture of saturated water vapor and entrained drops of liquid water in which the weight of the vapor constitutes 95 per cent of the total weight.

Boiling Point. When a liquid surface is exposed to a space in which the *total* gas pressure is less than the equilibrium vapor pressure of the liquid, a very rapid vaporization known as *boiling* takes place. Boiling results from the formation of tiny free spaces within the liquid itself. If the equilibrium vapor pressure is greater than the total pressure on the surface of the liquid, vaporization will take place in these free spaces which tend to form below the liquid surface. This vaporization will cause the formation of bubbles of vapor which crowd back the surrounding liquid and increase in size because of the greater pressure of the vapor. Such a bubble of vapor will rise to the surface of the liquid and join the main body of gas above it. Thus, when a liquid boils, vaporization takes place not only at the surface level but also at many interior surfaces of contact between the liquid and bubbles of vapor. The rising bubbles also break up the normal surface into more or less of a froth. The vapor once liberated from the liquid is at a higher pressure than the gas in which it finds itself and will immediately expand and flow away from the surface. These factors all contribute to make vaporization of a liquid relatively very rapid when boiling takes place. When the total pressure is such that boiling does not take place, vaporization will nevertheless continue, but at a slower rate, as long as the vapor pressure of the liquid exceeds the partial pressure of its vapor above the surface.

The temperature at which the equilibrium vapor pressure of a liquid equals the total pressure on the surface is known as the *boiling point*. The boiling point is dependent on the total pressure, increasing with an increase in pressure. Theoretically, any liquid may be made to boil at any desired temperature by sufficiently altering the total pressure on its surface. The temperature at which a liquid boils when under a total pressure of 1.0 atmosphere is termed the *normal boiling point*. This is the temperature at which the equilibrium vapor pressure equals 760 millimeters of mercury or 1.0 atmosphere.

Vapor Pressures of Solids. Solid substances possess a tendency to disperse directly into the vapor state and to exert a vapor pressure just as do liquids. The transition of a solid directly into the gaseous state is termed *sublimation,* a process entirely analogous to the vaporization of a liquid. A familiar example of sublimation is the disappearance of snow in sub-zero weather.

A solid exerts an equilibrium vapor pressure just as a liquid does; this is a function of the nature of the material and its temperature. Sublimation will take place whenever the partial pressure of the vapor in contact with a solid surface is less than the equilibrium vapor pressure of the solid. Conversely, if the equilibrium vapor pressure of the solid is exceeded by the partial pressure of its vapor, condensation directly from the gaseous to the solid state will result.

At the melting point the vapor pressures of a substance in the solid and liquid states are equal. At temperatures above the melting point the solid state cannot exist. However, by careful cooling a liquid can be caused to exist in an unstable, supercooled state at temperatures below its melting point. The vapor pressures of supercooled liquids are always greater than those of the solid state at the same temperature, and the liquid tends to change to the solid.

The vapor pressures of solids, even at their melting points, are generally small. However, in some cases these values become large and of considerable importance. For example, at its melting point of 114.5°C iodine crystals exert a vapor pressure of 90 millimeters of mercury.

Solid carbon dioxide at its melting point of $-56.7°C$ exerts a vapor pressure of 5.11 atmospheres and a pressure of 1.0 atmosphere at a temperature of $-78.5°C$. It is therefore impossible for liquid carbon dioxide to exist in a stable form at pressures less than 5.11 atmospheres.

Calculations dealing with the vapor pressures and sublimation of solids are analogous to those of the vaporization of liquids. The principles and methods outlined in the following sections are equally applicable to sublimation and to vaporization processes.

EFFECT OF TEMPERATURE ON VAPOR PRESSURE

The forces causing the vaporization of a liquid are entirely derived from the kinetic energy of translation of its molecules. It follows that an increase in kinetic energy of molecular translation should increase the rate of vaporization and therefore the vapor pressure. In Chapter II it was pointed out that the kinetic energy of translation is directly proportional to the absolute temperature. On the basis of this theory, an increase in temperature should cause an increased rate of vaporization and a higher equilibrium vapor pressure. This is found to be universally the case where vapor pressures have been experimentally investigated. It must be remembered that it is the temperature of the liquid surface which is effective in determining the rate of vaporization and the vapor pressure.

An exact thermodynamic relationship between vapor pressure and temperature is developed in Chapter XI as

$$\left(\frac{\partial p}{\partial T}\right)_v = \frac{\Lambda}{T(V_g - V_l)} \tag{1}$$

where

p = vapor pressure
T = absolute temperature
Λ = heat of vaporization at temperature T
V_g = volume of gas
V_l = volume of liquid

The above relationship is also referred to as the Clapeyron equation. It is entirely rigorous, universal, and applies to any vaporization equilibrium. Its use in this form is, however, greatly restricted because it presupposes a knowledge of the variation of Λ, V_g, and V_l with temperature.

The latent heat of vaporization, Λ, is the quantity of heat which must be added in order to transform a substance from the liquid to the vapor state at the same temperature. The heat of vaporization decreases as pressure increases, becoming zero at the critical point. This property is fully discussed in subsequent chapters. Values of the heats of vaporization at the normal boiling point of many compounds are listed in Tables XI and XIII, pages 234-237.

By neglecting the volume of liquid and assuming the applicability of the ideal gas law the above relation reduces to the Clausius-Clapeyron equation:

$$\frac{dp}{p} = \frac{\lambda dT}{RT^2} \text{ or } d\ln p = -\frac{\lambda d}{R}\left(\frac{1}{T}\right) \tag{2}$$

where

$$R = \text{gas law constant}$$

The Clausius-Clapeyron equation in the form written above is accurate only when the vapor pressure is relatively low, where it may be assumed that the vapor obeys the ideal gas law and that the volume in the liquid state is negligible as compared with that of the vapor state.

Where the temperature does not vary over wide limits it may be assumed that the molal latent heat of vaporization, λ is constant and Equation (2) may be integrated, between the limits p_0, T_0, and p, T, to give,

$$\ln \frac{p}{p_0} = \frac{\lambda}{R}\left(\frac{1}{T_0} - \frac{1}{T}\right) \tag{3}$$

or

$$\log \frac{p}{p_0} = \frac{\lambda}{2.303R}\left(\frac{1}{T_0} - \frac{1}{T}\right) \tag{4}$$

Equation (4) permits calculation of the vapor pressure of a substance at a temperature T if the vapor pressure p_0 at another temperature T_0 is known, together with the latent heat of vaporization λ. The results are accurate only over limited ranges of temperature in which it may be assumed that the latent heat of vaporization is constant and at such conditions that the ideal gas law is obeyed.

Illustration 1. The vapor pressure of ethyl ether is given in the International Critical Tables as 185 mm of Hg at 0°C. The latent heat of vaporization is 92.5 calories per gram at 0°C. Calculate the vapor pressure at 20°C and at 35°C.

Molecular weight................. 74.
λ.............................. 6850 calories per gram-mole.
R.............................. 1.99 calories per gram-mole per °K.
T_0.............................. 273°K.
p_0.............................. 185 mm Hg.

when $\qquad\qquad\qquad\qquad T = 293°K\ (20°C)$

$$\log \frac{p}{185} = \frac{6850}{2.30 \times 1.99}\left[\frac{1}{273} - \frac{1}{293}\right] = 1495\ (0.003663 - 0.003413) = 0.374$$

$$\frac{p}{185} = 2.36 \qquad\qquad\qquad p = 437\ \text{at}\ 20°C$$

when $\qquad\qquad\qquad\qquad T = 308°K\ (35°C)$

$$\log \frac{p}{185} = 1495\ (0.003663 - 0.003247) = 0.621$$

$$\frac{p}{185} = 4.18 \qquad\qquad\qquad p = 773\ \text{mm Hg at}\ 35°C$$

The values for the vapor pressure of ether which have been experimentally observed are 442 mm of Hg at 20°C and 775.5 mm of Hg at 35°C.

In the preceding illustration the Clausius-Clapeyron equation yields results which are satisfactory for many purposes. However, Equation (3) is only an approximation which may lead to considerable error in some cases. It should be used only in the absence of experimental data.

Tables of physical data contain experimentally determined values of the vapor pressures of many substances at various temperatures. Because of the frequent requirement of accurate values of the vapor pressure of water, extensive data are presented in Table I expressed in English units.

VAPOR-PRESSURE PLOTS

From experimental data various types of plots have been devised for relating vapor pressures to temperature. Use of an ordinary uniform scale of coordinates does not result in a satisfactory plot because of the wide ranges to be covered and the curvature encountered. A single chart cannot be used over a wide temperature range without sacrifice of accuracy at the lower temperatures and the rapidly changing slope makes both interpolation and extrapolation uncertain.

A better method which has been extensively used is to plot the logarithm of the vapor pressure (log p) against the reciprocal of the absolute temperature $(1/T)$. The resulting curves, while not straight, show much less curvature than a rectangular plot and may be read accurately over wide ranges of temperature. Another method is to plot the logarithm of the pressure against temperature on a uniform scale. These scales do not reduce the curvature of the vapor-pressure lines as much as the use of the reciprocal temperature scale but are more easy to construct and read.

TABLE I

Vapor Pressure of Water

English units

Pressure of aqueous vapor over *ice* in 10^{-3} inches of Hg from $-144°$ to $32°$F

Temp °F	0.0	2.0	4.0	6.0	8.0
−140	0.00095	0.00075	0.00063		
−130	0.00276	0.00224	0.00181	0.00146	0.00118
−120	0.00728	0.00595	0.00492	0.00409	0.00339
−110	0.0190	0.0157	0.0132	0.0111	0.00906
−100	0.0463	0.0387	0.0325	0.0274	0.0228
−90	0.106	0.0902	0.0764	0.0646	0.0543
−80	0.236	0.202	0.171	0.146	0.125
−70	0.496	0.429	0.370	0.318	0.275
−60	1.02	0.882	0.764	0.663	0.575
−50	2.00	1.75	1.53	1.33	1.16
−40	3.80	3.37	2.91	2.59	2.29
−30	7.047	6.268	5.539	4.882	4.315
−20	12.64	11.26	10.06	8.902	7.906
−10	22.13	19.80	17.72	15.83	14.17
−0	37.72	33.94	30.55	27.48	24.65
0	37.72	41.85	46.42	51.46	56.93
10	62.95	69.65	76.77	84.65	93.35
20	102.8	113.1	124.4	136.6	150.0
30	164.6	180.3			

Pressure of aqueous vapor over *water* in inches of Hg from $+4°$ to $212°$F

Temp °F	0.0	2.0	4.0	6.0	8.0
0			0.05402	0.05929	0.06480
10	0.07091	0.07760	0.08461	0.09228	0.1007
20	0.1097	0.1193	0.1299	0.1411	0.1532
30	0.1664	0.1803	0.1955	0.2118	0.2292
40	0.2478	0.2677	0.2891	0.3120	0.3364
50	0.3626	0.3906	0.4203	0.4520	0.4858
60	0.5218	0.5601	0.6009	0.6442	0.6903
70	0.7392	0.7912	0.8462	0.9046	0.9666
80	1.0321	1.1016	1.1750	1.2527	1.3347
90	1.4215	1.5131	1.6097	1.7117	1.8192
100	1.9325	2.0519	2.1775	2.3099	2.4491
110	2.5955	2.7494	2.9111	3.0806	3.2589
120	3.4458	3.6420	3.8475	4.0629	4.2887
130	4.5251	4.7725	5.0314	5.3022	5.5852
140	5.8812	6.1903	6.5132	6.850	7.202
150	7.569	7.952	8.351	8.767	9.200
160	9.652	10.122	10.611	11.120	11.649
170	12.199	12.772	13.366	13.983	14.625
180	15.291	15.982	16.699	17.443	18.214
190	19.014	19.843	20.703	21.593	22.515
200	23.467	24.455	25.475	26.531	27.625
210	28.755	29.922			

TABLE I — (*Continued*)

Pressure of aqueous vapor over *water* in lb/sq in. for temperatures 210–705.4°F

Temp °F	0.0	2.0	4.0	6.0	8.0
210	14.123	14.696	15.289	15.901	16.533
220	17.186	17.861	18.557	19.275	20.016
230	20.780	21.567	22.379	23.217	24.080
240	24.969	25.884	26.827	27.798	28.797
250	29.825	30.884	31.973	33.093	34.245
260	35.429	36.646	37.897	39.182	40.502
270	41.858	43.252	44.682	46.150	47.657
280	49.203	50.790	52.418	54.088	55.800
290	57.556	59.356	61.201	63.091	65.028
300	67.013	69.046	71.127	73.259	75.442
310	77.68	79.96	82.30	84.70	87.15
320	89.66	92.22	94.84	97.52	100.26
330	103.06	105.92	108.85	111.84	114.89
340	118.01	121.20	124.45	127.77	131.17
350	134.63	138.16	141.77	145.45	149.21
360	153.04	156.95	160.93	165.00	169.15
370	173.37	177.68	182.07	186.55	191.12
380	195.77	200.50	205.33	210.25	215.26
390	220.37	225.56	230.85	236.24	241.73
400	247.31	252.9	258.8	264.7	270.6
410	276.75	282.9	289.2	295.7	302.2
420	308.83	315.5	322.3	329.4	336.6
430	343.72	351.1	358.5	366.1	374.0
440	381.59	389.7	397.7	405.8	414.2
450	422.6	431.2	439.8	448.7	457.7
460	466.9	476.2	485.6	495.2	504.8
470	514.7	524.6	534.7	544.9	555.4
480	566.1	576.9	587.8	589.9	610.1
490	621.4	632.9	644.6	656.6	668.7
500	680.8	693.2	705.8	718.6	731.4
510	744.3	757.6	770.9	784.5	798.1
520	812.4	826.6	840.8	855.2	870.0
530	885.0	900.1	915.3	930.9	946.6
540	962.5	978.7	995.0	1011.5	1028.2
550	1045.2	1062.3	1079.6	1097.2	1115.1
560	1133.1	1151.3	1169.7	1188.5	1207.4
570	1226.5	1245.8	1265.3	1285.1	1305.3
580	1325.8	1346.4	1367.2	1388.1	1409.5
590	1431.2	1453.0	1475.0	1497.4	1520.0
600	1542.9	1566.2	1589.4	1613.2	1637.1
610	1661.2	1686.0	1710.7	1735.6	1761.0
620	1786.6	1812.3	1838.6	1865.2	1892.1
630	1919.3	1947.0	1974.5	2002.7	2031.1
640	2059.7	2088.8	2118.0	2147.7	2178.0
650	2208.2	2239.2	2270.1	2301.4	2333.3
660	2365.4	2398.1	2431.0	2464.2	2498.1
670	2531.8	2566.0	2601.0	2636.4	2672.1
680	2708.1	2745.0	2782.0	2819.1	2857.0
690	2895.1	2934.0	2973.5	3013.2	3053.2
700	3093.7	3134.9	3176.7	3206.2*	

* At 705.4°F, the critical temperature.

As a means of deriving consistent vapor-pressure data for homologous series of closely related compounds Coates and Brown developed a special method of plotting which has proved particularly valuable for the hydrocarbons.[1] For this plot rectangular coordinate paper is used with temperatures as abscissas and normal boiling points as ordinates. Curved lines of constant vapor pressure are then plotted from the experimental data available for the various members of the series. This method is particularly well adapted to extrapolating data obtained for the lower boiling homologs of a series in order to estimate vapor pressures for the higher boiling homologs.

Reference Substance Plots. The methods of plotting described above all result in lines having some degree of curvature, which makes necessary a considerable number of experimental data for the complete definition of the vapor pressure curve. Where only limited data are available there is great advantage to a method of plotting which will yield straight lines over a wide range of conditions. With such a method a complete curve can be established from only two experimental points and erratic data can be detected.

Where an accurate evaluation of a physical property has been developed over a wide range of conditions for one substance the resulting relationship frequently may be made the basis of empirical plots for other substances of not greatly different properties. This general method may be applied to vapor-pressure data by selecting a reference substance the temperature-vapor pressure relationship of which has been evaluated over a wide range. A function of the temperature at which some other substance exhibits a given vapor pressure may then be plotted against the same function of the temperature at which the reference substance has the same vapor pressure. Or, conversely, a function of the vapor pressure of the substance at a given temperature may be plotted against the same function of vapor pressure of the reference substance at the same temperature.

By proper selection of the reference substance and the functions of the properties plotted, curves which approximate straight lines over wide ranges of conditions are obtained. The best results are obtained with reference substances as similar as possible in chemical structure and physical properties to the compounds of interest.

Equal Pressure Reference Substance Plots. The first reference substance plot of vapor-pressure data was proposed by Dühring, who plotted the temperature at which the substance of interest has a given vapor pressure against the temperature at which the reference sub-

[1] Coates and Brown, " A Vapor Pressure Chart for Hydrocarbons," Dept. Engr. Research University of Michigan, Circular Series No. 2 (December, 1928).

stance has the same vapor pressure.   Dühring lines of sodium hydroxide solutions are plotted in Fig. 6, page 83, using water as the reference substance.   Each of these lines relates the temperature of the designated solution to the temperature at which water exerts the same vapor pressure.   Vapor-pressure data for water appear in Table I.

Equal Temperature Reference Substance Plots.   If Equation (2) is divided by a similar equation for a reference substance *at the same temperature* the following expression is obtained where the primed quantities indicate the reference substance:[3]

$$\frac{d \ln p}{d \ln p'} = \frac{\lambda}{\lambda'} \tag{5}$$

Integrating,[3]

$$\log p = \frac{\lambda}{\lambda'} \log p' + C \tag{6}$$

Thus, if the logarithm of the vapor pressure of a substance at a given temperature is plotted against the logarithm of the vapor pressure of a reference substance at the same temperature a line will be obtained the slope of which is equal to the ratio of the heat of vaporization of the substance of interest to that of the reference substance, provided that conditions are such that Equation (2) is applicable.   Since the ratio λ/λ' is found to be reasonably constant for most substances not close to their critical points, the lines should be straight, at least at conditions well removed from the critical point.

This method of plotting was introduced by Cox[2] and later more fully discussed by Othmer.[3]   Cox found that a wide variety of substances plotted as nearly straight lines by this method at conditions up to and including their critical points.   Since the ratio of the latent heats of vaporization of two substances varies rapidly as either one approaches its critical point, this behavior can be rationalized with Equation (6) only by assuming that the variations in the ratio of heats of vaporization are exactly compensated by deviations from the Clausius-Clapeyron equation from which Equation (6) was derived.   As was pointed out on page 60 this equation is valid only at conditions of relatively low pressures and large vapor volumes.

Figure 4 is a Cox chart from which, for simplicity in use, the logarithmic scale of pressures of the reference substance has been omitted and only the auxiliary temperature scale derived from it shows.   Such

---

[2] E. R. Cox, *Ind. Eng. Chem.* **15**, 592 (1923).   Reprinted with permission.

[3] D. F. Othmer, *Ind. Eng. Chem.* **32**, 841–856 (1940).   Reprinted with permission.

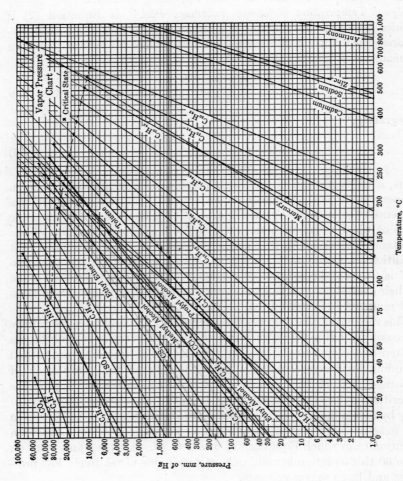

Fig. 4.   Cox vapor-pressure chart.

a chart may be constructed by plotting vapor pressures as ordinates against reference substance vapor pressures on multi-cycle double logarithmic paper. From the vapor-pressure data of the reference substance an auxiliary abscissa scale of temperatures is established. To extend the range of the chart to temperatures higher than the critical temperature of this reference substance a second higher boiling reference substance is selected and its vapor-pressure data plotted over the temperature range of the first reference substance. The vapor-pressure line of the second reference substance is then extended and from it the extension of the auxiliary temperature abscissa scale is established. Figure 4 was developed in this manner using water as the primary reference substance and mercury for temperatures above the critical of water.

The Cox method of plotting has been studied by Calingaert and Davis[4] who found that the data for widely varying types of materials yield lines with little curvature when plotted on such a chart. Furthermore, it was found that the curves of groups of closely related compounds converge at single points which are characteristic of the groups. For example, single points of convergence were found for each of the following groups: the paraffin hydrocarbons, the benzene mono-halides, the alcohols, the silicon hydride series, and the metals. For a member of a group of materials having convergent curves only one experimental point and the point of convergence of the group are necessary to establish a complete curve.

Calingaert and Davis also found that the method of Cox, when water is the reference substance, is equivalent to assuming that the vapor pressure of a substance is represented by the following equation:

$$\ln p = A - \frac{B}{T - 43} \tag{7}$$

where

$$p = \text{vapor pressure}$$
$$T = \text{temperature, } °K$$
$$A, B = \text{empirical constants}$$

Thus, by plotting log p against $1/(T - 43)$ a straight line should be obtained.

Illustration 2. The vapor pressure of chloroform is 61.0 mm of Hg at 0°C and 526 mm of Hg at 50°C. Estimate, from Fig. 4, the vapor pressure at 100°C.

Solution: The two experimental values of the vapor pressures at 0°C and 50°C are represented by points on Fig. 4. A straight line is projected through these two points to the abscissa representing 100°C. The ordinate at this point is approxi-

[4] *Ind. Eng. Chem.* **17**, 1287 (1925). Reprinted with permission.

mately 2450 mm of Hg, the estimated vapor pressure at 100°C. The experimentally observed value is 2430 mm of Hg.

Illustration 3. The vapor pressure of normal butyl alcohol at 40°C is 18.6 mm of Hg. Estimate the temperature at which the vapor pressure is 760 mm of Hg, the normal boiling point.

Solution: The experimental value of the vapor pressure at 40°C is represented by a point on Fig. 4. A straight line is drawn from this point to the point of convergence of the alcohol group. This point of convergence is located by extending the curves for methyl and propyl alcohols. The abscissa of the point at which this line crosses the 760-mm ordinate is about 117°C. The experimentally observed boiling point of normal butyl alcohol is 117.7°C.

Estimation of Critical Properties. By means of the Cox type of plot described above the vapor-pressure curve of a substance may be estimated with fair accuracy from only a few experimental points. However, to define completely the vapor-pressure characteristics of the substance, the terminus of the curve, represented by the critical point, must also be located. Once the vapor-pressure curve is established this point can be located from knowledge of only the critical temperature.

The critical temperature of a substance is also of great value in predicting its behavior in certain processes and in establishing relationships among other of its physical properties. However, the experimental determination of critical temperatures is difficult, and these data are not available for many substances. It is frequently desirable to predict the critical temperature of a substance from other more easily determined properties. Guldberg proposed a rule that the ratio of the boiling point, under a pressure of one atmosphere, to the critical temperature is a constant when both are expressed on an absolute scale of temperature. This rule is only a very rough approximation, which cannot be used where reliable results are required.

In a method proposed by Watson[5] the critical temperature of a *nonpolar* compound is predicted from its boiling point, molecular weight, and liquid density. A nonpolar compound is one having its atoms symmetrically arranged in the molecule so that there are no unbalanced electrical charges which tend to rotate the molecule when in an electrostatic field. Nonpolar compounds are, in general, chemically inactive and do not ionize or conduct electricity well. For example, the hydrocarbons of practically all series are relatively nonpolar, whereas water, alcohol, ammonia, and the like are highly polar. In general, compounds having symmetrical molecular arrangements such as methane (CH_4) or carbon tetrachloride (CCl_4) may be expected to have nonpolar characteristics. Compounds which do not have symmetrical molecular arrangements such as methyl chloride (CH_3Cl) or ethyl alcohol

[5] *Ind. Eng. Chem.* **23**, 360 (1931).

(C₂H₅OH) or acetic acid (CH₃COOH) may be expected to be polar. Polar compounds do not follow many of the generalizations which apply to the nonpolar group.

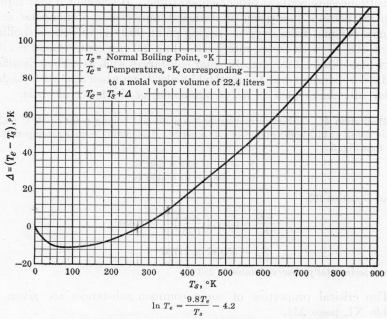

$$\ln T_e = \frac{9.8 T_e}{T_s} - 4.2$$

Fig. 5 Temperatures of constant vapor concentration.

The following empirical equation was proposed:

$$\frac{T_e}{T_c} = 0.283 \left(\frac{M}{\rho_B}\right)^{0.18} \tag{8}$$

where:

T_c = critical temperature (°K).

T_e = the temperature (°K) at which the substance is in equilibrium with its saturated vapor in a concentration of 1.0 gram-mole in 22.4 liters.

M = molecular weight.

ρ_B = density of the liquid in grams per cubic centimeter at its normal boiling point. By means of Fig. 109, Chapter XII, ρ_B can be estimated from density measurements at other temperatures.

The temperature T_e is a function of the normal boiling point of the substance. In Fig. 5, a curve is plotted relating $(T_e - T_s)$ to T_s, where

T_s is the normal boiling point in degrees Kelvin. From this curve T_e can be determined for any substance of known boiling point. The density at the boiling point may be determined or estimated from liquid density measurements at other temperatures. It was shown that Equation (8) permits prediction of critical temperatures of nonpolar substances ranging from oxygen (boiling point 90.1°K) to octane (boiling point 398°K) with errors rarely exceeding 2.0 per cent. Good results were also obtained for slightly polar substances such as carbon disulfide and chlorobenzene, but the method breaks down when applied to highly polar substances such as water, ammonia, or the lower alcohols.

Illustration 4. Carbon tetrachloride has a normal boiling point of 77°C and a liquid density at its boiling point of 1.48 g per cc. Calculate the critical temperature.

From Fig. 5

$$T_s = 77 + 273 = 350°K$$
$$T_e = 350 + 10 = 360°K$$

From Equation (8),

$$\frac{360}{T_c} = 0.283 \left(\frac{154}{1.48}\right)^{0.18} = 0.654$$

$$T_c = 550°K \text{ or } 277°C$$

The experimentally observed value is 283°C.

The critical properties of some common substances are given in Table XI, page 234.

H. P. Meissner and E. M. Redding[6] have developed more generally applicable methods for estimating all three critical constants. These methods apply to polar as well as nonpolar substances with the exception of water. For associated liquids the results are dependable for critical volume and temperature but not for critical pressure.

The *critical volume* is obtained from the equation:

$$v_c = (0.377\bar{P} + 11.0)^{1.25} \tag{9}$$

where

$$v_c = \text{critical volume, cc per gram-mole}$$
$$\bar{P} = \text{parachor}$$

The *parachor* is a measure of the molecular volume of a liquid at a standard surface tension:

$$\bar{P} = \frac{M\sigma^{0.25}}{\rho_l - \rho_g} \tag{10}$$

[6] *Ind. Eng. Chem.* **34**, 521–525 (1942). Reprinted with permission.

where M = molecular weight

                σ = surface tension, dynes per cm

                ρ_l = density of liquid, grams per cu cm

                ρ_g = density of gas, grams per cu cm

For organic liquids, the parachor can be estimated quite accurately from the structural formula by use of atomic and structural values listed in Table II. The value of the parachor of a compound is the sum of the contributions of its atomic and structural elements.

TABLE II
ATOMIC AND STRUCTURAL PARACHORS

C	4.8	Triple bond	46.6
H	17.1	Double bond	23.2
N	12.5	3-membered ring	16.7
P	37.7	4-membered ring	11.6
O	20.0	5-membered ring	8.5
S	48.2	6-membered ring	6.1
F	25.7	O_2 in esters	60.0
Cl	54.3		
Br	68.0		
I	91.0		

S. Sugden, *The Parachor and Valency*, Rutledge and Sons, London, 1930. Reprinted with permission.

For a few simple molecules, such as CO, CO_2, SO_3, the results of Table II are not accurate. For water and diphenyl Equation (9) does not apply. Otherwise, for a group of one hundred compounds selected at random the deviation never exceeded 5% from experimental values. Equation (9) is also applicable to associated liquids, the parachor being based upon the structure of the nonassociated liquid.

For the *critical temperature*, Equation (8) is preferable for nonpolar compounds. For polar substances or where the liquid density is not known the following empirical formulas have been developed by Meissner and Redding:

For compounds boiling below 235°K and for all elements.

$$T_c = 1.70T_B - 2.0 \tag{11}$$

For compounds boiling above 235°K

(a) *Containing halogens or sulfur:*

$$T_c = 1.41T_B + 66 - 11F \tag{12}$$

where   F = number of fluorine atoms in the molecule

(b) *Aromatic compounds and naphthenes free of halogens and sulfur:*

$$T_c = 1.41T_B + 66 - r(0.383T_B - 93) \tag{13}$$

where

r = ratio of noncyclic carbon atoms to the total number of carbon atoms in the compound

(c) *Other compounds* (boiling above 235°K, containing no aromatics, no naphthenes, no halogens, and no sulfur):

$$T_c = 1.027T_B + 159 \qquad (14)$$

The above equations for critical temperatures give agreement within 5% with experimental values of nearly all compounds regardless of degree of association with the exception of water. The equations have not been tried on substances normally boiling above 600°K.

The *critical pressure* in atmospheres may be predicted from the equation:

$$p_c = \frac{20.8T_c}{(v_c - 8)} = \frac{20.8T_c}{\left(\dfrac{M}{\rho_c} - 8\right)} \qquad (15)$$

With the exception of water, maximum errors of 15 per cent are encountered and the majority of the results are within 10% of the experimental values.

Meissner and Redding discuss a method of using the preceding equations when the normal boiling point is unknown but a vapor-pressure and a liquid density value are available at some other temperature.

Illustration 5. Estimate the critical properties of triethylamine $(C_2H_5)_3N$, normal boiling point 362.5°K.

$$\begin{aligned}
\text{Molecular weight} &= 101.1 \\
\text{Parachor: From Table II} &
\end{aligned}$$

$$\begin{array}{lll}
C_6 & 6 \times 4.8 = & 28.8 \\
H_{15} & 15 \times 17.1 = & 256.5 \\
N & 1 \times 12.5 = & 12.5 \\
\bar{P} & & = 297.8
\end{array}$$

From Equation (9):

$$v_c = [0.377(297.8) + 11.0]^{1.25} = 416 \text{ cc per gram-mole}$$
$$v_c \text{ (experimental)} = 403$$

Since triethylamine is neither aromatic nor naphthenic and contains no halogens or sulfur, Equation (14) may be used,

$$T_c = 1.027(362.5) + 159 = 531°K$$
$$T_c \text{ (experimental)} \qquad = 535.2$$

From Equation (15) using calculated values for T_c and v_c,

$$p_c = \frac{20.8(531)}{(416-8)} = 27.0 \text{ atm}$$

$$p_c \text{ (experimental)} = 30.0 \text{ atm}$$

Critical Pressures and Vapor Pressures of Organic Compounds. It was found by Gamson and Watson[7] that the vapor pressure data of all of over forty substances investigated may be represented by the following equation:

$$\log p = \frac{-A}{T_r} + B - e^{-20(T_r-b)^2} \tag{16}$$

where A, B, and b are constants characteristic of the substance. In applying this equation over vapor-pressure ranges from a few tenths of a millimeter of mercury to the critical point, deviations were found to be generally less than 3%.

Because of the difficulty of evaluating the constants, Equation (16) is not convenient for the extrapolation of fragmentary data. However, through generalized expressions for the constants of the equation, it may be used to predict a complete vapor-pressure curve from only a single measurement. The ranges of the constants are indicated in Table IIa.

TABLE IIa

VAPOR-PRESSURE CONSTANTS

p in mm of Hg

	A	B	b	p_c	$T_c°K$
Methane...........	2.3383	6.8800	0.000	34,810	190.7
Ethane............	2.5728	7.1411	0.088	37,090	305.5
Ethylene..........	2.5463	7.1269	0.098	38,680	282.8
Propane...........	2.6606	7.1819	0.125	33,210	370.0
n-Octane..........	3.2316	7.5034	0.236	18,700	569.4
Water.............	3.1423	8.3610	0.163	165,470	647.2
Methyl alcohol......	3.5876	8.3642	0.243	59,790	513.2
Diethyl ether.......	2.9726	7.4039	0.204	27,000	467.0
Acetone...........	3.0644	7.6173	0.180	35,720	508.2
Ammonia..........	2.9207	7.8519	0.163	85,350	406.1
Methylamine.......	2.9589	7.7066	0.239	55,940	430.1
Hydrogen cyanide...	3.2044	7.7761	0.000	37,300	456.7
Methyl chloride.....	2.7195	7.4185	0.052	50,010	416.4
Carbon tetrachloride.	2.7989	7.3329	0.158	34,200	556.3
Acetic acid.........	3.3908	8.0291	0.138	43,480	594.8

[7] B. W. Gamson and K. M. Watson, presented before Petroleum Div., Am. Chem. Soc., Pittsburgh meeting, Sept., 1943. To be published *Nat. Petr. News*, 1944.

Applying Equation (16) to the critical point,

$$\log p_c = B - A - e^{-20(1-b)^2}$$

Since the exponential correction term is found to be negligible at reduced temperatures above 0.8, $B = \log p_c + A$, and Equation (16) may be written:

$$\log p_r = \frac{-A(1 - T_r)}{T_r} - e^{-20(T_r - b)^2} \qquad (17)$$

If the methods previously described are used to calculate the critical temperatures it is possible to estimate a complete vapor pressure relationship from a single point by use of Equation (17) in conjunction with generalized expressions for the critical pressure and the constant b.

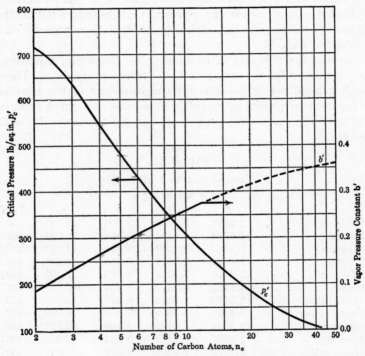

FIG. 5a.   Critical pressures and vapor-pressure constants of the paraffins.

(Reproduced in " C.P.P. Charts ")

It was found that as the number of carbon atoms in a homologous series of organic compounds is increased above 2, the critical pressure progressively diminishes while the constant b increases. These relationships are shown graphically for the paraffin hydrocarbons in Fig. 5a.

For series of compounds other than the paraffin hydrocarbons, Fig. 5a may be used with the following equations for estimating the critical pressures and constants b:

$$p_c = p_c' + \frac{\Delta p}{n_C + c} \tag{18}$$

$$b = b' + \Delta b \tag{19}$$

where p_c and b are the values for a compound containing n_C carbon atoms, p_c' and b' are the values read from Fig. 5a corresponding to n_C, and $\Delta p, c$, and Δb are constants characteristic of the homologous series, given in Table IIb.

TABLE IIb

CRITICAL AND VAPOR-PRESSURE CONSTANTS

	Δp lb/sq in.	c	Δb
Acids............................	300	0	0.05
Alcohols.........................	480	0	0.22
Aldehydes.......................	250	0(?)	0.0(?)
Amines, primary.................	69	−1.3	0.12
" secondary...............	80	0	0.12
Aromatic hydrocarbons (monocyclic)	830	−3.0	−0.02
Esters...........................	60	−1.6	0.09
Ethers...........................	0	0	0.04
Halogenated paraffins, mono........	47	−1.0	0.08
Ketones..........................	180	0	0.05
Naphthenes......................	970	0	−0.03
Nitriles..........................	0	0	0.02
Phenols..........................	1360	−3.0	0.0(?)
Olefins, mono....................	63	0	0.01

Illustration 6. The normal boiling point of n-propyl amine is 48.7°C and its critical temperature 223.8°C. Estimate the critical pressure and the vapor pressure at a temperature of 0°C.

$$n_C = 3$$

From Fig. 5a,

$$p_c' = 640 \qquad b' = 0.133$$

From Table IIb,

$$\Delta p = 69 \qquad c = -1.3 \qquad \Delta b = 0.12$$

Substituting in Equations (18) and (19),

$$p_c = 640 + \frac{69}{1.7} = 681 \text{ lb/sq in.}$$

(The experimentally observed value is 680.)

$$b = 0.133 + 0.12 = 0.253$$

At the normal boiling point,

$$T_r = \frac{322}{497} = 0.647; \quad p_r = \frac{14.7}{681} = 0.0216$$

Substituting in Equation (17),

$$-A\left(\frac{1 - 0.647}{0.647}\right) = \log 0.0216 + e^{-20(0.647-0.253)^2}$$

$$A = 2.9703$$

At 0°C,

$$T_R = \frac{273}{497} = 0.549$$

$$\log p_r = \frac{-2.9703(1 - 0.549)}{0.549} - e^{-20(0.549-0.253)^2} = -2.6139$$

$$p_r = 0.00243$$

$$p = 1.66 \text{ lb/sq in. or } 85.8 \text{ mm Hg}$$

It must be emphasized that the above generalizations are not rigorous and are not supported by experimental data beyond the first few members of any series. Where accuracy is important direct experimental measurements are desirable and generalized methods are to be used for approximations in the absence of experimental data. However, it appears that critical pressures derived by this method are somewhat more reliable than those from the more general method of Meissner previously described, and that the vapor pressure relations, particularly in the low ranges, are much better than those obtained from general reference substance plots or relations.

The most convenient method of using Equations (16) or (17) is through the derivation of curves on semi-logarithmic paper, plotting vapor pressures on the logarithmic scale and temperatures on the uniform scale.

MIXTURES OF IMMISCIBLE LIQUIDS

Two liquids which are immiscible in each other can exist together only in nonhomogeneous mixtures. In such systems, where intimate mixing is maintained, an exposed liquid surface will consist of areas of each of the component liquids. Each of these components will vaporize at the surface and tend to establish an equilibrium value of the partial pressure of its vapor above the surface. As has been pointed out the equilibrium vapor pressure of a liquid is independent of the relative proportions of liquid and vapor, but is determined by the temperature and the nature of the liquid. It follows from kinetic theory that the equilibrium vapor pressure of a liquid should be the same, whether it exists alone or as a part of a mixture, if a free surface of the pure liquid is exposed. In a nonhomogeneous mixture of immiscible liquids the vaporization and condensation of each component takes place at the respective surfaces of the pure liquids, independently of the natures or amounts of other components which may be present. Each component liquid actually exists in a pure state and as such exerts its normal equilibrium vapor pressure.

The total vapor pressure exerted by a mixture of immiscible liquids is the sum of the vapor pressures of the individual components at the existing temperature. When the vapor pressure of such a mixture equals the existing total pressure above its surface, it will boil, giving off a mixture of the vapors of its components. Since each component of the mixture adds its own vapor pressure depending only upon the temperature, it follows that *the boiling point of a nonhomogeneous mixture must be lower than that of any one of its components alone.* This fact is made use of in the important industrial process of *steam distillation* of materials which are insoluble in water. By mixing an immiscible material with water it can be distilled at a temperature always below the boiling point of water corresponding to the existing total pressure. In this manner it is possible to distill waxes, fatty acids of high molecular weight, petroleum fractions, and the like, at relatively low temperatures and with less danger of decomposition than by other methods of distillation.

The composition of the vapors in equilibrium with or rising from a mixture of immiscible liquids is determined by the vapor pressures of the liquids. The partial pressure of each component in the vapor is equal to its vapor pressure in the liquid state. The ratio of the partial pressure to the total pressure gives the mole fraction or percentage by volume, from which the weight percentage of the component in the vapor may be calculated. The total vapor pressure exerted by a mixture of immiscible liquids is easily calculated as the sum of the vapor pressures of the component liquids. Conversely, the boiling point of the mixture under a specified total pressure is the temperature at which the sum of the individual vapor pressures equals the total pressure. This temperature is best determined by trial or by a graphical method in which a plot of total vapor pressure against temperature is prepared.

Illustration 7. It is proposed to purify benzene from small amounts of nonvolatile solutes by subjecting it to distillation with saturated steam under atmospheric pressure of 745 mm of Hg. Calculate (a) the temperature at which the distillation will proceed and (b) the weight of steam accompanying 1 lb of benzene vapor.

Solution: This problem may be solved by trial, using the data of Fig. 4.

Temp.	v.p. C_6H_6	v.p. H_2O	Total v.p.
60°C	390 mm	150 mm	540 mm
70°C	550 mm	235 mm	785 mm
65°C	460 mm	190 mm	650 mm
68°C	510 mm	215 mm	725 mm
69°C	520 mm	225 mm	745 mm

The boiling point of the mixture will, therefore, be 69°C. This result could be obtained graphically by plotting a curve relating temperature to total vapor pressure.

Basis: 1 lb-mole of mixed vapor.

Benzene $= \dfrac{520}{745} = 0.70$ lb-mole or $0.70 \times 78 = \ldots\ldots\ldots\ldots\ldots$ 55 lb

Water $= 0.30$ lb-mole or$\ldots\ldots\ldots\ldots\ldots\ldots\ldots\ldots\ldots\ldots\ldots\ldots$ 5.4 lb

Steam per pound of benzene $= \dfrac{5.4}{55} = \ldots\ldots\ldots\ldots\ldots\ldots\ldots$ 0.099 lb

In the preceding illustration it has been assumed that the steam and the liquid being distilled leave the still in the proportions determined by their vapor pressures. This will be the case only when the liquids in the still are intimately mixed and when the steam which is introduced comes into intimate contact and equilibrium with the liquids. If these conditions are not realized the proportion of steam in the vapors will be higher than that corresponding to the theoretical equilibrium.

Illustration 8. It is desired to purify myristic acid ($C_{13}H_{27}COOH$) by distillation with steam under atmospheric pressure of 740 millimeters. Calculate the temperature at which the distillation will proceed and the number of pounds of steam accompanying each pound of acid distilled.

Vapor pressure of myristic acid at 99°C = 0.032 millimeter of mercury

The vapor pressure of myristic acid is negligible in its effect on the boiling point of the mixture, which may be assumed to be that of water at 740 millimeters of mercury, or 99°C.

Basis: 1 lb-mole of mixed vapors.

Myristic acid $= \dfrac{0.032}{740} = 4.3 \times 10^{-5}$ lb-mole or 4.3×10^{-5}

$\times 228 = \ldots\ldots\ldots\ldots\ldots\ldots\ldots\ldots\ldots\ldots\ldots\ldots\ldots\ldots\ldots\ldots$ 0.0098 lb

Water $= 1.0$ lb-mole $= \ldots\ldots\ldots\ldots\ldots\ldots\ldots\ldots\ldots\ldots\ldots$ 18 lb

Steam per pound of acid $= \dfrac{18}{0.0098} = \ldots\ldots\ldots\ldots\ldots\ldots\ldots$ 1840 lb

Vaporization with Superheated Steam. The preceding illustration deals with an organic compound having a high boiling point which cannot be subjected to ordinary direct distillation at atmospheric pressure. By distillation with saturated steam the boiling point of the mixture is reduced below 100°C, but as indicated by the results of the illustration, an enormous amount of steam must be used in order to obtain a small amount of product. An alternative method would be to conduct a direct distillation under a sufficiently reduced pressure to lower the boiling point to the desired temperature. However, the maintenance of high vacua in apparatus suitable for the vaporization of such materials is difficult and frequently impracticable.

These difficulties may be circumvented by maintaining the material to be vaporized at the highest permissible temperature and introducing superheated steam or some other inert gas. In this case there will be no liquid water in the system, and the superheated steam merely serves as a carrier which mixes with and removes the vapors of the material to be distilled. If the material being vaporized is allowed to reach equilibrium with its vapor, the partial pressure of the distillate vapor will be its equilibrium vapor pressure at the existing temperature. The partial pressure of the steam will be the difference between the existing total pressure and the partial pressure of the distillate vapor. The amount of steam required per unit quantity of distillate may, therefore, be diminished either by raising the temperature or lowering the total pressure. Distillation with superheated steam is frequently combined with reduced pressure in order to reduce the steam requirements for the distillation of high-boiling-point materials which will not withstand high temperatures.

Ordinarily the mixing of the steam with the material being vaporized will not be sufficiently intimate to result in equilibrium conditions. The steam will then leave the liquid without being completely saturated with distillate vapor.

Illustration 9. Myristic acid is to be distilled at a temperature of 200°C by use of superheated steam. It may be assumed that the relative saturation of the steam with acid vapors will be 80%.

(a) Calculate the weight of steam required per pound of acid vaporized if the distillation is conducted at an atmospheric pressure of 740 mm of Hg.

(b) Calculate the weight of steam per pound of acid if a vacuum of 26 in. of Hg is maintained in the apparatus.

Vapor pressure of myristic acid at 200°C = 14.5 mm of Hg

Basis: 1 lb-mole of mixed vapors.

(a) Partial pressure of acid = $14.5 \times 0.80 = 11.6$ mm of Hg

$$\text{Quantity of acid} = \frac{11.6}{740} = 0.0157 \text{ lb-mole or } 0.0157 \times 228 = \quad 3.58 \text{ lb}$$

Quantity of water = 0.9843 lb-mole or . . . . 17.7 lb

$$\text{Steam per pound of acid} = \frac{17.7}{3.58} = \text{. . . . .} \quad 4.95 \text{ lb}$$

(b) Total pressure = $740 - (26 \times 25.4) = 80$ mm

$$\text{Quantity of acid} = \frac{11.6}{80} = 0.145 \text{ lb-mole or . . . .} \quad 33.1 \text{ lb}$$

Quantity of water = 0.855 lb-mole or . . . . 15.4 lb

$$\text{Steam per pound of acid} = \frac{15.4}{33.1} = \text{. . . . .} \quad 0.465 \text{ lb}$$

SOLUTIONS

The surface of a homogeneous solution contains molecules of all its components, each of which has an opportunity to enter the vapor state. However, the number of molecules of any one component per unit area of surface will be less than if that component exposed the same area of surface in the pure liquid state. For this reason the rate of vaporization of a substance will be less per unit area of surface when in solution than when present as a pure liquid. However, any molecule from a homogeneous solution which is in the vapor state may strike the surface of the solution at any point and will be absorbed by it, re-entering the liquid state. Thus, although the opportunity for vaporization of any one component is diminished by the presence of the others, the opportunity for the condensation of its vapor molecules is unaffected. For this reason, the equilibrium vapor pressure which is exerted by a component in a solution will be, in general, less than that of the pure substance.

This situation is entirely different from that of a nonhomogeneous mixture. In a nonhomogeneous mixture the rate of vaporization of either component, per unit area of total surface, is diminished because the effective surface exposed is reduced by the presence of the other component. However, condensation of a component can take place only at the restricted areas where the vapor molecules impinge upon its own molecules. Thus, both the rate of vaporization and the rate of condensation are reduced in the same proportion and the equilibrium vapor pressure of each component is unaffected by the presence of the others.

Raoult's Law. The generalization known as *Raoult's law* states that the equilibrium vapor pressure which is exerted by a component in a solution is proportional to the mole fraction of that component. Thus,

$$p_A = P_A \left(\frac{n_A}{n_A + n_B + n_C + \cdots} \right) = N_A P_A \qquad (20)$$

where

p_A = vapor pressure of component A in solution with components $B, C, \ldots$

P_A = vapor pressure of A in the pure state

$n_A, n_B, n_C \ldots$ = moles of components $A, B, C, \ldots$

N_A = mole fraction of A

From the kinetic theory of equilibrium vapor pressures it would be expected that this generalization would be correct when the following conditions exist:

1. No chemical combination or molecular association takes place in the formation of the solution.

2. The dimensions of the component molecules are approximately equal.

3. The attractive forces between like and unlike molecules are approximately equal.

4. The component molecules are nonpolar and are not adsorbed at the surface of the solution.

Few combinations of liquids would be expected to fulfill all these conditions, and it is not surprising that Raoult's law represents only a more or less rough approximation to actual conditions. Where the conditions are fulfilled, a solution will be formed from its components without thermal change, and without change in total volume. A solution which exhibits these properties is termed an *ideal* or *perfect* solution. Solutions which approximate the ideal are formed only by liquids of closely related natures such as the homologs of a series of nonpolar organic compounds. For example, paraffin hydrocarbons of not too widely separated characteristics form almost ideal solutions in each other. The behavior of the ideal solution is useful as a criterion by which to judge solutions and also as a means of approximately predicting quantitative data for solutions which would not be expected to deviate widely from ideal behavior. For the accuracy required in the majority of industrial problems, a great many solutions of chemically similar materials may be included in this class.

Equilibrium Vapor Pressure and Composition. If the validity of Raoult's law is assumed, it is necessary to have only the vapor-pressure data for the pure components in order to predict the pressure and composition of the vapor in equilibrium with a solution. The total vapor pressure of the solution will be the sum of the vapor pressures of the components, each of which may be calculated from Equation (20). The partial pressure of each component in the equilibrium vapor will be equal to its vapor pressure in the solution, thus fixing the composition of the vapor.

Illustration 10. Calculate the total pressure and the composition of the vapors in contact with a solution at 100°C containing 35% benzene (C_6H_6), 40% toluene ($C_6H_5CH_3$), and 25% ortho-xylene ($C_6H_4(CH_3)_2$) by weight.

Vapor pressures at 100°C:

Benzene = ...	1340 mm Hg
Toluene = ...	560 mm Hg
o-Xylene = ...	210 mm Hg

Basis: 100 lb of solution:

$$\text{Benzene} = 35 \text{ lb or } \frac{35}{78} = \ldots\ldots\ldots\ldots\ldots\ldots \quad 0.449 \text{ lb-mole}$$

$$\text{Toluene} = 40 \text{ lb or } \frac{40}{92} = \ldots\ldots\ldots\ldots\ldots\ldots \quad 0.435 \text{ lb-mole}$$

$$o\text{-Xylene} = 25 \text{ lb or } \frac{25}{106} = \ldots\ldots\ldots\ldots\ldots\ldots \quad 0.236 \text{ lb-mole}$$

$$\text{Total} = 100 \text{ lb or } \quad \ldots\ldots\ldots\ldots\ldots\ldots \quad 1.120 \text{ lb-mole}$$

Vapor pressures:

$$\text{Benzene} = 1340 \times \frac{0.449}{1.120} = 1340 \times 0.401 = \ldots.. \quad 536 \text{ mm Hg}$$

$$\text{Toluene} = 560 \times \frac{0.435}{1.120} = 560 \times 0.388 = \ldots\ldots. \quad 217 \text{ mm Hg}$$

$$o\text{-Xylene} = 210 \times \frac{0.236}{1.120} = 210 \times 0.211 = \ldots\ldots. \quad 44 \text{ mm Hg}$$

$$\text{Total} = \ldots\ldots\ldots\ldots\ldots\ldots\ldots\ldots\ldots\ldots \quad 797 \text{ mm Hg}$$

Molal percentage compositions:

	Liquid	Vapor
Benzene	40.1%	536/797 = 67.3%
Toluene	38.8%	217/797 = 27.2%
o-Xylene	21.1%	44/797 = 5.5%
	100.0%	100.0%

In a similar manner the vapor pressure of the solution at any other temperature might be calculated and a curve plotted relating total vapor pressure to temperature. From such a curve the boiling point of the solution at any specified pressure may be predicted. It will be noted that the composition of the vapor may differ widely from that of the solution, depending on the relative volatilities. In the special case of a solution containing a nonvolatile component the vapor will contain none of this component but its presence in the liquid will diminish the vapor pressure of the other components in the same proportion that it reduces their mole fractions.

Nonvolatile Solutes. If one component of a binary solution has a negligible vapor pressure, its presence will have no effect on the composition of the vapor in equilibrium with the solution. The vapor will consist entirely of the volatile component but its equilibrium pressure will be less than that of the pure liquid at the same temperature. Thus, a nonvolatile solute produces a *vapor-pressure lowering* or a *boiling point elevation* in its solvent. If the components possess closely related characteristics the system may approach ideal behavior. In this case the total vapor pressure will be the product of the vapor pressure and the mole fraction of the solvent.   With ionizing or associating solutes the effective mole fraction of the solute is dependent upon the degree of ionization or association.   For these reasons, the theories of ideal behavior are of little assistance in the estimation of vapor-pressure data for many solutions, particularly those in which water is the solvent.

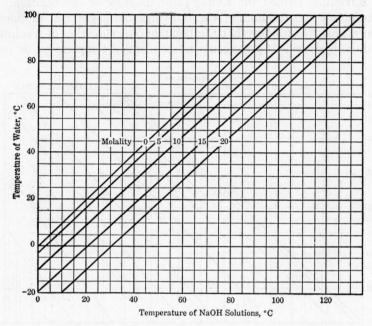

FIG. 6   Dühring lines of aqueous solutions of sodium hydroxide.

If the vapor pressure of a solution is known at two temperatures these data will establish a straight line on a reference substance chart prepared according to either the method of Cox or that of Dühring, pages 64 and 65.   In Fig. 6 are the Dühring lines corresponding to various concentrations of aqueous sodium hydroxide solutions.   Where sufficient data are available, it is advisable to plot the temperatures of the solutions against those of the pure solvent, in this case water. Using this method the curve representing zero concentration of solute will be a straight line of unit slope.   By interpolation between a set of Dühring lines the boiling point of a solution under any desired pressure or the vapor pressure at any temperature may be estimated.

Similarly, the Cox type of plot may be applied to solutions, preferably using one of the components as the reference substance.   Such a plot, developed by Othmer for sulfuric acid solutions is shown in Fig. 7.

This type of plot has the advantage over the Dühring plot of permitting estimation of thermal data from the slopes of the vapor pressure lines and Equation (6), and may or may not give closer approximation to straight line relationships, depending on the particular system involved.

The difference between the boiling point of a solution and that of the pure solvent is termed the *boiling-point elevation* of the solution. It will be noted that the lines of Fig. 6 diverge but slightly at the higher temperatures. It follows that the boiling-point elevation of a solution of sodium hydroxide is practically independent of temperature or pres-

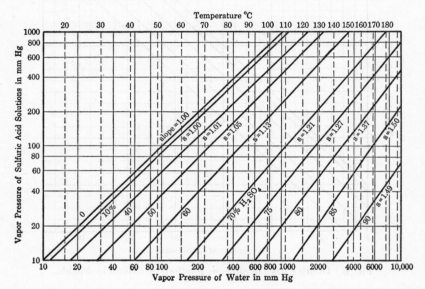

FIG. 7.   Vapor pressure of sulfuric acid solutions.   [From Othmer, *Ind. Eng. Chem.*, **32**, 847 (1940), with permission.]

sure. Although several systems exhibit this behavior it can by no means be considered general and the fact that the Dühring lines of sodium hydroxide solutions happen to be almost parallel is merely a characteristic of this system. It may be easily demonstrated that the boiling-point elevation of an ideal solution increases rapidly with an increase in temperature.

Relative Vapor Pressure. When it is desired to approximate the complete vapor-pressure data for a solution from only a single experimental observation, a modified form of Raoult's law will frequently give good results:

$$p = kp_0 \qquad\qquad (21)$$

where p = vapor pressure of solution.

              p_0 = vapor pressure of pure solvent.

              k = a factor, dependent on concentration.

For an ideal solution the factor k will equal the mole fraction of the solvent and will be independent of temperature or pressure. For non-ideal solutions k may differ widely from the mole fraction but in many cases it will be practically independent of temperature or pressure for a solution of a given composition. The factor k is sometimes termed the *relative vapor pressure* of a solution. The value of k for a solution may be obtained by a single determination of boiling point. Equation (21) may then be used to estimate the vapor pressures at other temperatures.

Illustration 11. An aqueous solution of sodium chloride contains 5 gram-moles of NaCl per 1000 g of water. The normal boiling point of this solution is 106° C. Estimate its vapor pressure at 25°C.

Mole fraction of water = $\left(1 - \dfrac{5}{55.5 + 5}\right)$ = 1 − 0.0826 = 0.9174

From Table I, p_0 at 106°C = ... 940. mm of Hg

p_0 at 25°C = ... 23.5 mm of Hg

$$k = \frac{760}{940} = 0.81$$

Vapor pressure of solution at 25°C = 23.5 × 0.81 = 19.0 mm Hg

Experimentally observed value = ... 18.97 mm Hg

In the preceding illustration it will be noted that the value of k is widely different from the mole fraction of the solvent. However, for this particular system it is, for all practical purposes, independent of temperature. In certain systems, notably aqueous solutions of strong bases, such constancy does not exist and the relative vapor pressure may vary considerably. For example, a solution of caustic soda having a molality of 10 has a relative vapor pressure, k, of 0.584 at 100°C and only 0.479 at 25°C. When the type of behavior of a system is unknown it is desirable to obtain at least two experimental points and to establish a Dühring line if it is desired to predict reliable values of vapor pressure.

PROBLEMS

1. (a) Obtaining the necessary data from a physical table, plot a curve relating the vapor pressure of acetic acid ($C_2H_4O_2$) in millimeters of Hg to temperature in degrees Centigrade. Plot the curve for the temperature range from 20° to 140°C, using vapor pressures as ordinates and temperatures as abscissas, both on uniform scales.

(b) Ethylene glycol (OHCH$_2$·CH$_2$OH) has a normal boiling point of 197°C. At a temperature of 120°C it exerts a vapor pressure of 39 mm of Hg. From these data construct a Dühring line for ethylene glycol using water as the reference substance. From this line estimate the vapor pressure at 160°C and the boiling point under a pressure of 100 mm of Hg.

(c) Ethyl bromide (C$_2$H$_5$Br) exerts a vapor pressure of 165 mm of Hg at 0°C and has a normal boiling point of 38.4°C. From Fig. 4 estimate its vapor pressure at 60°C.

(d) Nonane (C$_9$H$_{20}$) has a normal boiling point of 150.6°C. From Fig. 4 estimate its boiling point under a pressure of 100 mm of Hg.

2. For one of the substances from the following list, prepare a table of data and graphs as indicated below:

Substance	Temperature Range		Pressure Units
Ammonia	−30°C	to 24°C	atm
Carbon dioxide	−38°C	to Critical Temp.	atm
Carbon disulfide	0°C	to 46.3°C	mm Hg
Chlorine	−34.6°C	to 30°C	atm
Sulfur dioxide	0°C	to 50°C	atm
Acetone	10°C	to 60°C	mm Hg
Carbon tetrachloride	25°C	to 80°C	mm Hg
Chloroform	10°C	to 70°C	mm Hg
Ethyl alcohol	35°C	to 80°C	mm Hg
Ethyl ether	34.6°C	to 120°C	atm
Methyl chloride	−24.0°C	to 40°C	atm

(a) From a handbook of physical-chemical data, tabulate the following values. Indicate source of data.

           Column No. 1.   Temperature, °C (t)
           "   "   2.   Absolute temperature, °K (T)
           "   "   3. $1/T$
           "   "   4.   Vapor pressure (p)

(b) Using graph paper with uniform scales, plot vapor pressures as ordinates and temperature as abscissas. Draw a smooth curve through the plotted points.

(c) Using semi-log paper, plot vapor pressures as ordinates on the logarithmic scale versus $1/T$ as abscissas. Choose the scale for $1/T$ so that the curve will have a slope as close to 45° as feasible. Draw a smooth curve through the plotted points.

(d) Using semi-log paper plot vapor pressure as ordinates and temperature (°C) as abscissas.

3. Construct a Dühring chart and a Cox chart for ethyl alcohol, using water as the reference substance. Plot the curve using the following data:

Temperature	Vapor Pressure of Ethyl Alcohol
10°C	23.6 mm Hg
78.3°C	760.0 mm Hg

Using the Dühring line based on these figures, determine:

(a) The vapor pressure at 60°C.

(b) The boiling point under 500 mm pressure

4. Use the Cox chart (Fig. 4) to determine the boiling point at 2000 mm Hg of one of the following substances. Use the vapor pressure data that are given below to establish the line on the Cox chart.

Ethyl Acetate	Ethyl Formate	Sulfur
0°C 24.2 mm Hg	0°C 72.4 mm Hg	250°C 12 mm Hg
160°C 8.349 atm	200°C 28.0 atm	444.6°C 760.0 mm Hg

5. The boiling point of benzene is 80°C and its liquid density at the boiling point is 0.80 gram per cc. Estimate its critical temperature both by Equation (8) and Equations (11) to (14).

6. In the following tabulation, the normal boiling point, critical temperature, and liquid density at the normal boiling point are given for several hydrocarbons. For one of the materials assigned from this list, calculate the critical temperature, using Equation (8) and Fig. 5. A comparison of this computed value with the experimental value will give an indication of the accuracy that may be expected by this method.

Hydrocarbon	Normal Boiling Point °C	Critical Temperature °C	Liquid Density at Normal Boiling Point
n-Butane	−0.3	153	0.605 g/cc
n-Pentane	36.1	197	0.610
n-Hexane	69.0	235	0.613
n-Heptane	98.4	267	0.613
n-Octane	125.8	296	0.611
Cyclohexane	80.8	281	0.719

7. Using equations (9) to (15) calculate the critical properties of the following compounds and compare them with the tabulated experimental values:

Compound	T_B°K	T_C°K	p_c atm	V_C cc/gr mole
Acetone..............	329.0	508.2	47.0	216.5
Acetic acid..........	391.0	594.8	57.2	171.0
Ammonia.............	239.6	405.6	111.5	72.5
Chlorobenzene........	405.0	632.2	44.6	308.0
Trifluorotrichloroethane	322.0	487.0	34.0	.....
Ethyl alcohol.........	351.0	516.3	63.1	167.0

8. Using the data of Fig. 5a and Table IIb calculate the necessary data and plot vapor pressure curves extending from 0.01 lb per sq in. to the critical point for the following substances. Plot the curves on five cycle semi-logarithmic paper, using vapor pressures as ordinates on the logarithmic scale and temperatures in °F as abscissas.

  (a) Decane (t_c = 347°C)   (t_B = 174°C)
  (b) Toluene (t_c = 320.6°C)   (t_B = 110.8°C)
  (c) Propyl ethyl ether (t_c = 227.4°C)   (t_B = 64°C)

9. Using the data of Fig. 4 estimate the temperature required for the distillation of hexadecane ($C_{16}H_{34}$) at a pressure of 750 mm Hg in the presence of liquid water. Calculate the weight of steam evolved per pound of hexadecane distilled.

10. A fuel gas has the following analysis by volume (in the third column are the normal boiling points of the pure components):

Components	Percentage	Boiling Point
Ethane (C_2H_6)	4.0	−88°C
Propane (C_3H_8)	38.0	−44°C
Isobutane (C_4H_{10})	8.0	−10°C
Normal butane (C_4H_{10})	44.0	0°C
Pentanes (C_5H_{12})	6.0	+30°C (average)
	100.0	

It is proposed to liquefy this gas for sale in cylinders and tank cars.

(a) Calculate the vapor pressure of the liquid at 30°C and the composition of the vapor evolved. (The vapor pressures may be estimated from Fig. 4 and the normal boiling points.)

(b) Calculate the vapor pressure of the liquid at 30°C if all the ethane were removed.

11. Assuming that benzene (C_6H_6) and chlorobenzene (C_6H_5Cl) form ideal solutions, plot curves relating total and partial vapor pressures to mole percentages of benzene in the solution at temperatures of 90°, 100°, 110°, and 120°C. Also plot the curves relating total vapor pressure to the mole percentage of benzene in the vapor at 90 and 120°C. The normal boiling point of benzene is 79.6°C. Following are other vapor pressure data:

Vapor Pressure — mm Hg		
Temperature	Chlorobenzene	Benzene
90°C	208	1013
100	293	1340
110	403	1744
120	542	2235
132.1	760	2965

12. From the curves of Problem 11, plot the isobaric boiling-point curves of benzene-chlorobenzene solutions under a pressure of 760 mm of Hg. The points on both the liquid and vapor curves corresponding to temperatures of 90°, 100°, 110°, and 120°C should be used in establishing the curves.

13. An aqueous solution of $NaNO_3$ containing 10 gram moles of solute per 1000 g of water boils at a temperature of 108.7°C under a pressure of 760 mm of Hg. Assuming that the relative vapor pressure of the solution is independent of temperature, calculate the vapor pressure of the solution at 30°C and the boiling-point elevation produced at this pressure.

14. Obtaining the necessary data from the section on " Boiling Points of Mixtures " of Volume III of the International Critical Tables, plot the isobaric boiling-point curves for the system acetic acid (CH_3COOH) and water under a pressure of 760 mm of Hg. Both the liquid and vapor composition curves should be plotted using percentages of water by weight as abscissas and temperatures in degrees Centigrade as ordinates.

is saturated with vapor, the composition expressed by the first method is independent of both the nature of the gas and the total pressure but varies with the nature and concentration of the liquid. When the composition is expressed by the third method it varies with the nature of the liquid, the temperature, and the total pressure but is independent of the nature of the gas. At a given temperature if the equilibrium vapor pressure of the liquid, the composition of a vapor-saturated gas may be readily

CHAPTER IV

HUMIDITY AND SATURATION

When a gas or a gaseous mixture remains in contact with a liquid surface, it will acquire vapor from the liquid until the partial pressure of the vapor in the gas mixture equals the vapor pressure of the liquid at its existing temperature. When the vapor concentration reaches this equilibrium value the gas is said to be *saturated* with the vapor. It is not possible for the gas to contain a greater stable concentration of vapor, because as soon as the vapor pressure of the liquid is exceeded by the partial pressure of the vapor condensation takes place. The vapor content of a saturated gas is determined entirely by the vapor pressure of the liquid and may be predicted directly from vapor-pressure data.

The pure-component volume of the vapor in a saturated gas may be calculated from the relationships derived in Chapter II. Thus, if the ideal gas law is applicable:

$$V_v = V \frac{p_v}{p} \tag{1}$$

where

V_v = pure-component volume of vapor
p_v = partial pressure of vapor = the vapor pressure of
    the liquid at the existing temperature
V = total volume
p = total pressure

From Equation (1) the percentage composition by volume of a vapor-saturated gas may be calculated. When the ideal gas law is applicable, the composition by volume of a vapor-saturated gas is independent of the nature of the gas but is dependent on the nature and temperature of the liquid and on the total pressure. The composition by weight varies with the natures of both the gas and the liquid, the temperature, and the total pressure.

For certain types of engineering problems it is convenient to use special methods of expression for the vapor content of a gas. The weight of vapor per unit volume of vapor-gas mixture, the weight of vapor per unit weight of vapor-free gas, and the moles of vapor per mole of vapor-free gas are three common and useful methods of expression. When a gas is saturated with vapor, the composition expressed by the first method is independent of both the nature of the gas and the total pressure but varies with the nature and temperature of the liquid. When the composition is expressed by the third method it varies with the nature of the liquid, the temperature, and the pressure but is independent of the nature of the gas. From a knowledge of the equilibrium vapor pressure of the liquid the compositions of vapor-saturated gases may be readily calculated in any of these methods of expression, using the principles developed in Chapter III.

Illustration 1. Ethyl ether at a temperature of 20°C exerts a vapor pressure of 442 mm of Hg. Calculate the composition of a saturated mixture of nitrogen and ether vapor at a temperature of 20°C and a pressure of 745 mm of Hg expressed in the following terms:

(a) Percentage composition by volume.
(b) Percentage composition by weight.
(c) Pounds of vapor per cubic foot of mixture.
(d) Pounds of vapor per pound of vapor-free gas.
(e) Pound-moles of vapor per pound-mole of vapor-free gas.

(a) *Basis:* 1.0 cu ft of mixture.

Pure-component volume of vapor = $1.0 \times$

$\dfrac{442}{745} =$ .... 0.593 cu ft

Composition by volume:

Ether vapor.................................... 59.3%
Nitrogen....................................... 40.7%

(b) *Basis:* 1.0 lb-mole of the mixture.

Vapor present = 0.593 lb-mole or..................... 43.9 lb
Nitrogen present = 0.407 lb-mole or.................. 11.4 lb
    Total mixture................................. 55.3 lb

Composition by weight:

Ether vapor.................................... 79.4%
Nitrogen....................................... 20.6%

(c) *Basis:* Same as (b).

Volume = $359 \times \dfrac{760}{745} \times \dfrac{293}{273} =$ .... 393 cu ft

Weight of ether per cubic foot = $\dfrac{43.9}{393} =$ .... 0.112 lb

This result is independent of the total pressure. For example, an increase in the total pressure would decrease the volume per mole of mixture but would correspondingly decrease the weight of vapor per mole of mixture.

(d) *Basis:* Same as (b).

Weight of vapor per pound nitrogen = $\dfrac{43.9}{11.4} =$ .... 3.85 lb

(e) *Basis:* Same as (b).

$$\text{Moles of vapor per mole of nitrogen} = \frac{0.593}{0.407} = 1.455$$

Partial Saturation. If a gas contains a vapor in such proportions that its partial pressure is less than the vapor pressure of the liquid at the existing temperature, the mixture is but partially saturated. The *relative saturation* of such a mixture may be defined as the percentage ratio of the partial pressure of the vapor to the vapor pressure of the liquid at the existing temperature. The relative saturation is therefore a function of both the composition of the mixture and its temperature as well as of the nature of the vapor.

From its definition it follows that the relative saturation also represents the following ratios:

a. The ratio of the percentage of vapor by volume to the percentage by volume which would be present were the gas saturated at the existing temperature and total pressure.

b. The ratio of the weight of vapor per unit volume of mixture to the weight per unit volume present at saturation at the existing temperature and total pressure.

Another useful means for expressing the degree of saturation of a vapor-bearing gas may be termed the *percentage saturation*. The percentage saturation is defined as the percentage ratio of the existing weight of vapor per unit weight of vapor-free gas to the weight of vapor which would exist per unit weight of vapor-free gas if the mixture were saturated at the existing temperature and pressure. The percentage saturation also represents the ratio of the existing moles of vapor per mole of vapor-free gas to the moles of vapor which would be present per mole of vapor-free gas if the mixture were saturated at the existing temperature and pressure.

Care must be exercised that the *relative saturation* and the *percentage saturation* are not confused. They approach equality when the vapor concentrations approach zero but are different at all other conditions. The quantitative relationship between the two terms is readily derived from their definition. Thus,

$$\text{Relative saturation} = \frac{p_a}{p_s} \times 100 = s_r \qquad (2)$$

where

p_a = partial pressure of vapor actually present
p_s = partial pressure at saturation

$$\text{Percentage saturation} = \frac{n_a}{n_s} \times 100 = s_p \qquad (3)$$

where

$$n_a = \text{moles of vapor per mole of vapor-free gas actually present}$$
$$n_s = \text{moles of vapor per mole of vapor-free gas at saturation}$$

from Dalton's law,

$$\frac{n_a}{1} = \frac{p_a}{p - p_a} \quad \text{and} \quad \frac{n_s}{1} = \frac{p_s}{p - p_s} \tag{4}$$

or

$$\frac{n_a}{n_s} = \frac{p_a}{p_s}\left(\frac{p - p_s}{p - p_a}\right) \tag{5}$$

hence

$$s_p = s_r \left(\frac{p - p_s}{p - p_a}\right) \tag{6}$$

where

$$p = \text{total pressure}$$

Illustration 2. A mixture of acetone vapor and nitrogen contains 14.8% acetone by volume. Calculate the relative saturation and the percentage saturation of the mixture at a temperature of 20°C and a pressure of 745 mm of Hg.

The vapor pressure of acetone at 20°C is..............	184.8 mm of Hg
Partial pressure of acetone = 0.148 × 745............	110.0 mm of Hg
Relative saturation = 110/184.8....................	59.7%

Basis: 1.0 lb-mole of mixture.

Acetone...................................	0.148 lb-mole
Nitrogen...................................	0.852 lb-mole
Moles of acetone per mole of nitrogen = 0.148/0.852	0.174

Basis: 1.0 lb-mole of saturated mixture at 20°C and 745 mm of Hg.

Percentage by volume of acetone = 184.8/745.....	24.8%
Lb-moles of acetone......................	0.248
Lb-moles of nitrogen.......................	0.752
Moles of acetone per mole of nitrogen = 0.248/0.752	0.329
Percentage saturation = 0.174/0.329............	52.9%

As indicated by this illustration, the percentage saturation is always somewhat smaller than the relative saturation.

The composition of a partially saturated gas-vapor mixture is fixed if the relative or percentage saturation and the temperature and pressure are specified. From this information and a knowledge of the equilibrium vapor pressure at this temperature the composition may be expressed in any other terms. Conversely, the relative or percentage saturation may be calculated if the composition, pressure, and temperature are specified. The temperature required to produce a specified degree of saturation may be calculated if the composition at a specified pressure is known.

Illustration 3. Moist air is found to contain 8.1 grains of water vapor per cubic foot at a temperature of 30°C. Calculate the temperature to which it must be heated in order that its relative saturation shall be 15%.

Basis: 1 cu ft of moist air.

Water $= \dfrac{8.1}{7000} = 1.16 \times 10^{-3}$ lb or .... 6.42×10^{-5} lb-mole

Pure-component volume of water vapor $= 6.42 \times 10^{-5}$

$\times\ 359$ .... 0.0230 cu ft at S.C.

Partial pressure of water vapor $= 760 \times \dfrac{0.0230}{1.0} \times \dfrac{303}{273}$ .. 19.4 mm of Hg

Vapor pressure of water at temperature correspond-

ing to 15% relative saturation $= \dfrac{19.4}{0.15}$ .... 130 mm of Hg

From the vapor-pressure data for water it is found that this pressure corresponds to a temperature of 57°C.

Humidity. Because of the widespread occurrence of water vapor in gases of all kinds, special attention has been given to this case and a special terminology has been developed. The *humidity* of a gas is generally defined as the weight of water per unit weight of moisture-free gas. The *molal humidity* is the number of moles of water per mole of moisture-free gas. When the vapor under consideration is water the percentage saturation is termed the *percentage humidity*, H_p. The relative saturation becomes the *relative humidity*, H_r. The relation between these two humidities follows from Equation (6) as

$$H_p = H_r \left(\frac{p - p_s}{p - p_a} \right) \tag{7}$$

Considerable confusion exists in the literature in the use of these terms, and care must always be exercised to avoid misuse. The terminology recommended above is an extension of that proposed by Grosvenor.[1]

The Dew Point. If an unsaturated mixture of vapor and gas is cooled, the relative amounts of the components and the percentage composition by volume will at first remain unchanged. It follows that, if the total pressure is constant, the partial pressure of the vapor will be unchanged by the cooling. This will be the case until the temperature is lowered to such a value that the vapor pressure of the liquid at this temperature is equal to the existing partial pressure of the vapor in the mixture. The mixture will then be saturated, and any further cooling

[1] *Trans. Am. Inst. Chem. Eng.*, **1** (1908).

will result in condensation. The temperature at which the equilibrium vapor pressure of the liquid is equal to the existing partial pressure of the vapor is termed the *dew point* of the mixture.

The vapor content of a vapor-gas mixture may be calculated from dew-point data, or conversely, the dew point may be predicted from the composition of the mixture.

Illustration 4. A mixture of benzene vapor and air contains 10.1% benzene by volume.

(a) Calculate the dew point of the mixture when at a temperature of 25°C and a pressure of 750 mm of Hg.

(b) Calculate the dew point when the mixture is at a temperature of 30°C and a pressure of 750 mm of Hg.

(c) Calculate the dew point when the mixture is at a temperature of 30°C and a pressure of 700 mm of Hg.

Solution:

(a) Partial pressure of benzene = 0.101 × 750 = ... 75.7 mm Hg

From the vapor pressure data for benzene, Fig. 4, it is found that this pressure corresponds to a temperature of 20.0°C, the dew point.

(b) Partial pressure of benzene......................... 75.7 mm

   Dew point .... 20.0°C

(c) Partial pressure of benzene = 0.101 × 700............. 70.7 mm Hg

The temperature corresponding to a vapor pressure of 70.7 mm of Hg is found to be 18.7°C. From these results it is seen that the dew point does not depend on the temperature but does vary with the total pressure.

VAPORIZATION PROCESSES

The manufacturing operations of drying, air conditioning, and certain types of evaporation all involve the vaporization of a liquid into a stream of gases. In dealing with such operations it is of interest to calculate the relationships between the quantities and volumes of gases entering and leaving and the quantity of material evaporated. Such problems are of the general class which was discussed in Chapter II under the heading of "Volume Changes with Change in Composition." The concentrations of vapor in these problems are generally expressed in terms of the dew points, the relative saturations, or the moles of vapor per mole of vapor-free gas. The first two methods of expression are convenient because they are directly determined from dew point or wet- and dry-bulb temperature measurements. From such data the partial pressures of vapor may be readily calculated and the partial pressure method of solution might be used as described in Chapter II.

The vaporization processes all require the introduction of energy in the form of heat. The effective utilization of this heat is frequently the most important factor governing the operation of the process, and a knowledge of the relationships between the quantity of heat introduced and that dissipated in various ways is of great significance. The calculation of such an *energy balance* is greatly simplified if the quantities of all materials concerned are expressed in molal or weight units rather than in volumes. These units have the advantage of expressing quantity independent of change of temperature and pressure. The same desirability of weight or molal units arises when relationships are derived for the design of vaporization equipment. For these reasons it has become customary to express all data in either weight or molal units where thermal calculations are to be made or where design relationships are to be used. The molal units are preferable. From data expressed in molal units, volumes at any desired conditions may be readily obtained.

It will be noted that in any mixture following the ideal gas law the ratio of the number of moles of vapor to the number of moles of vapor-free gas is equal to the ratio of the partial pressure of the vapor to the partial pressure of the vapor-free gas. Accordingly, vapor concentration in moles of vapor per mole of vapor-free gas is obtained by dividing the partial pressure of the vapor by the partial pressure of the vapor-free gas.

Illustration 5. It is proposed to recover acetone, which is used as a solvent in an extraction process, by evaporation into a stream of nitrogen. The nitrogen enters the evaporator at a temperature of 30°C containing acetone such that its dew point is 10°C. It leaves at a temperature of 25°C with a dew point of 20°C. The barometric pressure is constant at 750 mm of Hg.

(a) Calculate the vapor concentrations of the gases entering and leaving the evaporator, expressed in moles of vapor per mole of vapor-free gas.

(b) Calculate the moles of acetone evaporated per mole of vapor-free gas passing through the evaporator.

(c) Calculate the weight of acetone evaporated per 1000 cu ft of gases entering the evaporator.

(d) Calculate the volume of gases leaving the evaporator per 1000 cu ft entering.

Solution: The vapor pressure of acetone is:

$$116 \text{ mm of Hg at } 10°C$$
$$185 \text{ mm of Hg at } 20°C$$

(a) Entering gases:

Partial pressure of acetone.....................	116 mm of Hg
Partial pressure of nitrogen = 750 − 116..........	634 mm of Hg
Moles of acetone per mole of nitrogen = 116/634...	0.183

Leaving gases:

Partial pressure of acetone.....................	185 mm of Hg
Partial pressure of nitrogen = 750 − 185..........	565 mm of Hg
Moles of acetone per mole of nitrogen = 185/565...	0.328

(b) *Basis:* 1.0 lb-mole of nitrogen.

Acetone leaving the process.....................	0.328 lb-mole
Acetone entering the process....................	0.183 lb-mole
Acetone evaporated............................	0.145 lb-mole

(c) *Basis:* 1.0 lb-mole of nitrogen.

Total gas entering the process = $1.0 + 0.183$....... 1.183 lb-moles

Volume of gas entering = $1.183 \times 359 \times \dfrac{760}{750} \times \dfrac{303}{273}$ 477 cu ft

Molecular weight of acetone..................... 58

Weight of acetone evaporated = 58×0.145....... 8.4 lb

Acetone evaporated per 1000 cu ft of gas entering =

$\dfrac{8.4}{477} \times 1000$................................. 17.6 lb

(d) *Basis:* 1.0 lb-mole of nitrogen.

Total gas leaving the process = $1.0 + 0.328$........ 1.328 lb-moles

Volume of gas leaving = $1.328 \times 359 \times \dfrac{760}{750} \times \dfrac{298}{273} =$ 526 cu ft

Volume of gas leaving per 1000 cu ft entering the

process = $\dfrac{526}{477} \times 1000$....................... 1102 cu ft

CONDENSATION

The relative saturation of a partially saturated mixture of vapor and gas may be increased in two ways without the introduction of additional vapor. If the temperature of the mixture is reduced, the vapor concentration corresponding to saturation is reduced, thereby increasing the relative saturation even though the existing partial pressure of the vapor is unchanged. If the total pressure is increased the existing partial pressure of the vapor is increased, again increasing the relative saturation. Thus, by sufficiently increasing the pressure or reducing the temperature of a vapor-gas mixture it is possible to cause it to become saturated, the existing partial pressure of vapor equaling the vapor pressure of the liquid at the existing temperature. Further reduction of the temperature or increase of the pressure will result in condensation, since the partial pressure of the vapor cannot exceed the vapor pressure of the liquid in a stable system.

In problems dealing with condensation processses, four interdependent factors are to be considered: the initial composition, the final temperature, the final pressure, and the quantity of condensate. It may be desired to calculate any one of these factors when the others are known or specified. Such calculations are readily carried out by selecting as a basis a definite quantity of vapor-free gas and calculating the quantities of vapor which are associated with it at the various stages of the process. In a condensation process the final conditions will be those of saturation at the final temperature and pressure. Any one of the three methods of calculation demonstrated in Chapter II under " Volume Changes with Change in Composition " may be used. However, for the reasons given in the preceding section it is generally desirable to express the vapor concentrations in moles of vapor per mole of vapor-free gas if any thermal or design calculations are to be carried out.

Illustration 6. Air at a temperature of 20°C and a pressure of 750 mm of Hg has a relative humidity of 80%.

(a) Calculate the molal humidity of the air.

(b) Calculate the molal humidity of this air if its temperature is reduced to 10°C and its pressure increased to 35 lb per sq in., condensing out some of the water.

(c) Calculate the weight of water condensed from 1000 cu ft of the original wet air in cooling and compressing to the conditions of part (b).

(d) Calculate the final volume of the wet air of part (c).

Vapor pressure of water:

$$17.5 \text{ mm of Hg at } 20°C$$
$$9.2 \text{ mm of Hg at } 10°C$$

Solution:

(a) Initial partial pressure of water $= 0.80 \times 17.5$ . . . . .   14.0 mm of **Hg**

Initial molal humidity $= \dfrac{14.0}{750 - 14.0}$ 0.0190

(b) Final partial pressure of water . . . .   9.2 mm of **Hg**

Final total pressure $= 35 \times \dfrac{760}{14.7}$ . . . .   1810 mm of **Hg**

Final molal humidity $= \dfrac{9.2}{1810 - 9.2}$ 0.0051

Basis: 1000 cu ft of original wet air.

(c) Partial pressure of dry air $= 750 - 14$ . . . .   736 mm of **Hg**

Partial volume of dry air at S.C. $= 1000 \times \dfrac{736}{760} \times \dfrac{273}{293} = 903$ cu ft

Moles of dry air $= 903/359$ . . . .   2.52 lb-moles

Water originally present $= 2.52 \times 0.0190$ . . . .   0.0478 lb-mole

Water finally present $= 2.52 \times 0.0051$ 0.0128

Water condensed $= 0.0350$ lb-mole or . . . .   0.630 lb

(d) Total wet air finally present $= 2.52 + 0.0128$ . . . . . .   2.53 lb-moles

Final volume of wet air $= 2.53 \times 359 \times \dfrac{760}{1810} \times \dfrac{283}{273}$ 396 cu ft

Illustration 7. A mixture of dry flue gases and acetone at a pressure of 750 mm of Hg and a temperature of 30°C has a dew point of 25°C. It is proposed to condense 90% of the acetone in this mixture by cooling to 5°C and compressing. Calculate the necessary pressure in pounds per square inch.

Vapor pressure of acetone:

$$\text{at } 30°C = 282.7 \text{ mm of Hg}$$
$$\text{at } 25°C = 229.2 \text{ mm of Hg}$$
$$\text{at } 5°C = 89.1 \text{ mm of Hg}$$

Solution:

Basis: 1.0 lb-mole of original mixture.

Partial pressure of acetone.........................	229.2 mm of Hg
Acetone present $= 229.2/750$	0.306 lb-mole
Flue gases present................................	0.694 lb-mole
Acetone present in final mixture $= 0.10 \times 0.306$......	0.0306 lb-mole
Final mixture of gas $= 0.694 + 0.0306$..............	0.725 lb-mole
Partial pressure of acetone in final mixture...........	89.1 mm
(the vapor pressure at 5°C)	
Mole percentage of acetone in final mixture $=$	
$0.0306/0.725$..................................	4.22%
Final pressure $= \dfrac{89.1}{0.0422} = 2110$ mm of Hg or........	40.8 lb per sq in.

WET- AND DRY-BULB THERMOMETRY

When a liquid evaporates into a large volume of gas at the same temperature, if no heat is supplied from an external source the liquid will be cooled as a result of the heat removed from it to supply the relatively large demand of the latent heat of vaporization. If the body of liquid has a large surface area in proportion to its mass, its temperature will quickly drop to an equilibrium value. This equilibrium temperature which is assumed by the evaporating liquid will be such that heat will be transferred to it from the warmer gas at a rate which is just adequate to supply the necessary latent heat for the vaporization. The equilibrium temperature of a liquid when vaporizing into a gas is termed the *wet-bulb* temperature and is always less than the actual *dry-bulb* temperature of the gas into which evaporation is taking place. If the gas is saturated there is neither vaporization nor depression of the wet-bulb temperature. It is therefore possible to use the depression of the wet-bulb temperature as a measure of the degree of unsaturation of the mixture of gas and vapor.

Although, by the evaluation of the required equations or charts, it is possible to extend wet- and dry-bulb thermometry to the determination of concentrations of different types of vapors, the only systems for which the method has been extensively used are those involving water vapor. This special application of wet- and dry-bulb thermometry is termed *hygrometry* or *psychrometry*. The United States Weather Bureau has worked out extensive psychrometric tables and charts from which humidities of air may be obtained with accuracy. For engineering purposes *humidity charts* have been developed which permit the determination of the humidities of any of the common gases from wet- and dry-bulb thermometric data. The limitation of these charts is that *each chart is applicable only to systems which exist under a single, specified total pressure.* For other total pressures it is necessary to develop other charts. The humidity charts are useful for rapid solutions of many of the problems of vaporization, condensation, and air-conditioning processes which occur at substantially constant atmospheric pressure.

Charts exactly similar to the humidity charts may be calculated for other systems of liquids and gases from existing data. The construction of such charts is justified only where a considerable amount of attention is to be devoted to a single system.

The Humidity Chart. In Fig. 8 is plotted a molal humidity chart covering a range of molal humidities from 0.00 to 0.34, and a range of dry-bulb temperatures from 80° to 210°F. A chart of this type was described in the literature by Hatta.[2] A lower range on a semilog scale is shown in Fig. 9. Values of molal humidities are plotted as ordinates and dry-bulb temperatures as abscissas. The chart is based on a total pressure of 1.0 atmosphere. Included on the same chart is curve C showing the relationship between the molal heat of vaporization of water as ordinates and temperature as abscissas.

Curve A of Fig. 8, termed the *saturation curve*, is a plot of molal humidity against temperature for any gas which is saturated with water vapor. Curves A' express similar relationships between molal humidities and temperatures corresponding to the specified values of *percentage* humidity. These curves are all independent of the nature of the gas under consideration.

Curves B are lines of constant wet-bulb temperature when the gas is composed of only diatomic components such as oxygen, nitrogen, carbon monoxide, hydrogen, air, and the like. One of these curves indicates the wet-bulb temperature of any mixture whose dry-bulb temperature and molal humidity determine a point falling on this curve. By interpolation between these curves it is possible to estimate any one of the properties of molal humidity, wet-bulb temperature, and dry-bulb temperature if the other two are known. It will be noted that where the wet-bulb temperature lines cross the saturation curve A the wet-bulb temperature must be equal to the dry-bulb temperature.

At wet-bulb temperatures of 100° and 150°F are groups of curves which indicate the effect on the wet-bulb temperature lines of the presence of triatomic gases such as carbon dioxide. The symbol x indicates the mole fraction of triatomic gases in the dry gas mixture. All wet-bulb lines on the chart for which no value of x is designated correspond to a composition of $x = 0$. The wet-bulb temperature lines which correspond to various mixtures of gases must converge at the saturation curve where the wet-bulb temperature is equal to the dry-bulb, regardless of

[2] *Chem. and Met. Eng.*, **37**, 137 (1930).

the dry gas composition. These lines permit application of the chart to mixtures of all the common gases.

A line on the chart which is parallel to the temperature axis represents a change of temperature without change in molal humidity. This fact

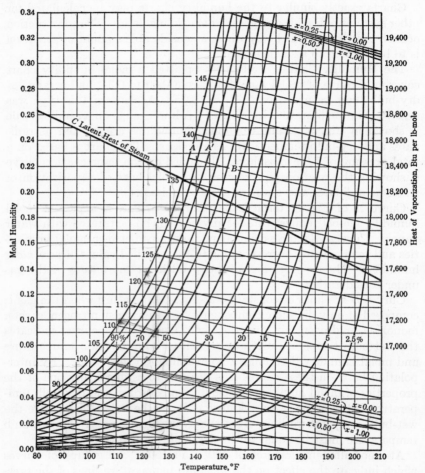

Fig. 8. Molal humidity chart (high temperature range). To obtain pounds of water per pound of dry air multiply molal humidity by 0.62.

(Reproduced in " C.P.P. Charts ")

may be used for estimation of the dew point of a mixture whose properties are represented by an established point on the chart. A line is projected through this point, parallel to the temperature axis, to the saturation curve A. The abscissa of the intersection of this line with the saturation curve is the dew point of the mixture.

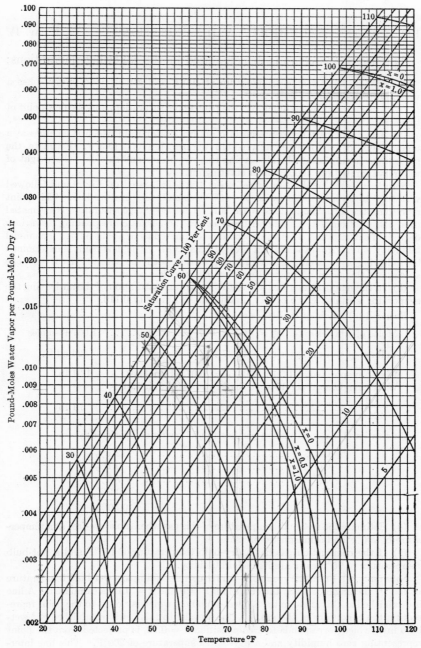

Pressure = 1 atmosphere = 29.92 in. of mercury
x = mole fraction of CO_2 in dry gas

Fig. 9. Molal humidity chart (low temperature range). To use either Fig. 8 or
9 in obtaining the pounds of water per pound of dry air multiply molal humidity,
ordinates of the charts, by 18/29 or 0.62.

(Reproduced in " C.P.P. Charts ")

101

The use of the chart is demonstrated in the following illustrations:

Illustration 8. Air at a temperature of 100°F and atmospheric pressure has a wet-bulb temperature of 85°F.

(a) Estimate the molal humidity, the percentage saturation, and the dew point of this air.

(b) The air of part (a) is passed into an evaporator from which it emerges at a temperature of 120°F with a wet-bulb temperature of 115.3°F. Estimate the percentage saturation of the air leaving the evaporator and calculate the weight of water evaporated per 1000 cu ft of entering air.

Solution: (a) The abscissa representing 100°F is located on Fig. 9 and followed vertically to its intersection with the 85°F wet-bulb temperature line. This point represents the initial conditions of the air. The percentage saturation is estimated from the position of this point with respect to the curves corresponding to various degrees of saturation. This value would be estimated to be about 52%. The molal humidity is read from the scale of ordinates as 0.037. The dew point is determined by following a horizontal line to its intersection with the saturation curve. The abscissa of this point of intersection is 80.5°F, the dew point.

(b) In the manner described above the percentage saturation of the air leaving the evaporator is estimated to be 84% and its molal humidity 0.110.

Basis: 1.0 lb-mole of moisture-free air.

Moles of wet air entering = 1.0 + 0.037 .... 1.037 lb-moles

Volume of wet air entering = $1.037 \times 359 \times \dfrac{560}{492}$ .... 424 cu ft

Water evaporated = 0.110 − 0.037 = 0.073 lb-mole or .. 1.31 lb
Water evaporated per 1000 cu ft of entering wet air 1.31 ×
$\dfrac{1000}{424}$ .... 3.1 lb

Illustration 9. A combustion gas has the following composition:

CO_2............................. 12.1%
CO............................. 0.1%
O_2............................. 7.6%
N_2............................. 80.2%
_____
100.0%

(a) Estimate the wet-bulb temperature of this gas when moisture-free at a temperature of 200°F and atmospheric pressure.

(b) If the combustion gas has a dry-bulb temperature of 140°F and a wet-bulb temperature of 95°F estimate its molal humidity.

Solution: (a) From the group of x-curves corresponding to a wet-bulb temperature of 100°F, the angle between the curves for $x = 0$ and $x = 0.12$ is estimated. A line is established at this angle to the 85° wet-bulb line, which is closest to the point representing the conditions of the gas. A line parallel to this newly established 85° wet-bulb line, which corresponds to a composition $x = 0.12$, is projected from the point representing zero humidity and a dry-bulb temperature of 200°F. This line intersects the saturation curve at a temperature of 87°F, which is the wet-bulb temperature. These projections may be carried out with the aid of two draftsman's triangles.

(b) In the manner described above, the angle between the curve $x = 0$ and $x = 0.12$ at a wet-bulb temperature of 100° is estimated. A line is established at this angle to the 95° wet-bulb line, determining the 95° wet-bulb line corresponding to $x = 0.12$. The intersection of this line with the 140° dry-bulb temperature line establishes the point representing the conditions of the mixture. The ordinate of this point is 0.040, the molal humidity of the mixture.

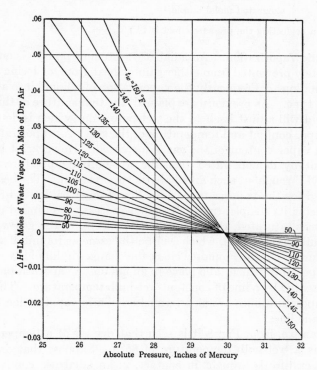

FIG. 9a.　Pressure correction to humidity charts.
(Reproduced in " C.P.P. Charts ")

If many calculations are to be made for a gas having an approximately constant value of x, it will be convenient to rule wet-bulb lines corresponding to this composition directly onto the chart.

Humidity charts, Figs. 8 and 9, are strictly applicable only to a barometric pressure of 29.92 in. Hg. Where wet-bulb temperature readings are taken at pressures other than atmospheric a correction should be applied to the chart readings to obtain the true humidity. In Fig. 9a the humidity correction to be added to the readings on the atmospheric pressure charts is plotted against total pressure for different wet-bulb readings. It will be observed that the correction increases with departure from atmospheric pressure and with increase in the wet-bulb temperature.

Illustration 9a. At 26.42 in. Hg the dry-bulb temperature of air is 150°F and its wet-bulb temperature is 120°F. Obtain the correct humidity from the charts.

From Fig. 8, molal humidity	= 0.116
From Fig. 9a, molal humidity correction =	.020
Corrected molal humidity	.136

The error in neglecting the pressure effect is 14.7%.

Adiabatic Vaporization. An adiabatic system or process is one which is completely prevented from either gaining heat from, or losing heat to, its surroundings. Frequently the vaporization of a liquid into a gas will be of this type. As previously explained, the temperature of the vaporizing liquid will adjust itself to the wet-bulb value which is determined by the vapor content and temperature of the gas. As vaporization proceeds the vapor content of the gas is increased. However, it has been experimentally demonstrated that the wet-bulb temperature remains unchanged throughout such an adiabatic vaporization. The heat which is required for the vaporization of the liquid must be derived from the gas, reducing its temperature. If such a process is continued until the gas becomes saturated, the dry-bulb temperature of the gas will equal its wet-bulb temperature, which will be the same as its initial wet-bulb temperature. On the humidity chart the changes in humidity and dry-bulb temperature which take place in an adiabatic vaporization process are represented by a line of constant wet-bulb temperature. These wet-bulb temperature lines are also termed adiabatic vaporization or *adiabatic cooling* lines.

For example, from Fig. 8 it is seen that dry air at a temperature of 197°F has a wet-bulb temperature of 85°F. Suppose that dry air at this temperature is brought in contact, in an adiabatic compartment, with water at a temperature of 85°F. As vaporization takes place the molal humidity of the air will increase, but if its wet-bulb temperature is to remain constant the dry-bulb temperature must correspondingly decrease along the 85°F wet-bulb temperature line. Thus, if vaporization continues until the molal humidity of the air becomes 0.020 the dry-bulb temperature of the air will be reduced to 144°F. If the vaporization is continued until the air is saturated, the molal humidity will then be 0.042 and the dry-bulb temperature will be 85°F. This molal humidity of 0.042 represents the maximum quantity of water which can be adiabatically evaporated into dry air at an initial temperature of 197°F.

Many types of industrial equipment for drying and evaporation are practically adiabatic in operation, the heat for the vaporization being almost entirely abstracted from the hot gases which enter the process. The humidity chart permits rapid calculation of the quantities of water which can be evaporated in such processes and of the temperatures and humidities throughout.

Illustration 10. Air enters a dryer at atmospheric pressure, a dry-bulb temperature of 190°F, and a wet-bulb temperature of 90°F. It is found that 0.028 lb-mole of water is evaporated in the dryer per pound-mole of dry air entering it. Assuming that the vaporization is adiabatic, estimate the dry-bulb temperature, the wet-bulb temperature, and the percentage saturation of the air leaving the dryer.

Solution: On Fig. 8 it is found that a dry-bulb temperature of 190° and a wet-bulb temperature of 90° correspond to a molal humidity of 0.011.

Molal humidity of air leaving = 0.011 + 0.028 = 0.039.

If the process is adiabatic the wet-bulb temperature will remain constant at 90°F.

A wet-bulb temperature of 90° and a molal humidity of 0.039 correspond to a dry-bulb temperature of 116°F, the temperature of the air leaving the dryer. The percentage saturation is estimated as 35%.

Illustration 11. Carbon dioxide is saturated with water vapor by passing it through a wetted chamber. The gas enters the chamber dry, at atmospheric pressure, and at a temperature of 120°F. It may be assumed that the vaporization in the saturator is adiabatic. Estimate the temperature and molal humidity of the saturated carbon dioxide leaving the chamber.

Solution: For pure carbon dioxide, wet-bulb curves corresponding to $x = 1.0$ must be used. In the group of 100° wet-bulb curves the angle is determined between the curves $x = 0$ and $x = 1.0$. A line is established at this angle to the 65° wet-bulb line, thus determining the 65° wet-bulb line for $x = 1.0$. A line parallel to this is projected from the point representing a dry-bulb temperature of 120° and zero humidity to the saturation curve. The intersection is at a temperature of 71°F, the wet-bulb temperature. The molal humidity at this point is 0.027.

PROBLEMS

Problems 1–19 inclusive should be solved without use of the humidity charts.

1. (*a*) Calculate the composition, by volume and by weight, of air which is saturated with water vapor at a pressure of 750 mm of Hg and a temperature of 70°F.

(*b*) Calculate the composition by volume and by weight of carbon dioxide which is saturated with water vapor at the conditions of part *a*. (The necessary data may be obtained from Table I.)

2. Nitrogen is saturated with benzene vapor at a temperature of 25°C and a pressure of 750 mm of Hg. Calculate the composition of the mixture, expressed in the following terms:

(*a*) Percentage by volume.

(*b*) Percentage by weight.

(*c*) Grains of benzene per cubic foot of mixture.

(*d*) Pounds of benzene per pound of nitrogen.

(*e*) Pound-moles of benzene per pound-mole of nitrogen.

3. Carbon dioxide contains 0.053 pound-mole of water vapor per pound-mole of dry CO_2 at a temperature of 35°C and a total pressure of 750 mm of Hg.

(*a*) Calculate the relative saturation of the mixture.

(*b*) Calculate the percentage saturation of the mixture.

(*c*) Calculate the temperature to which the mixture must be heated in order that the relative saturation shall be 30%.

4. A mixture of benzene and air at a temperature of 24°C and a pressure of 745 mm of Hg is found to have a dew point of 11°C.

(*a*) Calculate the percentage by volume of benzene.

(*b*) Calculate the moles of benzene per mole of air.

(*c*) Calculate the weight of benzene per unit weight of air.

5. An industrial heating gas is metered at a temperature of 69°F and a pressure of 752 mm of Hg. The relative humidity in the meter is found to be 47%.

The gas has a heating value of 500 Btu per cu ft measured at 60°F under a pressure of 30 in. of Hg and saturated with water vapor. Calculate the heating value per cubic foot measured in the above meter.

6. The heating value of an illuminating gas is best determined by means of the continuous-flow calorimeter. In such a determination a gas sample is burned which has a volume of 0.2 cu ft. This sample is measured, when saturated with water vapor, at a total pressure of 29.42 in. of Hg and a temperature of 78°F. What volume would be occupied by the same quantity of moisture-free gas as contained in this sample, were it at the gas-testers' standard conditions of 60°F, a pressure of 30.0 in. of Hg, and saturated with water vapor?

7. It is desired to construct a dryer for removing 100 lb of water per hour. Air is supplied to the drying chamber at a temperature of 66°C, a pressure of 760 mm of Hg, and a dew point of 4.5°C. If the air leaves the dryer at a temperature of 35°C, a pressure of 755 mm of Hg, and a dew-point of 24°C, calculate the volume of air, at the initial conditions, which must be supplied per hour.

8. Air, at a temperature of 60°C, a pressure of 745 mm of Hg, and a percentage humidity of 10%, is supplied to a dryer at a rate of 50,000 cu ft per hour. Water is evaporated in the dryer at a rate of 60 lb per hour. The air leaves at a temperature of 35°C, and a pressure of 742 mm of Hg.

(*a*) Calculate the percentage humidity of the air leaving the dryer.

(*b*) Calculate the volume of wet air leaving the dryer per hour.

9. For the best hygienic conditions, air in a living room should be at a temperature of 70°F and a relative humidity of 62%. It is also desirable that fresh air be admitted at such a rate that the air is completely renewed twice each hour. This requires that air be introduced at a volume rate, measured at the conditions of the room, equal to twice the volume of the room.

(*a*) Calculate the dew point of the air in the room.

(*b*) If air is taken in from the outside at a temperature of 10°F and saturated, calculate the weight of water which must be evaporated, per hour, in order to maintain the above specified conditions in a room having a volume of 3000 cu ft. (Vapor pressure of water at 10°F, over ice = 1.6 mm of Hg; barometric pressure = 743 mm of Hg.)

10. Illuminating gas at a temperature of 90°F and a pressure of 760 mm of Hg enters a gas holder carrying 14 grains of water vapor per cubic foot. If, in the holder, the gas is cooled to 35°F, calculate the weight of water condensed per 1000 cu ft of gas entering the holder. The pressure in the holder remains constant.

11. A gas mixture at a temperature of 27°C and a pressure of 750 mm of Hg contains carbon disulfide vapor such that the percentage saturation is 70%. Calculate the temperature to which the gas must be cooled, at constant pressure, in order to condense 40% of the CS_2 present.

12. A compressed-air tank having a volume of 15 cu ft is filled with air at a gauge pressure of 150 lb per sq in. and a temperature of 85°F and saturated with water vapor. The tank is filled by compressing atmospheric air at a pressure of 14.5 lb per sq in., a temperature of 75°F, and a percentage humidity of 60%. Calculate the amount of water vapor condensed in compressing enough air to fill the tank, assuming that it originally contained air at atmospheric conditions.

13. Air at a temperature of 30°C and a pressure of 750 mm of Hg has a percentage humidity of 60. Calculate the pressure to which this air must be compressed, at constant temperature, in order to remove 90% of the water present.

14. In a process in which benzene is used as a solvent it is evaporated into dry nitrogen. The resulting mixture at a temperature of 24°C and a pressure of 14.7 lb per sq in. has a percentage saturation of 60. It is desired to condense 80% of the benzene present by a cooling and compressing process. If the temperature is reduced to 10°C to what pressure must the gas be compressed?

15. Acetone is used as a solvent in a certain process. Recovery of the acetone is accomplished by evaporation in a stream of nitrogen, followed by cooling and compression of the gas-vapor mixture. In the solvent recovery unit, 50 lb of acetone are to be removed per hour. The nitrogen is admitted at a temperature of 100°F and 750 mm of mercury pressure, and the partial pressure of the acetone in the incoming nitrogen is 10.0 mm of mercury. The nitrogen leaves at 85°F, 740 mm of mercury, and a percentage saturation of 85%.

(a) How many cubic feet of the incoming gas-vapor mixture must be admitted per hour to obtain the required rate of evaporation of the acetone?

(b) How many cubic feet of the gas-vapor mixture leave the solvent recovery unit per hour?

16. The gas-vapor mixture in Problem 15 leaves the evaporator at 85°F, 740 mm of mercury, and a percentage saturation of 85%. The mixture is compressed and cooled to 32°F, after which the condensate is removed. The nitrogen, with the residual acetone vapor, is expanded to 750 mm of mercury pressure, heated to 100°F, and re-used in the solvent recovery system. The partial pressure of the acetone in the gas admitted to the solvent recovery system is 10 mm of mercury.

(a) To what pressure must the mixture be compressed at 32°F?

(b) On the basis of 1 cu ft of the gas-vapor mixture emerging from the solvent recovery unit, calculate:

(1) The volume of the gas-vapor mixture at 32°F after compression.

(2) The volume of the gas-vapor mixture entering the solvent recovery unit.

(3) The acetone condensed after compressing and cooling to 32°F.

17. A continuous dryer is operated under such conditions that 250 lb of water are removed per hour from the stock being dried. The air enters the dryer at 175°F, and a pressure of 765 mm of mercury. The dew point of the air is 40°F. The air emerges from the dryer at 95°F, a pressure of 755 mm of mercury, and at 90% relative humidity.

(a) How many cubic feet of the original air must be supplied per hour?

(b) How many cubic feet of air emerge from the dryer per hour?

18. Air at 25°C, 740 mm of mercury, and 55% relative humidity is compressed and cooled to 0°C.

(a) What pressure must be applied if 90% of the water is to be condensed?

(b) On the basis of 1 cu ft of original air, what will the volume be of the gas-residual vapor at 0°C in the compressed state?

19. Air at 25°C, 740 mm of mercury, and 55% relative humidity is compressed to 10 atm.

    (*a*) To what temperature must the gas-vapor mixture be cooled if 90% of the water is to be condensed?

    (*b*) On the basis of 1 cu ft of original air, what will be the volume of the gas-vapor mixture at 10 atmospheres after cooling to the final temperature?

In working out the following problems, the humidity charts in Figures 8 and 9 are to be used as far as possible. The use of the large-scale chart in Figure 9 is preferable within its range. Unless otherwise specified the pressure is assumed to be one atmosphere.

20. Air at atmospheric pressure has a wet-bulb temperature of 62°F and a dry-bulb temperature of 78°F.

    (*a*) Estimate the percentage saturation, molal humidity, and the dew point.

    (*b*) Calculate the weight of water contained in 100 cu ft of the air.

21. Hydrogen is saturated with water vapor at atmospheric pressure and a temperature of 90°F. The wet gas is passed through a cooler in which its temperature is reduced to 45°F and a part of the water vapor condensed. The gas after leaving the cooler is heated to a temperature of 70°F.

    (*a*) Estimate the weight of water condensed in the cooler per pound of moisture-free hydrogen.

    (*b*) Estimate the percentage humidity, wet-bulb temperature, and molal humidity of the gas in its final conditions.

22. It is desired to maintain the air entering a building at a constant temperature of 75°F and a percentage humidity of 40%. This is accomplished by passing the air through a series of water sprays in which it is cooled and saturated with water. The air leaving the spray chamber is then heated to 75°F.

    (*a*) Assume that the water and air leave the spray chamber at the same temperature. What is the temperature of the water leaving?

    (*b*) Estimate the water content of the air in the building in pounds of water per pound of dry air.

    (*c*) If the air entering the spray chamber has a temperature of 90°F and a percentage humidity of 65%, how much water will be evaporated or condensed in the spray chamber per pound of moisture-free air?

23. In a direct-fired evaporator the hot combustion gases are passed over the surface of the evaporating liquid. The gases leaving the evaporator have a dry-bulb temperature of 190°F and a wet-bulb temperature of 145°F. The dry combustion gases contain 11% CO_2. Estimate the percentage saturation of the gases leaving the evaporator, their molal humidity, and their dew point.

24. An air-conditioning system takes warm summer air at 95°F, 85% humidity, and 760 mm of mercury. This air is passed through a cold-water spray, and emerges saturated. The air is then heated to 70°F.

    (*a*) If it is required that the final percentage humidity be 35%, what must be the temperature of the air emerging from the spray?

    (*b*) On the basis of 1000 cu ft of entering air, calculate the weight of water condensed by the spray.

25. Cold winter air at 20°F, 760 mm pressure, and 70% humidity is conditioned by passing through a bank of steam-heated coils, through a water spray, and finally through a second set of steam-heated coils. In passing through the first bank of steam-heated coils, the air is heated to 75°F. The water supplied to the spray chamber is adjusted to the wet-bulb temperature of the air admitted to the chamber, hence the humidifying unit may be assumed to operate adiabatically. It is required that the air emerging from the conditioning unit be at 70°F and 35% humidity.

(a) What should be the temperature of the water supplied to the spray chamber?

(b) In order to secure air at the required final conditions, what must be the percentage humidity of the air emerging from the spray chamber?

(c) What is the dry-bulb temperature of the air emerging from the spray chamber?

(d) On the basis of 1 cu ft of outside air, calculate the volume at each step of the process.

(e) Calculate the pounds of water evaporated per cubic foot of original air.

26. At the top of a chimney the flue gas from a gas-fired furnace has a temperature of 180°F and a wet-bulb temperature of 133°F. The composition of the moisture-free flue gas is as follows:

$$
\begin{array}{ll}
CO_2\ldots\ldots\ldots\ldots\ldots\ldots\ldots & 14.1\% \\
O_2\ldots\ldots\ldots\ldots\ldots\ldots\ldots & 6.0\% \\
N_2\ldots\ldots\ldots\ldots\ldots\ldots\ldots & \underline{79.9\%} \\
& 100.0\%
\end{array}
$$

(a) If the temperature of the gases in the stack were reduced to 90°F, calculate the weight of water in pounds which would be condensed per pound-mole of dry gas.

(b) The gas burned in the furnace is estimated to contain 7% CH_4, 27% CO, and 3% CO_2 by volume; it does not contain any carbon compounds other than these. The gas is burned at a rate of 4000 cu ft per 24 hours measured at a temperature of 65°F, saturated with water vapor. Estimate the weight of water condensed in the chimney per day under the conditions of part (a).

27. A fuel gas has the following analysis:

$$
\begin{array}{llll}
\text{Carbon dioxide}\ldots\ldots & CO_2 & = & 3.2\% \\
\text{Ethylene}\ldots\ldots\ldots\ldots & C_2H_4 & = & 6.1 \\
\text{Benzene}\ldots\ldots\ldots\ldots & C_6H_6 & = & 1.0 \\
\text{Oxygen}\ldots\ldots\ldots\ldots & O_2 & = & 0.8 \\
\text{Hydrogen}\ldots\ldots\ldots\ldots & H_2 & = & 40.2 \\
\text{Carbon monoxide}\ldots\ldots & CO & = & 33.1 \\
\text{Paraffins}\ldots\ldots\ldots\ldots & C_{1.20}H_{4.40} & = & 10.2 \\
\text{Nitrogen}\ldots\ldots\ldots\ldots & N_2 & = & \underline{5.4} \\
& & & 100.0\%
\end{array}
$$

Assume that this gas is metered at 65°F, 100% saturation, and is burned with air supplied at 70°F, 1 atm pressure, and 60% humidity. Assume 20% excess air is supplied and that combustion is complete.

(a) What is the dew point of the stack gases?

(b) If the gases leave the stack at 190°F, what is the wet-bulb temperature?

(c) If the gases are cooled to 85°F before leaving the chimney, how many pounds of condensate drain down the chimney on the basis of 100 cu ft of fuel gas metered?

28. Air at 185°F and a dew point of 40°F is supplied to a dryer, which operates under adiabatic conditions.

(a) What is the minimum temperature to which the air will cool in the dryer?

(b) What is the maximum evaporation in pounds that can be obtained on the basis of 1000 cu ft of entering air?

(c) If maximum evaporation is obtained, what will be the volume of the emerging air on the basis of 1000 cu ft of entering air?

29. Air enters a dryer at a temperature of 240°F with a dew point of 55°F. The dryer may be assumed to approach adiabatic vaporization in its operation, all the heat being supplied by the air. If the air leaves the dryer saturated with water vapor, how much water can be evaporated per 1000 cu ft of entering wet air?

30. A dryer for the drying of textiles may be assumed to operate adiabatically. The air enters the dryer at a temperature of 160°F with a dew point of 68°F. It is found that 0.82 lb of water is evaporated per 1000 cu ft of wet air entering the dryer. Estimate the percentage saturation and temperature of the air leaving the dryer.

<u>Natural Sodium Salts</u>

Sea water, salt domes, natural brines, etc

Solar and vacuum evaporation.

<u>Natural Potassium Salts</u>

Natural deposits.

Fractional crystallization

Sodium Carbonate

Le Blanc Process

$2NaCl + H_2SO_4 \rightarrow Na_2SO_4 + 2HCl$　　　(furnace)

$Na_2SO_4 + 4C \rightarrow Na_2S + 4CO$　　　(another furnace)

$Na_2S + CaCO_3 \rightarrow Na_2CO_3 + CaS$　　　(tanks)

　Liquor is carbonated.

　　$2NaOH + CO_2 \rightarrow Na_2CO_3 + H_2O$

　　$Na_2S + H_2O + CO_2 \rightarrow Na_2CO_3 + H_2S$

Recovery of HCl

　Weldon process: HCl is dissolved.

　　$4HCl + MnO_2 \rightarrow MnCl_2 + 2H_2O + Cl_2$

　　$MnCl_2 + Ca(OH)_2 \rightarrow Mn(OH)_2 + CaCl_2$

　　$Mn(OH)_2 + \frac{1}{2}O_2 \rightarrow MnO_2 + H_2O$

　Deacon process

　　$4HCl + O_2 \xrightarrow{CuCl_2} 2H_2O + 2Cl_2 + \Delta$

CHAPTER V

SOLUBILITY AND SORPTION

Industrial problems involving phase equilibria among solids, liquids, and gases involve all the mutual equilibrium relations among these three phases, such as the dissolution of a solid by a liquid, the crystallization of a solid from a liquid, the absorption and desorption of gases by liquids, the distribution of a solute between immiscible liquids, and the adsorption and desorption of gases by solids.

DISSOLUTION AND CRYSTALLIZATION

A substance which is a solvent for a solid has an entirely specific effect upon the distribution of particles between the solid and its dispersed state. Thus a substance which is an excellent solvent for one solid may exert no appreciable influence on another. The solvent action of a liquid is presumed to result from a high affinity or attractive force between the liquid and the solid particles. When a solvent and a solid are brought into contact with each other, the attractive forces of the liquid aided by the thermal motion of the solid particles tend to break apart the structure of the solid and to disperse molecules or ions from its surface in a manner somewhat analogous to the vaporization of a liquid into a gas. As a result the ions or molecules enter the liquid as individually mobile units, thus forming what is termed a *solution* of the solid in the liquid. The dispersed material in a solution is termed the *solute*. This distinction between the solvent and solute in a solution is entirely arbitrary since either component may be considered as either solute or solvent. For example, a solution of naphthalene in benzene may equally well be considered as a solution of benzene in naphthalene.

The solute particles of a solution are free to move about as a result of the kinetic energy of translation which is possessed by each. By bombarding the confining walls these particles exert an *osmotic pressure* which is entirely analogous to the partial pressure exerted by each component of a gaseous mixture. Thus, when a solution is in contact with the solid from which it was formed, there will be a continual return of dissolved particles to the solid surface. The dispersion of a solid into a liquid is termed *dissolution*. The reverse process is known

111

as *crystallization*. As dissolution proceeds, the concentration of dispersed particles is increased, resulting in an increased osmotic pressure and a greater rate of crystallization. When the solute concentration becomes sufficiently high, the rate of crystallization will equal the rate of dissolution, and a dynamic equilibrium will be established, the concentration of the solution then remaining constant. This equilibrium is analogous to that between a liquid and its vapor. Under these conditions the solution is said to be *saturated* with the solute and is incapable of dissolving greater quantities of that particular solute.

The concentration of solute in a saturated solution is termed the *solubility* of the solute in the solvent. Solubilities are dependent on the nature of the solute, the nature of the solvent, and the existing temperature. Pressure also has a small effect on solubility, which ordinarily may be neglected. This effect is such that, in a solution whose volume is less than the sum of the volumes of its components in their pure states, solubility is increased by an increase in pressure. In the opposite case, in which the total volume is increased in the formation of a solution, an increase in pressure lowers the solubility.

The simple generalizations which apply to equilibria between liquids and vapor-gas mixtures are not applicable to liquid solutions. Little is known regarding the relationships between solubilities and the specific properties of the solute and solvent. The properties of each particular system must be individually determined by experimental means, and it is impossible to predict quantitatively the behavior of one system from that of another. This results from the fact that what may be termed the *solution pressure* of a solid is entirely dependent on the nature of the solvent with which it is in contact. In many cases it is impossible to predict, even qualitatively, the effect of temperature on solubility.

Solubilities of Solids Which Do Not Form Compounds with the Solvent. In general, where no true chemical compounds are formed between a solute and solvent, the solubility increases with increasing temperature. The solubility curves of such systems resemble vapor-pressure curves in general form. In Fig. 10 are plotted the solubility data of a typical system, naphthalene in benzene.

Curve *EB* of Fig. 10 represents the conditions of temperature and concentration which correspond to saturation with *naphthalene* of a solution of naphthalene in benzene. A solution whose concentration and temperature fix a point on this curve will remain in dynamic equilibrium with solid crystals of naphthalene. If the temperature is lowered or the concentration increased by the removal of solvent, naphthalene crystals will be formed in the solution. The area to the right of curve *EB* represents conditions of homogeneous, partially saturated solutions.

The area between curves EB and ED represents conditions of non-homogeneous mixtures of crystals of pure naphthalene in solutions of naphthalene in benzene. If a solution whose concentration and temperature are represented by point x_1 is cooled without change in composition, its conditions will vary along the dotted line parallel to the temperature axis. The temperature at which this dotted line intersects the solubility curve EB is the temperature at which pure naphthalene will begin to crystallize from the solution. As the temperature is reduced further, more naphthalene will crystallize and the remaining saturated solution will diminish in concentration, its condition always

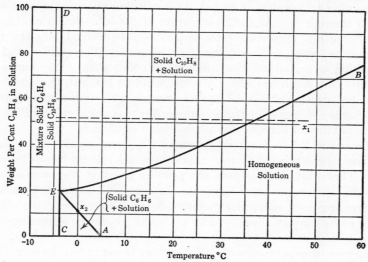

Fig. 10.   Solubility of naphthalene in benzene.

represented by a point on curve EB. When the temperature corresponding to line CED is reached the system will consist of pure naphthalene crystals and a saturated solution whose concentration corresponds to point E. Further decrease in temperature will cause this remaining solution to solidify to a mixture of crystals of pure benzene and pure naphthalene. The line CED represents the completion of solidification, and the area to the left of it corresponds to completely solid systems of naphthalene and benzene crystals.

The point E is termed the *eutectic* point, and the corresponding temperature and composition are, respectively, the *eutectic temperature* and the *eutectic composition*. If a solution of eutectic composition is cooled it will undergo no change until reaching the eutectic temperature, when it will completely solidify without further change in temperature. A solution of eutectic composition solidifies completely at one definite temperature which is also the lowest solidification point possible for the system.

The curve EA represents the conditions of temperature and composition which correspond to saturation with benzene of a solution of naphthalene in benzene. Whereas curve EB represents the solubility of naphthalene in benzene, curve EA represents the solubility of benzene in naphthalene. The area AEC represents conditions of nonhomogeneous mixtures of benzene crystals in saturated solutions of naphthalene in benzene. If a solution whose composition and temperature are represented by point x_2 is cooled, its conditions will vary along a horizontal through x_2. At the temperature of the intersection of this line with curve EA, crystals of pure benzene will be formed. As the temperature is further reduced, more pure benzene will crystallize, and the remaining saturated solution will vary in composition along curve EA towards E. When the eutectic temperature is reached, the remaining saturated solution will be of eutectic composition, represented by point E. Further decrease in temperature will cause complete solidification into a mixture of benzene and naphthalene crystals.

The curve EA is frequently termed the *freezing-point curve* of the solution representing the temperatures at which solvent begins to freeze out. The point A represents the freezing point of the pure solvent, benzene. From the same viewpoint the curve EB might be considered as a freezing-point curve along which solute begins to freeze out. Thus, either curve may be considered as a solubility curve or a freezing-point curve.

The *percentage saturation* of a solution may be defined as the percentage ratio of the existing weight of solute per unit weight of solvent to the weight of solute which would exist per unit weight of solvent if the solution were saturated at the existing temperature. The percentage saturation of a solution may be varied by changing either its temperature or composition. The effects of such changes on the percentage saturation of a solution may be predicted by locating the point representing the conditions of the solution on a solubility chart such as Fig. 10. A change in temperature will move this point along a line parallel to the temperature axis. A change in composition will move it along a line parallel to the concentration axis. A process in which both composition and temperature are changed simultaneously is best considered as proceeding in two steps: a change in composition at constant temperature and a change in temperature with constant composition.

Solubilities of Solids Which Form Solvates with Congruent Points. Many solutes possess the property of forming definite chemical compounds with their solvents. Such compounds of definite proportions between solutes and solvents are termed *solvates,* or if the solvent is water, *hydrates.* Several solvates of different compositions may be formed by a single system, each a stable form, under certain conditions of temperature and composition. The presence of such solvates greatly complicates the solubility relationships of the system.

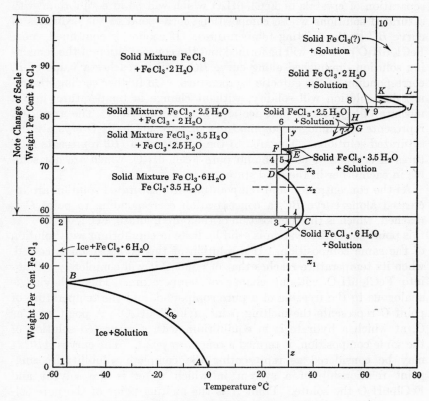

FIG. 11. Solubility of ferric chloride in water.

The system ferric chloride and water is an excellent illustration of the effects of the formation of many hydrates. In Fig. 11 are plotted the solubility curves of this system. Point A represents the freezing point of the pure solvent, water, and curve AB the conditions of equilibrium in a solution which is saturated with the solid solvent, ice. This curve is analogous to curve AE of Fig. 10 and represents the solubility of water in ferric chloride. The area $AB1$ represents nonhomogeneous mixtures of pure ice in equilibrium with saturated solutions.

Point B of Fig. 11 is a eutectic point analogous to point E of Fig. 10. Curve BC represents the conditions of saturation of a solution of ferric chloride in water with the solid hydrate, $FeCl_3 \cdot 6H_2O$. If a solution, whose conditions are represented by point x_1, is cooled, it will become saturated with $FeCl_3 \cdot 6H_2O$ when conditions corresponding to this concentration on curve BC are reached. Further cooling will result in separation of crystals of $FeCl_3 \cdot 6H_2O$ which will be in equilibrium with saturated solutions whose compositions correspond to the ordinates of curve BC at the existing temperatures. If cooling is continued, more $FeCl_3 \cdot 6H_2O$ crystals will be formed and the concentration of the remaining solution diminished along curve BC until it reaches a value of B, corresponding to the eutectic temperature. On further cooling, the remaining solution will solidify, without change in temperature, into a mixture of crystals of pure ice and pure $FeCl_3 \cdot 6H_2O$. The area $BC2$ represents nonhomogeneous mixtures of pure $FeCl_3 \cdot 6H_2O$ crystals and saturated solutions. The area to the left of curve $1B2$ represents mixtures of crystals of pure ice and pure $FeCl_3 \cdot 6H_2O$, which are not soluble in each other in the solid state.

As the concentration and temperature of a saturated solution are increased along curve BC, a concentration corresponding to point C is reached, which is the composition of the pure hydrate, $FeCl_3 \cdot 6H_2O$. At this point the entire system is solid hydrate in equilibrium with solution of the same composition. If a solution of this composition is cooled, when its temperature reaches that of point C it will completely solidify into $FeCl_3 \cdot 6H_2O$ without change of temperature. This behavior is analogous to the freezing of a pure compound, and the temperature of point C represents the melting point of $FeCl_3 \cdot 6H_2O$. A point such as C, at which a hydrate is in equilibrium with a saturated solution of the same composition, is termed a *congruent point*. The curves $ABC21$ may be considered as representing the complete solubility-freezing-point relationship of a system in which water is the solvent and $FeCl_3 \cdot 6H_2O$ the solute. Point A is the melting point of the pure solvent and C that of the pure solute; B represents the eutectic point of this system.

From the negative slope of curve CD it follows that at concentrations higher than that of point C the concentration of $FeCl_3$ in a saturated aqueous solution may be increased by *lowering* the temperature. This behavior may be readily understood by considering that the curves $3CDE4$ represent the solubility-freezing-point data of a new system in which the solvent is $FeCl_3 \cdot 6H_2O$ and the solute is $FeCl_3 \cdot 3.5H_2O$. In this system curve CD is analogous to AB and represents the conditions of equilibrium between pure, solid solvent, $FeCl_3 \cdot 6H_2O$, and a solution of $FeCl_3 \cdot 3.5H_2O$ in $FeCl_3 \cdot 6H_2O$. If a solution at conditions represented by point x_2 is cooled, pure $FeCl_3 \cdot 6H_2O$ will crystallize out when a temperature corresponding to this concentration on curve CD is reached. Further cooling will result in the formation of more pure solvent crystals, $FeCl_3 \cdot 6H_2O$, and the remaining saturated solution will increase in concentration, along curve CD. When the temperature of curve $4D3$ is reached this remaining solution will have the composition of point D. Further cooling will result in complete solidification, without further change in temperature, into a mixture of crystals of pure $FeCl_3 \cdot 6H_2O$ and pure $FeCl_3 \cdot 3.5H_2O$. Point D represents the eutectic point of this system and is analogous to point B. The area $CD3$ represents conditions of nonhomogeneous mixtures of pure crystals of $FeCl_3 \cdot 6H_2O$ and saturated solutions of $FeCl_3 \cdot 3.5H_2O$ in $FeCl_3 \cdot 6H_2O$.

Curve DE represents the conditions of equilibrium between crystals of pure solute, $FeCl_3 \cdot 3.5H_2O$, and a saturated solution of $FeCl_3 \cdot 3.5H_2O$ in $FeCl_3 \cdot 6H_2O$. As the concentration and temperature of a saturated solution are increased along curve DE, a concentration corresponding to point E is reached which is the composition of the hydrate $FeCl_3 \cdot 3.5H_2O$. At this point the entire system must be solid $FeCl_3 \cdot 3.5H_2O$ in equilibrium with solution of the same concentration. Point E is, therefore, a second congruent point, representing the melting point of $FeCl_3 \cdot 3.5H_2O$. Area $DE4$ represents conditions of nonhomogeneous mixtures of crystals of pure $FeCl_3 \cdot 3.5H_2O$ and saturated solutions. The area to the left of curve $3D4$ represents entirely solid systems composed of mixtures of crystals of pure $FeCl_3 \cdot 6H_2O$ and $FeCl_3 \cdot 3.5H_2O$.

By similar reasoning, the curves $5EFG6$ may be considered as representing the solubility-freezing-point data of a system in which the solute is $FeCl_3 \cdot 2.5H_2O$ and the solvent $FeCl_3 \cdot 3.5H_2O$. Curves $7GHJ8$ represent the system of $FeCl_3 \cdot 2H_2O$, solute, and $FeCl_3 \cdot 2.5H_2O$, solvent. Curves $9JKL$ represent the system of $FeCl_3$, solute, and $FeCl_3 \cdot 2H_2O$, solvent. Points F, H, and K are the eutectic points of these systems. Points G and J are the congruent melting points of $FeCl_3 \cdot 2.5H_2O$ and $FeCl_3 \cdot 2H_2O$, respectively.

By means of a chart such as Fig. 11 the changes taking place in even very complicated systems may be readily predicted or explained. An illustration of the peculiar phenomena which may take place in complex systems is furnished by the isothermal concentration of a dilute aqueous solution of ferric chloride along the line zy. Such concentration may be carried out by evaporation at constant temperature. At the intersection of zy with BC, crystals of $FeCl_3 \cdot 6H_2O$ will begin to form. At line $C2$ the system will be entirely solid $FeCl_6 \cdot 6H_2O$. Further concentration results in the appearance of liquid solution, and at curve CD the system will be once more entirely liquid. As concentration is continued to curve DE, solidification again begins, becoming complete at line $E5$. Further concentration causes the reappearance of liquid, and liquefaction is completed at curve EF. At curve FG solidification begins and becomes complete at $G6$. This surprising alternation of the liquid and solid states while concentration is progressing may be easily demonstrated.

Many other systems, both organic and inorganic, behave in a manner similar to that of aqueous ferric chloride and form one or more solvates and congruent points.

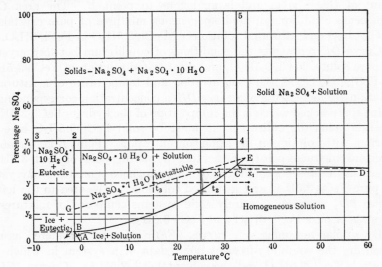

FIG. 12. Solubility of sodium sulfate in water.

Solubilities of Solids Which Form Solvates without Congruent Points. In certain systems solvates are formed which are not stable and which decompose before the temperature of a congruent point is reached. Such solvates undergo direct transition from the solid state into other chemical compounds. These transitions take place at sharply defined temperatures which are termed *transition points*. The system of sodium sulfate and water illustrates this type of behavior. In Fig. 12 are plotted the solubility and freezing-point data of this system.

Curve AB of Fig. 12 represents conditions of equilibrium between ice and aqueous solutions of sodium sulfate. Point B is the eutectic point of the system of $Na_2SO_4 \cdot 10H_2O$ (solute) and water (solvent). Curve BC represents the equilibrium between crystals of $Na_2SO_4 \cdot 10H_2O$ and saturated solution. As the temperature of the system is increased the concentration of the saturated solution increases along curve BC. Normally this increase would continue until a congruent point was reached. However, at a temperature of 32.384°C, $Na_2SO_4 \cdot 10H_2O$ becomes unstable and is decomposed into Na_2SO_4 and water. The solubility of the anhydrous Na_2SO_4 as indicated by curve CD diminishes with increasing temperature. Points on this curve represent conditions of equilibrium between crystals of anhydrous Na_2SO_4 and saturated solutions. A solution whose conditions are represented by point x_1 will become saturated if either heated or cooled sufficiently. If cooled, crystals of $Na_2SO_4 \cdot 10H_2O$ will form when the conditions of curve BC are reached. If heated, crystals of anhydrous Na_2SO_4 will form at the temperature corresponding to composition x_1 on curve CD.

The significance of the areas of Fig. 12 is similar to that of the other diagrams which have been discussed. The area of the small triangle to the left of line AB represents a region of equilibrium between crystals of pure ice and saturated solution. Area $BC42$ is a region of equilibrium between crystals of pure $Na_2SO_4 \cdot 10H_2O$ and saturated solution. The area to the left of line $B2$ represents a solid mixture of crystals of ice and $Na_2SO_4 \cdot 10H_2O$. Line $C45$ indicates the transition temperature at which the decahydrate decomposes to form the anhydrous salt. Line 324 indicates the composition of the pure decahydrate, $Na_2SO_4 \cdot 10H_2O$. The area to the right of curve $54CD$ represents conditions of equilibria between crystals of pure Na_2SO_4 and saturated solutions. The area above and to the left of curves 3245 is a region of entirely solid mixtures of Na_2SO_4 and $Na_2SO_4 \cdot 10H_2O$. Line 45 therefore represents a temperature of complete solidification.

Many systems which form several solvates show solubility relationships both with and without congruent points and transition points in the same system. For example, zinc chloride forms five hydrates. Four of these hydrates decompose, as does $Na_2SO_4 \cdot 10H_2O$, exhibiting transition points before congruent concentrations are reached. The fifth hydrate, $ZnCl_2 \cdot 2.5H_2O$, exhibits a true congruent point, as do the hydrates of ferric chloride. The significance of such solubility relationships may be understood from the principles discussed in the preceding sections.

Effect of Particle Size on Solubility. The solution pressure and solubility of a solid is affected by its particle size in a manner entirely analogous to the effect of particle size on vapor pressure. The solubility of a substance is increased with increase in its degree of subdivision. This increasing solubility with diminishing particle size is demonstrated by the behavior of crystals which are in equilibrium with their saturated solutions. Where such an equilibrium exists the total amounts of solid and liquid must remain unchanged. However, the equilibrium is dynamic, resulting from equality between the rates of dissolution and crystallization. As a result of the effect of particle size on solubility, the small crystals in such a solution will possess higher solution pressures than the large ones and will tend to disappear with corresponding increase in size of the large crystals. This growth of large crystals at the expense of the small ones in a saturated solution is a familiar phenomenon of considerable industrial importance. For the same reasons, an irregular crystalline mass will change its shape in a saturated solution. The sharp corners and points will exert higher solution pressures than the plane surfaces and will disappear, building up the plane surfaces and tending to produce a regular shape. Like vapor pressures, solubilities are not noticeably affected by particle size until submicroscopic dimensions are approached.

Supersaturation. Just as spontaneous condensation of a vapor is made difficult because of the high vapor pressure of small drops, spontaneous crystallization is interfered with by the high solubility of small crystals. In order to produce spontaneous crystallization the concentration of a solution must be sufficiently high that the small crystalline nuclei which are formed by simultaneous molecular or ionic impacts do not immediately redissolve. Such a concentration will be much greater than that which is in equilibrium with large crystals of the same solid. Once crystallization is started and nuclei are formed it will continue and the nuclei will grow until the normal equilibrium conditions are reached. For these reasons it is relatively easy to obtain solutions whose concentrations are higher than the values normally corresponding to saturation. Such solutions are *supersaturated* with respect to large crystals but are only partially saturated with respect to the tiny nuclei which tend to form in them. Supersaturated solutions may be formed by careful exclusion of all crystalline particles of solute and by slow changes in temperature or concentration without agitation.

Because of the phenomenon of supersaturation, it is possible to extend the curves of solubility diagrams, such as Fig. 12 into regions where the equilibria which they represent would not normally be stable. The dotted curves of Fig. 12 represent such equilibria which have been experimentally observed in supersaturated solutions of this system. Such equilibria are termed *metastable* and possess the continual tendency to revert to the normal, stable state corresponding to their conditions of temperature and concentration.

The dotted curve GE of Fig. 12 represents the metastable equilibrium between crystals of $Na_2SO_4 \cdot 7H_2O$ and its saturated solutions. If a solution at conditions x_1 is carefully cooled, normal crystallization of

$Na_2SO_4 \cdot 10H_2O$ may be prevented and a supersaturated solution produced at conditions x_1'. If cooling is continued the crystallization of $Na_2SO_4 \cdot 7H_2O$ may be induced at a temperature corresponding to composition x_1 on curve GE. A supersaturated solution at conditions x_1' is capable of dissolving $Na_2SO_4 \cdot 7H_2O$ and is unsaturated with respect to this compound. On the other hand, its conditions are unstable, and any disturbance such as agitation, a sudden temperature change, or the introduction of a crystal of $Na_2SO_4 \cdot 10H_2O$, will cause it to assume its normal equilibrium conditions with the crystallization of $Na_2SO_4 \cdot 10H_2O$.

Metastable equilibria are of little industrial significance, but supersaturation is very commonly encountered. It may be produced in two ways. By the exclusion of all particles of solid solute the formation of crystalline nuclei may be entirely prevented as described above. Another type of supersaturation may result from sudden changes of the conditions of a saturated solution even though crystallization has started and crystalline solute is present. This results from the fact that crystallization, especially in certain types of viscous solutions, is a slow process. Sudden cooling of such a solution will produce temporary conditions of supersaturation simply because the system is slow in adjusting itself to its equilibrium conditions. Agitation of the solution will hasten this adjustment.

In the majority of crystallization operations it is desirable to avoid supersaturation. Supersaturation of the type resulting from the absence of crystalline nuclei is prevented by *seeding* saturated solutions with crystals of solute. Spontaneous nucleus formation is also favored by the presence of rough, adsorbent surfaces. The crystallization of sugar on pieces of string to form rock candy and the scratching of the wall of a beaker to cause crystallization of an analytical precipitate are familiar illustrations of this principle. Supersaturation due to slow crystallization rates is avoided by using correspondingly slow rates of change of the conditions which promote supersaturation and in some cases by agitation.

DISSOLUTION

Problems arise in which it is required to calculate the amount of solute which can be dissolved in a specified quantity of solvent or solution, or, conversely, the quantity of solvent required to dissolve a given amount of solute to produce a solution of specified degree of saturation. Where solvates are not formed in the system such calculations introduce no new difficulties. From the solubility data is determined the quantity of solute which may be dissolved in a unit quantity of solvent to form a saturated solution at the existing temperature. The amount of solute which may be dissolved in a solution is then the difference between the amount already present and the amount which may be present if the solution is saturated at the specified conditions, both quantities being based on the same quantity of solvent.

Illustration 1. A solution of sodium chloride in water is saturated at a temperature of 15°C. Calculate the weight of NaCl which can be dissolved by 100 lb of this solution if it is heated to a temperature of 65°C.

Solubility of NaCl at 15°C = 6.12 lb-moles per 1000 lb of H_2O

Solubility of NaCl at 65°C = 6.37 lb-moles per 1000 lb of H_2O

Basis: 1000 lb of water.

NaCl in saturated solution at 15°C = 6.12 × 58.5 = ... 358 lb

Percentage of NaCl by weight = 358/1358 = ... 26.4%

NaCl in saturated solution at 65°C = 6.37 × 58.5 = ... 373 lb

NaCl which may be dissolved per 1000 lb of H_2O = 373 − 358 = 15 lb

Water present in 100 lb of original solution = 100 × (1.0 − 0.264) = 73.6 lb

NaCl dissolved per 100 lb of original solution = $\dfrac{15 \times 73.6}{1000}$ = ... 1.1 lb

It will be noted that the solubility of NaCl changes but little with change in temperature.

Where the substance to be dissolved is a solvated compound the problem is complicated by the fact that both solute and solvent are added to the solution. Such calculations are carried out by equating the total quantities of solute entering and leaving the process. Algebraic expressions are formed for the sum of the quantity of solute to be dissolved plus that originally present in the solution, and for the quantity of solute in the final solution. Since the total quantity of solute must be constant, these two expressions are equal and may be equated and solved. This method is demonstrated in the following illustration:

Illustration 2. After a crystallization process a solution of calcium chloride in water contains 62 lb of $CaCl_2$ per 100 lb of water. Calculate the weight of this solution necessary to dissolve 250 lb of $CaCl_2 \cdot 6H_2O$ at a temperature of 25°C. Solubility at 25°C = 7.38 lb-moles of $CaCl_2$ per 1000 lb of H_2O.

Basis: x = weight of water in the required quantity of solution.

$CaCl_2 \cdot 6H_2O$ to be dissolved = $\dfrac{250}{219}$ = ... 1.14 lb-moles

Total $CaCl_2$ entering process = $1.14 + \dfrac{62x}{111 \times 100}$ = $1.14 + \dfrac{0.559x}{100}$ lb-moles

Total water entering process = $x + (1.14 \times 6 \times 18)$ = $x + 123$ lb

Total $CaCl_2$ leaving process = ... $7.38 \dfrac{x + 123}{1000}$ lb-moles

Equating:

$$1.14 + \frac{0.559x}{100} = 7.38 \frac{x + 123}{1000}$$
$$1140 + 5.59x = 7.38x + 908$$
$$1.79x = 232$$
$$x = 130 \text{ lb}$$

Total weight of solution required $= 130 + (130 \times 0.62) = \dots\dots 211$ lb

CRYSTALLIZATION

The crystallization of a solute from a solution may be brought about in three different ways. The composition of the solution may be changed by the removal of pure solvent, as by evaporation, until the remaining solution becomes supersaturated and crystallization takes place. The second method involves a change of temperature to produce conditions of lower solubility and consequent supersaturation and crystallization. A third method by which crystallization may be produced is through a change in the nature of the system. For example, inorganic salts may be caused to crystallize from aqueous solutions by the addition of alcohol. Other industrial processes involve the *salting out* of a solute by the addition of a more soluble material which possesses an ion in common with the original solute. The calculations which are involved in this third type of crystallization processes are frequently very complicated and require a large number of data regarding the particular systems involved. Such systems involve more than two components and require application of the principles of complex equilibria which are discussed in a later section.

Where Solvates Are Not Formed. The most important crystallization processes of industry are those which combine the effect of increasing the concentration by the removal of solvent with the effect of change of temperature. Where crystallization is brought about only through change in temperature, the yields of crystals and the necessary conditions may be calculated on the basis of the quantity of solvent, which remains constant throughout the process. From the solubility data may be obtained the quantity of solute which will be dissolved in this quantity of solvent in the saturated solution which will remain after crystallization. The difference between the quantity of solute originally present and that remaining in solution will be the quantity of crystals formed. Such problems may be of two types: one in which it is desired to calculate the yield of crystals produced by a specified temperature change; and the converse, in which the amount of temperature change necessary to produce a specified yield is desired. The percentage yield of a crystallization process is the percentage which the yield of crystallized solute forms of the total quantity of solute originally present.

Illustration 3. A solution of sodium nitrate in water at a temperature of 40°C contains 49% NaNO₃ by weight.

(a) Calculate the percentage saturation of this solution.

(b) Calculate the weight of NaNO₃ which may be crystallized from 1000 lb of this solution by reducing the temperature to 10°C.

(c) Calculate the percentage yield of the process.

Solubility of NaNO₃ at 40°C = 51.4% by weight
Solubility of NaNO₃ at 10°C = 44.5% by weight

Basis: 1000 lb of original solution.

(a) Percentage saturation $= \dfrac{49}{51} \times \dfrac{48.6}{51.4} = 91.0\%$

(b) Yield of NaNO₃ crystals $= x$ lb

From a NaNO₃ balance

$$1000(0.49) = (1000 - x)(0.445) + x$$

Hence $\qquad\qquad\qquad\qquad x = 81$ lb

(c) Percentage yield $= \dfrac{81}{490} = 16.5\%$

Illustration 4. A solution of sodium bicarbonate in water is saturated at 60°C. Calculate the temperature to which this solution must be cooled in order to crystallize 40% of the NaHCO₃.

Solubility of NaHCO₃ at 60°C = 1.96 lb-moles per 1000 lb H₂O

Basis: 1000 lb of H₂O.

NaHCO₃ in original solution = ...   1.96 lb-moles
NaHCO₃ in final solution = 1.96 × 0.60 = ...   1.18 lb-moles

From the solubility data of NaHCO₃ it is found that a saturated solution containing 1.18 lb-moles per 1000 lb of H₂O has a temperature of 23°C. The solution must be cooled to this temperature to produce the specified percentage yield.

Calculations of the yields and necessary conditions of crystallization by concentration may be carried out by consideration of the quantity of solvent remaining after concentration has taken place. The quantity of solute which will be dissolved in this quantity of solvent in the saturated solution remaining after crystallization may be calculated from solubility data. The quantity of crystals formed in the process will be the difference between the quantity of solute originally present and that finally remaining in solution. If the concentration is accompanied or followed by a temperature change the problem is unchanged. It is only necessary to consider the final temperature in order to determine the quantity of solute remaining in solution. In such processes three variable factors are present: the yield, the temperature change, and the degree of concentration. Problems arise in which it is necessary to evaluate any one of these factors if the other two are specified.

Illustration 5. A solution of potassium dichromate in water contains 13% $K_2Cr_2O_7$ by weight. From 1000 lb of this solution are evaporated 640 lb of water. The remaining solution is cooled to 20°C. Calculate the amount and the percentage yield of $K_2Cr_2O_7$ crystals produced.

Solubility of $K_2Cr_2O_7$ at 20°C = 0.390 lb-mole per 1000 lb H_2O

Basis: 1000 lb of original solution.

Water = ...	870 lb
$K_2Cr_2O_7$ = ...	130 lb
Water remaining after concentration = 870 − 640 = ...	230 lb
$K_2Cr_2O_7$ in solution after crystallization at 20°C = ...	
$\dfrac{230}{1000} \times 0.390 = 0.090$ lb-mole or $0.090 \times 294 = $	26.4 lb
Yield of $K_2Cr_2O_7$ crystals = 130 − 26.4 = ...	103.6 lb
Percentage yield = $\dfrac{103.4}{130} = $	79.7%

Care must always be exercised that the true solvent in a crystallizing system is recognized. For example, in Fig. 10, curve *EB* represents conditions under which naphthalene is the solute and benzene the solvent. However, curve *EA* represents the solubility of benzene as *solute* in naphthalene as solvent. If a solution having a concentration of naphthalene less than that of point *E* is cooled to produce crystallization, pure benzene will crystallize as the solute and the naphthalene will be the solvent, remaining constant in quantity throughout the process. Similarly, in aqueous solutions of salts, if the concentration is less than the eutectic value, cooling will produce the crystallization of water as pure ice and the system may be treated as a solution of water in salt.

Illustration 6. A solution of sodium nitrate in water contains 100 grams of $NaNO_3$ per 1000 grams of water. Calculate the amount of ice formed in cooling 1000 grams of this solution to a temperature of −15°C.

Concentration of saturated, water-in-$NaNO_3$ solution at −15°C = 6.2 gram-moles of $NaNO_3$ per 1000 grams of H_2O

Basis: 1000 grams H_2O.

$NaNO_3$ in original solution = ...	100 grams
Per cent $NaNO_3$ by weight = $\dfrac{100}{1100} = $	9.1%

Basis: 1000 grams of original solution.

$NaNO_3$ = 91 grams or 91/85 = ...	1.07 gram-moles
Water = ...	909 grams
Water dissolved in $NaNO_3$ in residual solution = ...	
$\dfrac{1000}{6.2} \times 1.07 = $	173 grams
Weight of ice formed = 909 − 173 = ...	736 grams

Where Solvates Are Present. Where solvates are involved it becomes necessary to consider the solvent chemically combined with the solute which is removed from solution when solvated crystals are precipitated or which is added to the solution when solvated crystals are dissolved. The calculations involved are most easily performed by establishing a material balance for either component. A binary system of weight W containing y per cent component A and $(100 - y)$ per cent component B will be considered. It is assumed that this solution separates under a given temperature change into two phases, phase 1 having a weight w_1 and a composition of y_1 per cent component A, and phase 2 having a weight $W - w_1$, and a composition of y_2 per cent component A. A material balance of component A gives

$$yW = y_1w_1 + y_2(W - w_1) \tag{1}$$

or

$$\frac{w_1}{W} = \frac{y - y_2}{y_1 - y_2} \tag{2}$$

If separation results from evaporation of a weight w_2 of pure component B, the material balance of component A becomes

$$yW = y_1w_1 + y_2(W - w_1 - w_2) \tag{3}$$

Illustration 7. An aqueous solution of sodium sulfate is saturated at 32.5°C. Calculate the temperature to which this solution must be cooled in order to crystallize 60% of the solute as $Na_2SO_4 \cdot 10H_2O$.

From Fig. 12 the solubility at 32.5° is seen to be 32.5% Na_2SO_4.

Basis: 1000 lb of initial solution.

Na_2SO_4 crystallized = 325 × 0.6 = 195 lb

$Na_2SO_4 \cdot 10H_2O$ crystallized = 195/0.441 = 442 lb

Water in these crystals = 442 − 195 = 247 lb

Water left in solution = 675 − 247 = 428 lb

Na_2SO_4 left in solution = 325 × 0.4 = 130 lb

Composition of final solution = 130/(130 + 428) = 23.3% Na_2SO_4

From Fig. 12 it is found that this concentration corresponds to a temperature of 27°C, the required crystallizing temperature.

Illustration 8. A solution of ferric chloride in water contains 64.1% $FeCl_3$ by weight. Calculate the composition and yield of the material crystallized from 1000 lb of this solution if it is so cooled as to produce the maximum amount of crystallization from a residual liquid.

From Fig. 11 it is seen that, if a solution of this composition is cooled, the hydrate $FeCl_3 \cdot 6H_2O$ will crystallize. The maximum crystallization from a liquid residue will result from cooling to the eutectic temperature, 27°C. Further cooling would cause complete solidification of the system. From Fig. 11 the solubility of $FeCl_3$ at the eutectic temperature is 68.3% by weight.

Basis: 1000 lb of original solution.

Percentage $FeCl_3$ in $FeCl_3 \cdot 6H_2O = \dfrac{162.2}{162.2 + 108} = 60.0\%$

Let x = pounds of $FeCl_3 \cdot 6H_2O$ crystallized.

Material balance of $FeCl_3$.

Original solution		Final solution		Crystals
(1000)(0.641)	=	(1000 − x)(0.683)	+	0.600x

or $x = 511$ lb $FeCl_3 \cdot 6H_2O$ crystals

Illustration 9. A solution of sodium sulfate in water is saturated at a temperature of 40°C. Calculate the weight of crystals and the percentage yield obtained by cooling 100 lb of this solution to a temperature of 5°C.

From Fig. 12 it is seen that at a temperature of 5°C the decahydrate will be the stable crystalline form. The solubilities read from Fig. 12 are as follows:

at 40°C: 32.6% Na_2SO_4

at 5°C: 5.75% Na_2SO_4.

Basis: 100 lb of original solution, saturated at 40°C.

Percentage Na_2SO_4 in $Na_2SO_4 \cdot 10H_2O$ crystals $= \dfrac{142}{142 + 180} = 44.1\%$

Let x = pounds of $Na_2SO_4 \cdot 10H_2O$ crystals formed.

Material balance of Na_2SO_4.

Original solution		Final solution		Crystals
0.326(100)	=	0.057(100 − x)	+	0.441x

or $x = 69.5$ lb $Na_2SO_4 \cdot 10H_2O$ formed

Weight of $Na_2SO_4 \cdot 10H_2O$ in original solution $= 32.6\dfrac{322}{142} = 74$ lb

Percentage yield $= \dfrac{69.5}{74} = 94\%$

Calculations from Line Segments of Equilibrium Diagrams.

In the separation of crystals from a solution the weight ratio of the crystals to the weight of the original solution is given by Equation (2), page 126. The ratios of Equation (2) can be obtained directly from the line segments on a binary equilibrium diagram when compositions are plotted in weight percentage. This method is illustrated by Fig. 12 for the system Na_2SO_4 and H_2O. Starting with a homogeneous solution at a temperature t_1 containing y per cent Na_2SO_4 (component A), and cooling, crystals of $Na_2SO_4 \cdot 10H_2O$ (phase 1) will start to separate at a temperature t_2. As the temperature drops crystallization will continue, the crystals having a uniform composition of 44.1% Na_2SO_4 corresponding to the decahydrate as represented by the upper boundary of this two-phase field. The percentage of Na_2SO_4 in the remaining liquid phase becomes progressively less as crystallization proceeds, and its decreasing composition will be represented by points on the solubility curve corresponding to the existing temperature. When the final temperature t_3 is reached the composition of the residual liquid will be represented by ordinate y_2, where y_2 represents the percentage of component A (Na_2SO_4) in the liquid phase. The weight of the $Na_2SO_4 \cdot 10H_2O$ crystals relative to the entire weight of the system as given by Equation (2) will be seen to be equal to the ratio of the line segments $y - y_2$ to $y_1 - y_2$.

In general, the percentage by weight of any phase in a two-phase equilibrium mixture at a given temperature may be obtained from the equilibrium diagram by extending the concentration line across the field showing the phases present. The weight ratio of phase 1 to the total weight will then be the ratio of the line segment extending between the composition of the entire system and the composition of phase 2 to the line segment extending between the compositions of phases 1 and 2. Phase 1, represented by segment $y - y_2$, always corresponds to that phase which is richest in component A. For example, in the above system phase 1 represents the $Na_2SO_4 \cdot 10H_2O$ crystal phase for compositions above the eutectic and represents the liquid phase below the eutectic composition where ice separates as the solid phase.

Illustration 9 will be solved from line segments on the equilibrium diagram. By referring to Fig. 12 the value of y corresponding to saturation at 40°C is 32.6 per cent. At 5°C, the composition y_1 of the solid phase is 44.1 per cent, and y_2 of the liquid phase is 5.75 per cent. Applying the above rule of segments the percentage weight of crystals separated is $\dfrac{32.6 - 5.75}{44.1 - 5.75} = 69.5$ per cent. By this ratio of segment lines the composition of any equilibrium mixture of two phases can be estimated from the equilibrium diagram.

It must be emphasized that the line-segment rule can be directly applied only when compositions are expressed in weight percentages and never when they are expressed in molality or mole fraction.

SOLUBILITIES IN COMPLEX SYSTEMS

FRACTIONAL CRYSTALLIZATION

All the systems thus far discussed have contained only two primary components. Where three or more components are involved the solubility relationships become very complex because of the effect which the presence of each solute has on the solubility of the others. These mutual effects cannot be predicted without experimental data on the particular case under consideration. For salts which are ionized in solution the following qualitative generalization applies: Where the same ion is formed from each of two ionized solute compounds, the solubility of each compound is diminished by the presence of the other. For example, the solubility of NaCl in water may be diminished by the addition of another solute which forms the chloride ion. This principle is applied in the crystallization of pure NaCl by bubbling HCl into concentrated brine. Similarly, the solubility of NaCl in water is reduced by the addition of sodium hydroxide. This accounts for the recovery of nearly pure caustic soda by evaporation of cathode liquor in the electrolysis of sodium chloride.

The mutual solubility relationships in complex systems are of great importance in many industrial processes. Soluble substances frequently may be purified or separated from other substances by properly conducted *fractional crystallization* processes. In such processes conditions are so adjusted that only certain of the total group of solutes will crystallize, the others remaining in solution. For the production of very pure materials it is frequently necessary to employ several successive fractional crystallizations. In such a scheme of recrystallization the crystal yield from one step is redissolved in a pure solvent and again fractionally crystallized to produce further purification.

Complete data have been developed for solubilities in many complex systems. The presentation and use by such data for more complicated cases are beyond the scope of a general discussion of industrial calculations. In an excellent monograph[1] by Blasdale the scientific principles involved in such problems are thoroughly discussed. Data for many systems are included in the International Critical Tables.

In Fig. 13 are the solubility data of Caspari[2] for the system of sodium sulfate and sodium carbonate in water, presented in an isometric, three-coordinate diagram. A diagram of this type permits visualization of the relationship existing in such systems of only three components. The solubility data determine a series of surfaces which, with the axial planes, form an irregular-shaped enclosure in space. Any conditions of concentration and temperature which fix a point lying within this enclosure correspond to a homogeneous solution. Points lying outside the enclosure represent conditions under which at least one solid substance is present. The surfaces themselves represent conditions of equilibria between the indicated solids and saturated solutions. The line formed by the intersection of two of these surfaces represents con-

[1] W. G. Blasdale, *Equilibria in Saturated Salt Solutions*, Chemical Catalog Co., New York (1927).

[2] W. A. Caspari, *J. Chem. Soc.*, **125**, 2381 (1924).

ditions of equilibria between a mixture of two solid substances and
saturated solution.

For example, curves $OABC$ of Fig. 13 represent the solubility in water
of Na_2SO_4 alone, corresponding to Fig. 12. The addition of Na_2CO_3 to
such a system lowers the solubility of the Na_2SO_4, as indicated by the
surfaces $ABRQP$ and $BCUR$. The former surface corresponds to
equilibria between crystals of pure $Na_2SO_4 \cdot 10H_2O$ and solutions containing both Na_2SO_4 and Na_2CO_3. Surface $BCUR$ represents similar
equilibria with anhydrous Na_2SO_4 as the solid. Similarly, surface
$GFSQP$ represents equilibria between crystals of pure $Na_2CO_3 \cdot 10H_2O$

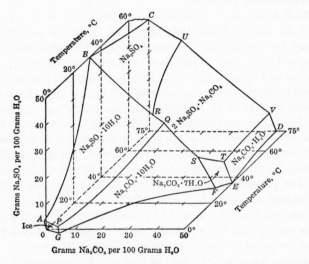

Fig. 13. Solubility of sodium carbonate-sodium sulfate in water.

and solutions containing both Na_2CO_3 and Na_2SO_4. These solutions
are, therefore, saturated with respect to Na_2CO_3 but unsaturated with
respect to Na_2SO_4. If Na_2SO_4 were added to a solution whose conditions correspond to a point on surface $GFSQP$, the salt would dissolve
and $Na_2CO_3 \cdot 10H_2O$ would crystallize.

The line PQ represents solutions which are saturated with both
Na_2CO_3 and Na_2SO_4. Such solutions are incapable of dissolving greater
quantities of either salt and are in equilibrium with crystals of each.
It so happens that when crystals of $Na_2CO_3 \cdot 10H_2O$ and $Na_2SO_4 \cdot 10H_2O$
are formed together in a solution, each solid compound is slightly soluble
in the other, forming what are termed *solid solutions*. Therefore, line
PQ represents equilibria between crystals of $Na_2CO_3 \cdot 10H_2O$ containing

small amounts of $Na_2SO_4 \cdot 10H_2O$ in solid solution, crystals of $Na_2SO_4 \cdot 10H_2O$ containing small amounts of $Na_2CO_3 \cdot 10H_2O$ in solid solution, and liquid solution containing both Na_2SO_4 and Na_2CO_3.

At the higher temperatures the system is complicated by the decomposition of the hydrates and by the formation of a stable double salt of definite composition, $2Na_2SO_4 \cdot Na_2CO_3$. The surface $RQSTVU$ represents equilibria between pure crystals of this double salt and liquid solutions. The other surfaces and lines of the diagram may be interpreted in a similar manner from the composition of the equilibrium solid which is marked on each surface.

An isometric diagram such as Fig. 13, though valuable as an aid to visualization, is not suitable as a basis for quantitative calculations. Data for calculation purposes are better presented as a family of isothermal solubility curves on a triangular diagram as shown in Fig. 14. The coordinate scales represent weight percentages of Na_2SO_4, Na_2CO_3 and H_2O respectively. Each line of Fig. 14 represents the solubility relationships at one indicated temperature. By interpolation between these lines solubilities at any desired temperature may be obtained. Points along curve AB represent conditions of saturation with both solutes. The curves running upward to the left from this curve represent solutions which are saturated with sodium carbonate but only partially saturated with sodium sulfate. Similarly, the curves running upward to the right represent saturation with sodium sulfate. The hydrates, $Na_2SO_4 \cdot 10H_2O$ (44.1% Na_2SO_4) and $Na_2CO_3 \cdot 10H_2O$ (37.1% Na_2CO_3), are represented by points C and D, respectively.

Point x_1 on Fig. 14 represents a composition which at a temperature of 25°C corresponds to a solution saturated with respect to sodium carbonate but unsaturated with respect to sodium sulfate. If such a solution is cooled, pure $Na_2CO_3 \cdot 10H_2O$ will crystallize and the composition of the residual solution will change along the dotted line Dx_1. At a temperature of 22.5°C, corresponding to the intersection of this dotted line with curve AB, the remaining solution will become saturated with both sodium carbonate and sulfate. Further cooling will result in crystallization of both decahydrates in crystals of the solid solutions which were previously discussed. The concentration of the residual solution will then diminish along curve AB. If cooling were stopped at 22.5°C a yield of pure $Na_2CO_3 \cdot 10H_2O$ crystals would be obtained, and a separation of one solute from the other would result.

Point x_2 on Fig. 14 represents a composition which at a temperature of 25°C corresponds to a solution unsaturated with respect to both solutes. If water is evaporated from such a solution at a temperature of 25°C the concentration of each solute will be increased. Since the relative proportions of the two solutes will remain unchanged, the composition of the solution will vary along a straight line passing through point G of the diagram. If evaporation is continued a composition x_2' will be reached at which the solution is saturated with respect to sodium sulfate. Further evaporation will result in crystallization of pure $Na_2SO_4\cdot$ $10H_2O$, and the composition of the residual solution will vary along the 25°C isothermal line from x_2' to B. When the residual solution reaches a composition B it will be saturated with both solutes. Further evaporation will produce crystallization of both solutes, and the com-

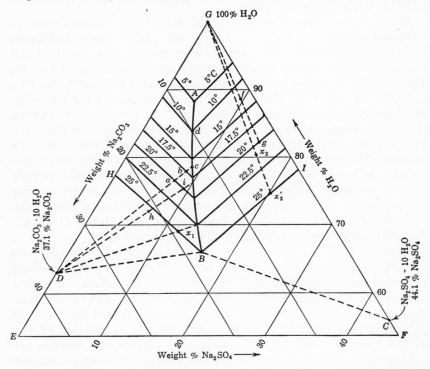

Fig. 14. Solubility of the system $Na_2CO_3 - Na_2SO_4 - H_2O$ at low temperatures.

position of the residual solution will remain unchanged at B. By such isothermal evaporation processes it is possible to separate the original solution into pure $Na_2SO_4\cdot10H_2O$ and a solution of composition B if the evaporation is not carried beyond the point of initial saturation with both solutes.

Points lying below a solubility isotherm on Fig. 14 represent the compositions of systems which at that temperature are heterogeneous mixtures. For example, at 25°C, any point lying in the field IBC will represent a mixture of solid $Na_2SO_4 \cdot 10H_2O$ and a solution saturated with sodium sulfate but not with sodium carbonate. Similarly, a point in the field HBD represents a system of solid $Na_2CO_3 \cdot 10H_2O$ and solution. A point in the field $EDBCF$ corresponds to a system saturated with both decahydrates. It will consist of a solution of composition B and a mixture of crystals of solid solutions of the two decahydrates.

With the aid of a chart of the type of Fig. 14 it is possible to carry out by combined graphical and arithmetical methods the calculations involved in crystallization problems.

Illustration 10. An aqueous solution at a temperature of 22.5°C contains 21 grams of Na_2CO_3 and 10 grams of Na_2SO_4 per 100 grams of water.

(a) Calculate the composition and weight of the crystals formed by cooling 1000 grams of this solution to a temperature of 17.5°C.

(b) Repeat the calculation of part (a) to correspond to a final temperature of 10°C.

Basis: 1000 grams of original solution.

Initial composition, point a on Fig. 14:

$$Na_2SO_4 \;=\; \frac{10}{131} \;=\; 7.64\%$$

$$Na_2CO_3 \;=\; \frac{21}{131} \;=\; 16.02\%$$

$$H_2O \qquad\quad = 76.34\%$$

(a) Upon cooling, $Na_2CO_3 \cdot 10H_2O$ will crystallize out along line aD, giving a composition at 17.5°C corresponding to point b.

$$Na_2SO_4 \;=\; 8.5\%$$
$$Na_2CO_3 \;=\; 13.8\%$$

Let x = weight of $Na_2CO_3 \cdot 10H_2O$ crystallized.
Then a material balance for Na_2CO_3 gives

$$(0.1602)(1000) = 0.138(1000 - x) + 0.371(x)$$
$$x = 98 \text{ g } Na_2CO_3 \cdot 10H_2O$$

(b) When the original solution cools to 10°C, crystallization of $Na_2CO_3 \cdot 10H_2O$ will proceed along Da until it strikes line AB, after which crystallization of both decahydrates will follow line AB.

Composition of final solution at d is

$$Na_2SO_4 \;=\; 6.2\%$$
$$Na_2CO_3 \;=\; 10.0\%$$
$$H_2O \qquad = 83.8\%$$

Let x = weight of $Na_2CO_3 \cdot 10H_2O$ crystallized,
y = weight of $Na_2SO_4 \cdot 10H_2O$ crystallized.

From a Na_2CO_3 balance

$$(0.1602)(1000) = 0.371(x) + (0.10)(1000 - x - y)$$

From a Na_2SO_4 balance

$$(0.0764)(1000) = 0.441(y) + (0.062)(1000 - x - y)$$

Hence $x = $ 251 grams
               $y = $ 79 grams

Where evaporation of solvent is involved in a crystallization process two types of problems may arise. It may be desired to calculate the quantity of solvent which must be removed in order to produce a specified yield of crystals or it may be desired to calculate the yield of crystals resulting from the removal of a specified quantity of solvent. In the first type of problem it is necessary to determine the composition of the solution remaining after the process. The quantity of solvent to be evaporated will then be the difference between that in the initial and final solutions. If both solutes crystallize in the process the residual solution will have a composition corresponding to saturation with both solutes at the final temperature. If only one solute crystallizes, the composition of the final solution may be determined graphically.

Illustration 11. An aqueous solution contains 9.8% Na_2SO_4 and 1.8% Na_2CO_3. How much water must be evaporated from 100 lb of this solution to saturate the solution with one solute at 20°C without crystallization?

Upon evaporation the composition of the solution will follow along a line fG, starting from the original composition corresponding to f on Fig. 14.

The line fG crosses the 20°C line at a composition of 15.1% Na_2SO_4 and 2.8% Na_2CO_3.

Let $x = $ water evaporated. Then a water balance gives

$$100(0.884) = x + (100 - x)(0.821)$$

or $$x = 35.2 \text{ lb}$$

If a specified quantity of solvent is to be evaporated from a solution and it is desired to calculate the resultant yield of crystals, the composition of the entire final mixture of crystals and solution is readily determined by subtraction of the quantity of evaporated solvent. If, in the final mixture, the concentration of only one of the solutes is greater than that corresponding to saturation, only this one, A, will be crystallized. The entire quantity of the other solute, B, must then be in the residual solution, fixing its composition with respect to this solute. The complete composition of the residual solution will be that corresponding to this concentration of solute B and saturation with solute A at the existing temperature. The quantity of solute A which will be crystallized will be the difference between the total quantity and that remaining in solution.

Illustration 12. A solution contains 19.8% Na_2CO_3 and 5.9% Na_2SO_4 by weight. Calculate the weight and composition of crystals formed from 100 lb of this solution when

(a) 10 lb of water is evaporated and the solution cooled to 20°C.

(b) 20 lb of water is evaporated and the solution cooled to 20°C.

Basis: 100 lb of solution.

(a) The residue after evaporation consists of

$$\frac{19.8}{90} = 22\% \ Na_2CO_3$$

$$\frac{5.9}{90} = 6.55\% \ Na_2SO_4$$

This composition corresponds to h on Fig. 14. Upon cooling, $Na_2CO_3 \cdot 10H_2O$ will crystallize out along line Dh reaching at 20°C a composition corresponding to point i, 9.4% Na_2SO_4 and 15.3% Na_2CO_3.

Let x = lb $Na_2CO_3 \cdot 10H_2O$ crystallized.
Then

$$90(0.22) = x(0.371) + (90 - x)(0.153)$$

or

$$x = 27.5 \text{ lb.}$$

(b) After evaporation of 20 lb water the residue consists of

$$\frac{19.8}{80} = 24.8\% \ Na_2CO_3$$

$$\frac{5.9}{80} = 7.4\% \ Na_2SO_4$$

At 20°C the solution becomes saturated with respect to both solutes, corresponding to a composition of 14.8% Na_2CO_3 and 11.4% Na_2SO_4.

Let x = lb $Na_2CO_3 \cdot 10H_2O$ crystallized,

y = lb $Na_2SO_4 \cdot 10H_2O$ crystallized.

From a Na_2CO_3 balance

$$0.248(80) = 0.371x + 0.148(80 - x - y)$$

From a Na_2SO_4 balance

$$0.074(80) = 0.44y + 0.114(80 - x - y)$$

or

$$x = 38.1 \text{ lb } Na_2CO_3 \cdot 10H_2O \text{ crystallized}$$
$$y = 3.45 \text{ lb } Na_2SO_4 \cdot 10H_2O \text{ crystallized}$$

The system of sodium sulfate, sodium carbonate, and water has been selected as an illustration, not because of its industrial importance but because it exhibits most of the phenomena to be found in such ternary systems. Many other systems involve similar formations and decompositions of hydrates, double salts, and solid solutions and may be dealt with by means of similar diagrams and methods. Several systems of industrial importance show peculiarities of individual behavior which form the bases of important processes. For example, in the system NaNO₃–NaCl–H₂O lowering the temperature decreases the solubility of NaNO₃ but *increases* that of NaCl in solutions which are saturated with both salts. This peculiarity makes it possible to crystallize pure NaNO₃ by cooling a solution saturated with both salts. This principle is used on a large scale for the commercial production of Chile saltpetre. In the system NaOH–NaCl–H₂O the solubility of NaCl in solutions containing high concentrations of NaOH becomes very small. This fact is taken advantage of in the separation of NaCl from electrolytically produced NaOH solutions. The solution is merely concentrated by evaporation at a relatively high temperature where the solubility of NaOH is great. As the concentration increases, NaCl crystallizes and solutions may be produced containing only traces of NaCl.

VAPOR PRESSURE AND RELATIVE HUMIDITY ABOVE SOLUTIONS

Certain aqueous solutions are used for drying gases, dehumidification of air, and control of humidity in air conditioning. Solutions of lithium chloride, calcium chloride, glycerol, and triethylene glycol are used for this purpose in large industrial applications, and in addition sulfuric acid, phosphoric acid, and caustic soda solutions are used as laboratory desiccants. To simplify calculations it is convenient to show in a single diagram the relation of vapor pressure, relative humidity, temperature, and composition of the water-desiccant system.

Such a diagram is shown for the calcium chloride-water system in Fig. 15. The area to the left of curve *abcde* represents unsaturated solutions of calcium chloride in equilibrium with water vapor. The curved lines are isotherms showing the decrease in vapor pressure with increase in concentration. The nearly vertical dotted lines are constant relative humidity. Curve *abc* represents the solubility of the hexahydrate in water with corresponding changes in temperature, vapor pressure, and relative humidity. Line *cf* represents pure hexahydrate crystals containing 50.7% CaCl₂. In area *abcf*, crystals of the hexahydrate and its saturated solution are present. The horizontal lines inside this area represent the vapor pressure of the saturated solution at various temperatures, with corresponding relative humidities.

The vapor pressure of the hexahydrate is lower than that of the saturated solution but any tendency for condensation on the surface of the hexahydrate is immediately reversed by the formation of a saturated solution having a vapor pressure equal to that of the system. At point *c*, 86°F, crystals of hexahydrate are transformed to the tetrahydrate, and crystals of the hexahydrate do not exist above this temperature. At

point d, 112°F, crystals of tetrahydrate are converted to the dihydrate
and crystals of the tetrahydrate do not exist above this temperature.   In
area $fchg$, crystals of both hexa- and tetrahydrates are present.   The
horizontal lines represent vapor pressures of the hexahydrate.   Three
phases exist in equilibrium and in mutual contact in this area.   The
vapor pressure of the hexahydrate is higher than that of the tetrahydrate

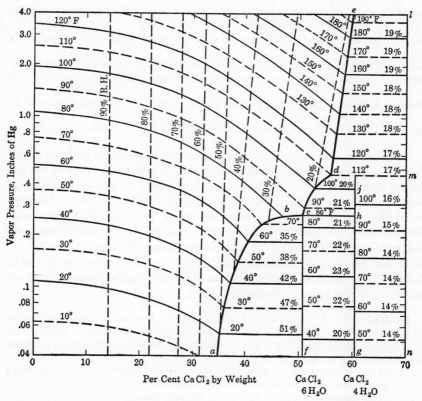

FIG. 15.   Vapor pressure and relative humidity over calcium chloride solutions.

but any tendency to condense water on the tetrahydrate is immediately
followed by hexahydrate formation with a vapor pressure equal to that
of the system.   In the area $cdjh$, crystals of tetrahydrate are in equilibrium with a saturated solution of tetrahydrate.   The horizontal lines
represent vapor pressures of the saturated solutions of tetrahydrate.   In
area $delm$, crystals of dihydrate are in equilibrium with a saturated solution of dihydrate.   The horizontal vapor pressure lines represent the
vapor pressure of the saturated solution at various temperatures.   In

the area *gjmn* two solid phases of tetrahydrate and dihydrate are present. The horizontal lines represent the vapor pressure of the tetrahydrate at various temperatures. The vapor pressure of the dihydrate is lower but the tendency for water to condense on the dihydrate is reversed by formation of the tetrahydrate with vapor pressure equal to that of the system.

AREA	PHASES PRESENT BESIDES WATER VAPOR	VAPOR PRESSURE LINES REPRESENT
abcf	Saturated solution and crystals of hexahydrate	Saturated solution of hexahydrate
fchg	Crystals of hexahydrate and of tetrahydrate	Crystals of hexahydrate
cdjh	Saturated solution and crystals of tetrahydrate	Saturated solution of tetrahydrate
delm	Saturated solution and crystals of dihydrate	Saturated solution of dihydrate
gjmn	Crystals of tetrahydrate and dihydrate	Crystals of tetrahydrate

In using the calcium-chloride-water system in a drying and regenerating cycle, during the drying stage the vapor pressure of water in the gas to be dried must exceed the equilibrium value of the system and in regeneration the vapor pressure of water in the hot air blown through must be less than the equilibrium vapor pressure of the system. Regeneration of calcium chloride from solution is frequently not considered feasible because of the relative cheapness of the salt.

It will be observed that in using a homogeneous solution of $CaCl_2$ below 40°F the lowest relative humidity obtainable is 45% whereas above 90°F the lowest attainable is 19%.

Illustration 13. Ninety pounds of anhydrous $CaCl_2$ is placed in a room of 90,000 cu ft capacity. The air is initially at 80°F and 90% relative humidity. Estimate the conditions of air and desiccant when equilibrium has been attained at 80°F, assuming a pressure of 1.0 atmosphere.

It will be estimated that enough water is removed to liquefy the $CaCl_2$, leaving a saturated solution of hexahydrate and air at a relative humidity read from Fig. 15 as 27%. This assumption will be verified by the calculations.

Weight of dry air in room	= 6560 lb
Initial humidity lb/lb	= 0.0198
Final humidity	= 0.0061
Water removed = 6560(0.0198 − 0.0061)	= 89.9 lb.

Let x = crystals of hexahydrate remaining.

Then, $(179.9 - x)$ = weight of saturated $CaCl_2$ solution.

From a $CaCl_2$ balance

$$90 = 0.507(x) + 0.46(179.9 - x)$$
$$x = 153 \text{ pounds of } CaCl_2 \cdot 6H_2O$$
$$(179.9 - 153) = 26.9 \text{ pounds of saturated solution.}$$

DISTRIBUTION OF A SOLUTE BETWEEN IMMISCIBLE LIQUIDS

When a solute is added to a system of two immiscible liquids, the solute is distributed between the liquids in such proportions that a definite equilibrium ratio exists between its concentrations in the two phases. The equilibrium between the solute in the two phases is of a dynamic nature with solute particles continually diffusing across the interface from one liquid to the other. At equilibrium conditions the concentrations adjust themselves so that the rate of loss of solute particles by each phase is compensated by its rate of gain of particles from the other phase.

The equilibrium distribution of a solute between immiscible solvents is expressed by the *distribution coefficient*, K, which is the ratio of the concentrations in the two phases. Thus,

$$K = \frac{C_B}{C_A} \tag{4}$$

where C_A, C_B = concentrations of solute in phases A and B, respectively. If sufficient solute is present to saturate the system completely, each phase must contain solute in the concentration corresponding to its normal saturation conditions. Therefore, the distribution coefficient at saturation is merely the ratio of the solubilities of the solute in the two liquids.

TABLE III

DISTRIBUTION OF PICRIC ACID BETWEEN BENZENE AND WATER

C_A = concentration of picric acid in water, gram-moles per liter of solution
C_B = concentration of picric acid in benzene, gram-moles per liter of solution
K = distribution coefficient at 15–18°C = C_B/C_A

C_B	K
0.000932	2.23
0.00225	1.45
0.01	0.705
0.02	0.505
0.05	0.320
0.10	0.240
0.18	0.187

Effect of Concentration. In ideal systems in which dissociation and association are absent the distribution coefficient is independent of concentration. Ordinarily, however, it shows marked variation with concentration, as indicated by the values in Table III for the distribution of picric acid, $HOC_6H_2(NO_2)_3$, between water and benzene, C_6H_6. Similar data for many other systems may be found in the International Critical Tables, Vol. III, page 418.

Effect of Temperature.   The effect of temperature on the distribution coefficient is generally small if the temperature coefficients of solubility are approximately equal in the two phases. Specific data are necessary in order to predict the effects of a temperature change. For many industrial calculations the effects of temperature changes of only a few degrees may be disregarded.

Distribution Calculations.   The distribution of a solute between two immiscible liquids is of considerable industrial importance in the separation and purification of organic compounds. Ordinarily one liquid will be water or an aqueous solution and the other some immiscible organic solvent. Equilibrium concentrations of a solute in such systems may be varied by the addition of a second solute which is soluble in only one of the liquids. The addition of such a solute is, in effect, a change in the nature of one of the liquids.

From values of distribution coefficients equilibrium conditions are readily calculated.

Illustration 14.   Picric acid exists in aqueous solution at 17°C in the presence of small amounts of inorganic impurities whose effects on its solubility may be neglected. The picric acid is to be extracted with benzene in which the inorganic materials are insoluble.

(*a*) If the aqueous solution contains 0.20 gram-mole of picric acid per liter, calculate the volume of benzene with which 1 liter of the solution must be extracted in order to form a benzene solution containing 0.02 gram-mole of picric acid per liter. (Neglect the difference between the volume of a solution and that of the pure solvent.)

(*b*) Calculate the percentage recovery of picric acid from the aqueous solution.

Basis: 1 liter of original aqueous solution.

Picric acid = ...   0.20 gram-mole

From Table III, $K = 0.505$ in final system = ... C_B/C_A

Final concentration of picric acid in aqueous solution = 0.02/0.505 = .   0.0396 gram-mole per liter

Picric acid in final benzene solution = 0.20 − 0.0396 = ...   0.16 gram-mole

Benzene required $= \dfrac{0.16}{0.02} = $ ....   8.0 liters per liter of aqueous solution

Percentage extraction of picric acid $= \dfrac{0.16}{0.20} = 80\%$

In calculations involving concentrated solutions the differences between the volume of a solution and that of the pure solvent cannot be neglected as was done in Illustration 14. In such cases it is convenient to express the concentrations and distribution coefficients in terms of the weight of solute per unit weight of solvent. The units in which distribution data are ordinarily expressed, as in Table III, may be readily converted into these terms if density-concentration data are available for both solutions.

If definite quantities of two immiscible solvents and a solute are mixed together, the final concentration of solute in either solution will be unknown.   If the distribution coefficient varies considerably with concentration it will also be unknown.   The distribution of the solute in such a case is best estimated by a method of successive approximations. A reasonable value of the final distribution coefficient is assumed as a first approximation.   On the basis of this assumed value a first approximation to the final concentrations is calculated.   The distribution coefficient is then corrected to correspond to these concentrations.   On the basis of the second approximation to the distribution coefficient, a second approximation to the final concentrations is calculated.   Unless the variation of distribution coefficient with concentration is very marked, two or three successive approximations of this type will yield results satisfactory for ordinary purposes.

Illustration 15.   One liter of a benzene solution containing 0.10 gram-mole of picric acid per liter is agitated with 1.0 liter of water.   Estimate the final concentration of picric acid in each solvent.

Solution: As a first approximation, assume from Table III that the final distribution coefficient will be 0.5.

Let x_1 = gram-moles of picric acid in final benzene solution.

Picric acid in final aqueous solution = $0.10 - x_1$.

$$\frac{x_1}{0.10 - x_1} = 0.5$$

$$x_1 = 0.033$$

The distribution coefficient corresponding to this concentration is taken from the data of Table III as a second approximation.

$$K_2 = 0.39$$

$$\frac{x_2}{0.10 - x_2} = 0.39$$

$$x_2 = 0.028$$

As a third approximation:

$$K_3 = 0.42 \text{ (corresponding to } C_B = 0.028)$$

$$\frac{x_3}{0.10 - x_3} = 0.42$$

$$x_3 = 0.029$$

This result may be taken as the final concentration of picric acid in the benzene solution.   If greater accuracy is desired, more approximations should be carried out.

Picric acid in final benzene solution = ...   0.029 gram-mole

Picric acid in final aqueous solution = ...   0.071 gram-mole

PARTIALLY MISCIBLE LIQUIDS

When two partially miscible liquids are brought together, a range of compositions occurs in which two liquid phases exist in equilibrium with each other. This behavior is typified by the water-phenol system illustrated in Fig. 16. At a temperature of 50°C mixtures containing more than 11.8% and less than 62.6% phenol by weight will separate into two layers, the upper containing 11.8% phenol and the lower 62.6%. Addition of either component to such a two-phase system will not change the

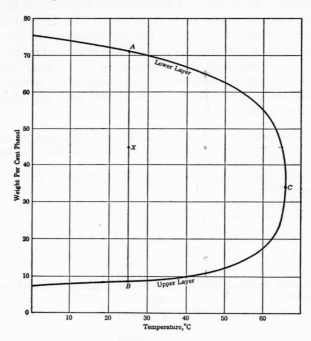

FIG. 16. Solubility of phenol in water.

composition of either phase but will merely shift the proportions in which they are present. Two liquid phases existing in equilibrium with each other in this manner with their compositions independent of the composition of the total two-phase mixture are termed *conjugate solutions*.

If the temperature of the phenol-water mixture is increased above 50°C the compositions of the conjugate solutions are changed, the phenol content of the upper layer increasing and that of the lower layer decreasing. When a temperature of 66°C is reached the compositions of the conjugate phases become equal and the mixture becomes homogeneous.

The composition at which equality of composition of the two phases is reached is termed the *critical solution composition* and the temperature at which miscibility becomes complete is the *critical solution temperature.* In Fig. 16, point C corresponds to a critical solution composition of 34% phenol and a critical solution temperature of 66°C.

The weights of the two phases in a heterogeneous binary system are readily calculated by component material balances if the composition of the entire mixture is known, together with compositions of the conjugate solutions as shown in Fig. 16. It may be demonstrated that if point X represents the composition of the entire mixture, composed of phases of compositions A and B, the weight of phase A is proportional to line segment BX and the weight of phase B is proportional to segment AX.

Ternary Liquid Mixtures. Mixtures of three or more partially miscible liquids may separate into two phases in equilibrium with each other in a manner analogous to that of the binary mixtures described above. In Figs. 17 and 18 are plotted the isothermal solubility data of a representative ternary system, tetrachlorethylene, isopropyl alcohol and water at 77°F, as determined by Bergelin, Lockhart and Brown.[3] At this temperature water is completely miscible with isopropyl alcohol. Similarly, tetrachlorethylene is completely miscible with isopropyl alcohol. However, water and tetrachlorethylene are almost completely immiscible with each other.

When the three components are mixed together, homogeneous solutions are formed only at compositions lying above curve *acb* of Fig. 17. Mixtures having compositions lying below this curve separate into two layers, the upper rich in water and the lower rich in tetrachlorethylene. Curve *acb*, bounding the two-phase region, is termed an *isothermal solubility curve.* Similar curves for other temperatures might be plotted on the same diagram to define completely the solubility characteristics of the system.

A mixture having a composition x in the two-phase region separates into two conjugate phases experimentally found to have compositions corresponding to points d and e. Thus, an upper layer of composition d must always be in equilibrium with a lower layer of composition e. As pointed out in Chapter I it is characteristic of the ternary diagram that mixtures of two solutions must have compositions corresponding to compositions falling on a straight line connecting the composition points of the original solutions. Thus, all mixtures of the conjugate phases having compositions d and e must lie along line de. This line is termed a *tie-line,* connecting the composition points of conjugate phases. It may be demonstrated from the geometry of the diagram that the weights of the

[3] Bergelin, Lockhart, and Brown, *Trans. Am. Inst. Chem. Eng.*, **39**, 173 (1943).

two phases into which a mixture of composition x separates are proportional to the lengths of the line segments dx and ex. Thus, the weight fraction of the upper-layer solution is equal to segment ex divided by segment de.

In order to define completely the solubility and equilibrium relationships of such a ternary system, a family of tie-lines such as fg and hi must be drawn in, linking together the compositions of conjugate solutions on the solubility curves. Tie-lines lying higher in the diagram than hi would be progressively shorter until at point c the length of the tie-

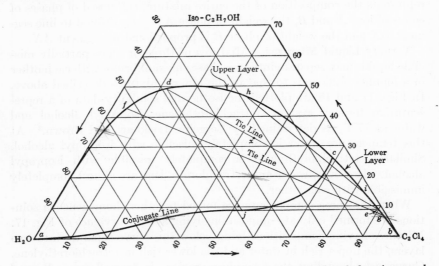

FIG. 17. Solubility curve and conjugate line for tetrachlorethylene-isopropyl alcohol-water system at 77°F in weight per cent. From Bergelin, Lockhart and Brown, *Trans. Am. Inst. Chem. Eng.*, **39**, 173 (1943).

line becomes zero. This point is termed the *critical solution point* or *plait point*, at which the compositions of the upper and lower layers of the two-phase system become equal.

The locations of all tie-lines can be defined without confusing the diagram by constructing them through use of curve ac, termed the *conjugate line*. This line is the locus of the points of intersection of lines drawn through conjugate compositions points parallel to the base and right side of the diagram respectively. Thus, point j is located as the intersection of lines dj and ej. Conversely, point j defines the location of tie-line de. In a similar manner any other tie-lines on the diagram may be located from the conjugate line.

It will be noted that for this system the critical solution point does not correspond to the maximum solubility of alcohol. Accordingly, the concentration of isopropyl alcohol must pass through a maximum in the upper layer as the isopropyl alcohol content of the system is increased. This behavior is shown more clearly in Fig. 18 where the concentration of isopropyl alcohol in the upper layer is plotted as a function of its concentration in the conjugate lower layer. This behavior is typical of many ternary systems and clearly indicates the extreme deviations

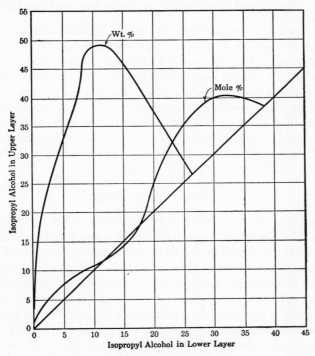

FIG. 18. Equilibrium distribution of isopropyl alcohol in water and tetrachlorethylene at 77°F in weight per cent. From Bergelin, Lockhart, and Brown, *Trans. Am. Inst. Chem. Eng.*, **39**, 173 (1943).

which may be encountered from the constancy of distribution factor assumed in Equation (4).

Methods for the correlation of solubility and equilibrium data and the prediction of tie-line locations have been developed by Othmer and Tobias[4] for systems of two slightly miscible components A and B in admixture with a third component C which is miscible with both pure A and pure B. For a large number of such systems the following em-

[4] Othmer and Tobias, *Ind. Eng. Chem.*, **34**, 690, 693, 696 (1942).

pirical equation was found to relate the compositions of the conjugate phases:

$$\frac{1 - a_1}{a_1} = u \left(\frac{1 - b_2}{b_2}\right)^v \tag{4a}$$

where a_1 = weight fraction of A in the phase rich in A

b_2 = weight fraction of B in the phase rich in B

u, v = constants characteristic of the system and temperature

Since all conjugate compositions lie on the solubility curve knowledge of a_1 and b_2 completely specifies the compositions of two conjugate solutions if the solubility curve is known. Knowledge of the compositions of two sets of conjugate solutions permits evaluation of the constants u and v in Equation (4a). Once u and v are established values of a_1 corresponding to assumed values of b_2 are readily calculated from Equation (4a) and the entire system of tie-lines and the conjugate line are established. These investigators also discuss methods for developing partial vapor-pressure isotherms which may be plotted directly on solubility charts such as Fig. 17.

SOLUBILITY OF GASES

. Sorption comprises the general phenomenon of the assimilation of a gas by a solid or liquid. When the sorbed gas forms a homogeneous solution with the liquid or forms a new chemical compound with the solid, the transformation is called *absorption*. When the gas is taken on only at the surface or in the capillaries of the solid to form a surface compound or condensate the phenomenon is designated as *adsorption*.

When a gas is brought into contact with the surface of a liquid, some of the molecules of the gas striking the liquid surface will dissolve. These dissolved molecules of gas will continue in motion in the dissolved state, some returning to the surface and re-entering the gaseous state. The dissolution of the gas in the liquid will continue until the rate at which gas molecules leave the liquid is equal to the rate at which they enter the liquid. Thus, a state of dynamic equilibrium is established and no further change will take place in the concentration of gas molecules in either the gaseous or liquid phases. The concentration of gas which is dissolved in a liquid is determined by the partial pressure of the gas above the surface.

Henry's Law. For many gases the relationship between the concentration of gas dissolved in a liquid and the equilibrium partial pressure of the gas above the liquid surface may be expressed by Henry's law. The ordinary statement of this law is that *the equilibrium value of*

the mole fraction of gas dissolved in a liquid is directly proportional to the partial pressure of that gas above the liquid surface, or

$$N_1 = \frac{1}{H} p_1 \qquad (5)$$

where

p_1 = equilibrium partial pressure of gas in contact with liquid

N_1 = mole fraction of gas in liquid

H = Henry's constant, characteristic of the system

This relationship has been found to be satisfactory at low concentrations, corresponding to low partial pressures of gas and high values of H. The factor H is a function of the specific nature of the gas and liquid and of the temperature, in general increasing with increase in temperature.

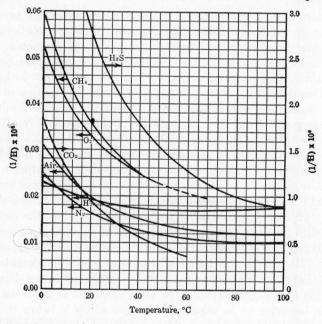

FIG. 19.   Solubilities of gases in water.   Variation of Henry's constant, H, with temperature (*pressures in millimeters of mercury*).

When pressures and concentrations are low the solubility data of a gas-liquid system are completely expressed by data relating values of Henry's constant, H, to temperature.   In Fig. 19 are curves expressing this relationship for several common gases in water, the numerical values of $1/H$ corresponding to pressures in millimeters of mercury.   The data for hydrogen sulfide and carbon dioxide will lead to considerable error if used for pressures above about 1 atmosphere.

Illustration 16. Calculate the volume of oxygen, in cubic inches, which may be dissolved in 10 lb of water at a temperature of 20°C and under an oxygen pressure of 1 atmosphere.

Solution:

From Fig. 19, for oxygen and water at 20°C, $1/H = $ . . . . 0.033×10^{-6}

Mole fraction of $O_2 = 760 \times 0.033 \times 10^{-6} = $ . . . . 25.1×10^{-6}

Mole fraction of water $= $ 1.00

Pound-moles of dissolved $O_2 = \dfrac{10}{18} \times 25.1 \times 10^{-6} = $ . . . . . 13.9×10^{-6}

Volume of dissolved $O_2 = 13.9 \times 10^{-6} \times 359 \times \dfrac{293}{273} = $ . . 5.37×10^{-3} cu ft

or 9.3 cu in. measured at 20°C, and a pressure of 1 atmosphere

Deviations from Henry's Law. Under conditions of high pressure or for a gas of relatively high solubility the direct proportionality of Henry's law breaks down. Aqueous solutions of ammonia, carbon dioxide, and hydrochloric acid are examples of systems whose behaviors deviate widely from that predicted by Henry's law except at low pressures. This deviation results in part from chemical reaction of the gas with the liquid and subsequent ionization of the dissolved molecules. Nernst has pointed out that Henry's law holds closely even for these cases when applied to the same single species in both phases, that is, to only the uncombined and unassociated molecules in the liquid phase. However, in general, experimentally determined data relating temperatures, pressures, and solubilities over the entire desired range are necessary in order to predict solubilities in such systems. These data are ordinarily expressed by *solubility isotherms* which relate the concentration of a dissolved gas to its partial pressure at a constant temperature. In Fig. 20 are solubility isotherms of ammonia in water at various temperatures. It will be noted that all the lines have considerable curvature, and that those corresponding to the lower temperatures show points of inflection at high pressures. The form of the 20°C isotherm is typical of the behavior to be expected of a gas which is below its critical temperature and dissolved in a solvent with which it is miscible when in the liquid state. If measurements were continued at higher pressures the slope of the curve would be expected to increase, becoming asymptotic to the abscissa corresponding to a pressure of 6420 millimeters, the vapor pressure of liquid ammonia at 20°C. Gaseous ammonia at 20°C could not exist at higher pressures, and at this pressure an infinite amount of ammonia could be condensed and dissolved in 1 gram of water. The isotherms corresponding to higher temperatures should exhibit similar points of inflection if extended to higher pressures.

From the data of Fig. 20 it is apparent that Henry's law should not be used when dealing with a gas with a considerable affinity for the solvent or at high pressures.   At high pressures the solubility calculated from Henry's law will, in general, be higher than the correct value.

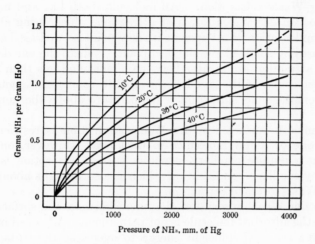

FIG. 20.   Solubility of ammonia in water.

In such cases specific experimental data such as those of Fig. 20 are necessary for dependable calculations.   Data for many systems, both in aqueous and organic solvents, are contained in the International Critical Tables (Vol. III, pages 255–283).   Data are included on the solubilities of gases in other solutions as well as in pure liquids.   In general, the solubility of a gas in a liquid is diminished by the addition of a nonvolatile solute with which it does not react chemically.

ADSORPTION OF GASES

The adsorption of gases by solids is of importance in many industrial operations such as the purification and drying of gases, the recovery of solvent vapors, the production of casing-head gasoline, and the operation of gas masks.   An application of increasing importance is in air conditioning, where dehumidification is often accomplished by adsorption of water vapor on solid desiccants.

Two types of adsorption should be recognized, one caused by the same type of intermolecular forces of attraction that produce normal condensation to the liquid state, called van der Waals forces, and the other caused by specific chemical bonds between the atoms on the surface of the solid and of the gas adsorbed, termed activated adsorption. Activated adsorption is of interest in connection with catalytic phenomena but of little interest in processes depending upon readily reversible adsorption and desorption such as discussed in this chapter.

van der Waals Adsorption. All molecules both like and unlike are subject to van der Waals forces of attraction, which bring about the normal condensation of a vapor at its dew point as well as the adsorption of gases by solids above the dew point. Where the van der Waals forces of attraction are greater between a solid and gas than between the like molecules of the gas, it is possible to condense the gas by adsorption on the solid surface at a temperature above the normal dew point.

On a smooth surface van der Waals adsorption is restricted to a layer of one or a few molecules in thickness. However, on a solid possessing a minute capillary structure surface adsorption is supplemented by capillary condensation which is also brought about by the van der Waals forces of attraction.

In van der Waals adsorption the union between the surface of the solid and the adsorbed molecule is not permanent. Adsorbed molecules which have acquired sufficient energy to overcome the surface forces continually evaporate while other molecules are being adsorbed. When a gas is brought into contact with an adsorbent surface, adsorption takes place until the rate at which gas molecules strike the surface and are adsorbed is equal to the rate of evaporation of adsorbed molecules. No further change will then take place in the concentration of the gas in either the gaseous or adsorbed phases and a condition of dynamic equilibrium will exist.

The equilibrium between a gas and a solid adsorbent is similar to that existing between a pure liquid and its vapor or between a gas and its solutions. The amount of gas which is adsorbed at equilibrium always increases with increase in partial pressure and decreases with increase in temperature.

Capillary Condensation. Adsorbents of industrial importance are generally substances of highly porous structure, which expose enormous interior surfaces. Activated alumina, charcoal, and silica gel are familiar examples. In silica gel, for instance, the capillary pores are about 4×10^{-7} centimeters in diameter, only 10 times the diameter of simple molecules, and comprise about 50% of the total volume. The total interior area is about one acre per cubic inch. In such materials capillary condensation is of great importance.

Capillary condensation can occur only when the solid is wetted by the condensate, resulting in concave surfaces of the condensed liquid. The equilibrium vapor pressure of a liquid having a concave surface is less than the normal value by an amount depending upon the radius of curvature.   In the submicroscopic capillary pores of a wetted solid the condensation of liquid will produce concave liquid surfaces of extremely small radii of curvature and correspondingly low vapor pressures.   For this reason vapors which are at partial pressures much less than the normal saturation value are condensed, augmenting the adsorption normally taking place on flat surfaces.   The adsorbing capacity of a material possessing submicroscopic capillarity is considerably greater than for one having the same surface area but having no capillary structure.

As the partial pressure of a gas is increased adsorption progressively increases.   At low partial pressures capillary condensation does not take place.   When a pressure sufficient to produce condensation in the smallest capillaries is reached, capillary condensation begins and the capillaries fill to levels of greater diameter at higher vapor pressures. The relationship between pressure and amount adsorbed is, therefore, dependent on the size distribution of the capillary pores as well as on the area of exposed surface and the nature of both adsorbent and gas.

Equilibrium Adsorption.   The amount of gas adsorbed by a solid under equilibrium conditions may be expressed either as percentage by weight, or as the mass or gaseous volume of adsorbate per unit mass of gas-free adsorbent.   The percentage water adsorbed by a solid can be plotted against partial pressure of water vapor in a family of isotherms, or against temperature in a family of isobars.   All the data presented in such extensive isotherm and isobar plots can be approximately represented by a single line for a given substance when moisture content is plotted against relative humidity.   In Fig. 21 are plotted equilibrium moisture contents against relative humidity for a number of common materials.

For silica gel the commercially dry basis is used in expressing moisture content.   The commercially dry gel contains 5% water in chemical combination which is not removed in desorption or regeneration operations and hence is not included in reporting moisture content.   If this chemically adsorbed water were removed, the adsorptive capacity of the gel would disappear.   For a particular silica gel the useful range of moisture content for temperatures up to 100°F may be represented by the simple equation:

$$w_e = 55H_r = 55\frac{p_a}{p_s} \tag{6}$$

where

H_r = relative humidity

w_e = equilibrium moisture content, pounds per 100 pounds of commercially dry gel

p_a = partial pressure of water vapor in air

p_s = vapor pressure of pure water at temperature of air

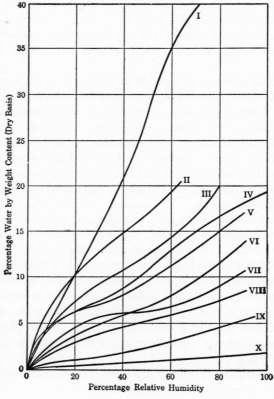

I. Silica gel.
II. Leather, chrome tanned.
III. Wool, worsted.
IV. Activated alumina.
V. Viscose

VI. Cotton cloth.
VII. Sulfite pulp, fresh, unbleached.
VIII. Bond paper.
IX. Cellulose acetate silk, fibrous.
X. Kaolin (Florida).

FIG. 21.   Equilibrium moisture content of various substances, at 77°F.

The simplicity of presenting equilibrium adsorption against relative saturation instead of as isotherms or isobars is at once apparent.   It should be noted that relative saturation should be employed and not percentage saturation, since the former is quite independent of total pressure, as shown on page 92.

Illustration 17. Air at atmospheric pressure and 70°F leaves a bed of silica gel which contains 5% water (commercially dry basis), at the exit end. Assuming equilibrium conditions, what is the relative humidity of the exit air? It is assumed that the operation is isothermal.

From Fig. 21 the relative humidity of air leaving at equilibrium with silica gel of 5% water content is 9% or from Equation (6) the relative humidity is

$$\frac{5}{55} \times 100 = 9.1\%$$

Equilibrium moisture content is of importance in the drying of materials which exhibit adsorptive characteristics. It is evident that water will not evaporate from an adsorbent solid into a gas the relative humidity of which is higher than that corresponding to the equilibrium moisture content of the solid. Thus, the equilibrium moisture content represents the minimum moisture content to which a material can be dried by a gas of a given relative humidity. Continued passage of gas will not result in further drying even though the gas is far from normal saturation with water.

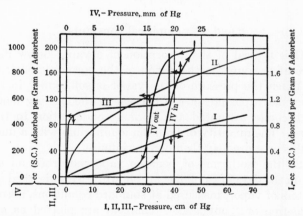

I. CO_2 on silica gel at 0°C. Patrick, Preston and Owens, *J. Phys. Chem.*, **29**, 421 (1925).

II. SO_2 on silica gel at 30°C. McGavack and Patrick, *J. Am. Chem. 'Soc.*, **42**, 946 (1920).

III. C_6H_6 on cocoanut charcoal at 59.5°C. Coolidge, *J. Am. Chem. Soc.*, **46**, 596 (1924).

IV. H_2O on pine charcoal at 25°C. Allmand, Hand, Manning, and Shiels, *J. Phys. Chem.*, **33**, 1682 (1929).

FIG. 22.   Adsorption isotherms.

Adsorption Isotherms.   A group of typical adsorption isotherms is shown in Fig. 22.   The quantity of adsorbate is expressed as the volume of gas measured in millimeters at 0°C and 760 millimeters of mercury, per gram of gas free solid.

It will be noted from these curves that adsorption is a specific property depending upon the nature of the system. Curves I and II are typical of the adsorption of a gas which is above its critical temperature or far removed from conditions of normal saturation and are similar in shape to the curves characteristic of activated adsorption. The adsorbed quantity increases with increased pressure but at a continually diminishing rate. The adsorption isotherm for this case may be expressed by the Taylor equation:

$$w = \frac{ap}{1 + ap} \tag{7}$$

where

w = the mass adsorbed per unit weight of adsorbent
p = the partial pressure of the adsorbate in the gas phase
a = a constant

In many cases the adsorption isotherm is satisfactorily represented by the empirical equation proposed by Freundlich:

$$w = kp^n \tag{8}$$

where

k and n = empirical constants

Curve III is typical of the adsorption of a vapor below its critical temperature at pressures in the region of the normal vapor saturation pressure. When the pressure of the vapor is sufficiently increased to equal the normal vapor pressure of benzene at the existing temperature the vapor becomes saturated and normal condensation to the liquid state results. The quantity of liquid in equilibrium with a unit weight of solid at this pressure may then become infinite.

In Fig. 23 the isotherms of benzene vapor adsorbed on activated charcoal from the data of Coolidge[5] are shown. The adsorbed quantity, x, in milliliters of vapor, measured at standard conditions, adsorbed per gram of gas-free or "outgassed" charcoal are plotted as ordinates. Logarithms of the pressures of benzene vapor in millimeters of mercury are plotted as abscissas. The logarithmic scale is desirable because of the wide range of pressures required. The characteristic shape of these curves when plotted in linear coordinates is indicated by Curve III of Fig. 22. The coconut charcoal used in the experiments on which Fig. 23 is based had been outgassed by heating to 550°C at reduced pressure.

[5] *J. Am. Chem. Soc.* **46**, 596 (1924).

It will be noted from Fig. 23 that, like the systems of Curves III and IV of Fig. 22, when the pressure is sufficiently increased to equal the normal vapor pressures of benzene at the existing temperature, the vapor becomes saturated and normal condensation to the liquid state results.

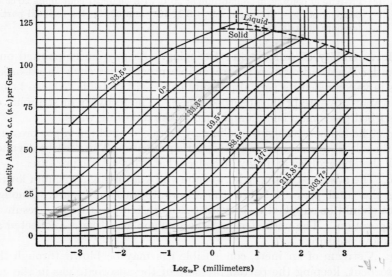

FIG. 23.   Adsorption of benzene on activated cocoanut charcoal
(outgassed at 550°C.)

The data of Fig. 23 are rigorously applicable only to the particular charcoal for which they were determined.   The quantity adsorbed is dependent on the source of charcoal, the method of its preparation, and its subsequent treatment.   The same limitation applies to adsorption data for other materials.

Illustration 18.   (a) Estimate the number of pounds of benzene which may be absorbed by 1 lb of the activated charcoal of Fig. 23 from a gas mixture at 20°C in which the partial pressure of benzene is 30 mm of Hg.

(b) Calculate the percentage of the adsorbed vapor of part (a) which would be recovered by passing superheated steam at a pressure of 5 lb per sq in. and a temperature of 200°C through the adsorbent until the partial pressure of benzene in the steam leaving is reduced to 10.0 mm of Hg.

(c) Calculate the residual partial pressure of benzene in a gas mixture treated with the freshly stripped charcoal of part (b) at a temperature of 20°C.

Solution: From Fig. 23,

(a) C_6H_6 adsorbed at 20°C, 30 mm of Hg = ... 110 cc per g

$$\text{or } \frac{110}{22,400} \times 78 = 0.382 \text{ g per g or lb per lb}$$

(b) C_6H_6 adsorbed at 200°C, 10.0 mm of Hg = ...   22 cc per g

     Benzene recovered = 110 − 20 = ...   90 cc per g

     Percentage recovery = $\dfrac{90}{110}$ = ...   82%

(c) Pressure of benzene in equilibrium at 20°C with charcoal containing 20 cc of benzene per g = antilog of $\overline{4}.9 = 8 \times 10^{-4}$ mm of Hg, the residual partial pressure of benzene.

Reversibility of Adsorption: Stripping. As previously pointed out, van der Waals adsorption is a reversible process and such an adsorbed gas is vaporized if its partial pressure in the gas phase is reduced below its vapor pressure in the adsorbed phase. The recovery or stripping of adsorbed gases may be accomplished in the following ways:

a. The temperature of the solid may be raised until the vapor pressure of the adsorbed gas exceeds atmospheric pressure. The adsorbate vapor will then be evolved and may be collected at atmospheric pressure.

b. The adsorbed gas may be withdrawn by applying a vacuum lowering the total pressure below the adsorbate vapor pressure. Enough heat should be supplied to prevent a drop in temperature as a result of evaporation. The undiluted adsorbate vapor may then be collected at this low pressure.

c. A stream of an inert, condensible gas may be blown through the adsorbent, keeping the partial pressure of the adsorbate gas in the gas stream below the equilibrium pressure of the adsorbate in the solid. The adsorbate vapor will be evolved in admixture with the inert gas. By using an easily condensible vapor for stripping, such as superheated steam, the adsorbed material may be easily recovered by condensing the stripping vapor only or by condensing the entire mixture and separating by decantation provided the two condensed vapors are immiscible.

d. The adsorbed vapors may be displaced by treatment with some other vapor which is preferentially adsorbed.

The various desiccants used for drying gases are usually regenerated by blowing hot air through the spent adsorbent. For example, silica gel is regenerated by hot gases at 250°F to 350°F and activated alumina by hot gases from 350° to 600°F.

Adsorption Hysteresis. Curve IV on Fig. 22 indicates a behavior which is typical of certain adsorbents which possess a high degree of capillarity. As indicated by the arrows on the double section of the curve, the quantity adsorbed in equilibrium with a selected partial pressure is dependent on the direction from which equilibrium is approached. If equilibrium is reached by the evolution of adsorbate the upper or " out " curve is applicable. If the equilibrium conditions are reached by adsorption, with a continually increasing concentration, the lower or

" in " curve applies. This behavior is known as *adsorption hysteresis* and is exhibited by many vapors when adsorbed in large quantities on charcoal and other capillary adsorbents. In calculations dealing with such systems data for both " in " and " out " curves are required.

This phenomenon can be explained by assuming a particular geometric shape of pore spaces within the solid. In Fig. 24, is shown a pore space inside a solid undergoing desorption in one case and adsorption in the other. If desorption is started with a full pore space and unsaturated gas is passed over the surface with a partial pressure of adsorbate gas of p_a, evaporation continues until the curvature attained in the upper capillary corresponds to the equilibrium vapor pressure at the level of liquid indicated. In adsorption, starting with an adsorbate-free solid, adsorption stops when the liquid level has reached the point indicated in *b*, corresponding to the same curvature as in *a* and exerting a vapor pressure equal to p_a. In each case equilibrium is attained, in desorption with a high moisture content and in ad-

sorption with a low moisture content as indicated by shaded portions in the two figures. The situation is exaggerated in the illustration. With interconnecting pore spaces open to the atmosphere the forces produced by small capillaries will cause liquid to flow from large openings into the finer capillaries, thus emptying the large pore spaces. A

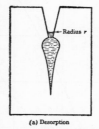
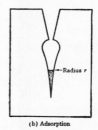

(a) Desorption   (b) Adsorption

FIG. 24.   Adsorption hysteresis.

discussion of these forces is given by Ceaglske and Hougen.[6] Other types of adsorption hysteresis are explained by Emmett and Dewitt.[7]

Preferential Adsorption. Preferential adsorption has already been referred to as one of the methods of removing an adsorbate gas. In Illustration 18 it was assumed that the presence of the gases from which the benzene was adsorbed had no effect on the equilibrium between the benzene vapor and the charcoal. This assumption is not necessarily true because charcoal adsorbs considerable quantities of all the ordinary gases as well as benzene vapors. It would be expected that when charcoal is exposed to a mixture of gases a complicated equilibrium would be reached between each of the gases and its adsorbed quantity. Few quantitative data are available on the adsorption of mixtures of gases and vapors, but it is apparent that when several gases are adsorbed the presence of each must affect the equilibrium concentration of others.

It is an experimentally observed fact that, in general for van der

[6] *Trans. Am. Inst. Chem. Eng.*, **33**, 283 (1937).
[7] *J. Am. Chem. Soc.*, **65**, 1258 (1943).

Waals adsorption, a gas of high molecular weight, high critical temperature, and low volatility is adsorbed in preference to a gas of low molecular weight, low critical temperature, and high volatility. Such a preferentially adsorbed gas or vapor will displace other gases which have already been adsorbed. The chemical nature of the gas also plays an

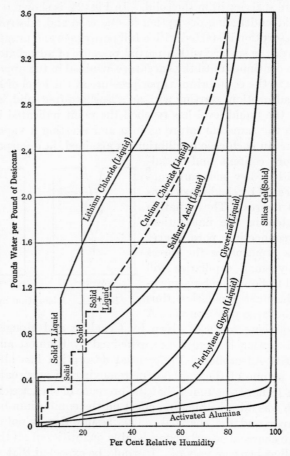

FIG. 25. Equilibrium moisture content of various desiccants.

important part, but ordinarily it may be assumed that a heavy vapor of low volatility will almost completely displace a light gas of high volatility and similar chemical type. In the experiments of Coolidge it was found that exposure of the outgassed charcoal to air before treatment with benzene vapors had no apparent effect on the final equilibrium.

In the absence of definite data it may ordinarily be assumed that when an adsorbent is treated with a vapor of low volatility such as water or benzene in admixture with a very volatile gas such as air, the adsorption of the gas will exert only a negligible influence on the normal equilibrium between the vapor and the adsorbent. When a gaseous mixture having several components of similar volatility is treated with an adsorbent,

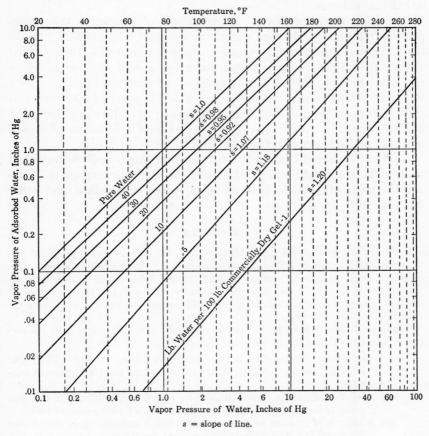

FIG. 26.   Vapor pressure of water adsorbed on silica gel.

equilibrium quantities of these components are adsorbed, and general predictions of the equilibrium conditions cannot be made without specific data.

The preferential adsorptive properties which many substances exhibit are of great industrial importance in the selective separation of components of gaseous mixtures. An important application of adsorbents is in the drying of gases. In Fig. 25 are the equilibrium moisture contents of silica gel and activated alumina compared with various solid and liquid chemical desiccants.   It will be observed that the solid desiccants compare unfavorably with the liquid desiccants in adsorption capacity. However, both silica gel and activated alumina are capable of drying air to much lower dew points than homogeneous solutions of lithium chloride or calcium chloride.

Cox Chart for Adsorbents. A useful way of presenting the vapor-pressure relations of adsorbed water is by means of the Cox chart wherein for the same temperature the vapor pressure of adsorbed water is plotted against the vapor pressure of pure water for lines of constant moisture content.   Straight lines generally result on a logarithmic plot, as shown in Fig. 26 for silica gel.   The relative humidity at any temperature and composition is obtained directly from the ratio p/p_s, where p_s is the vapor pressure of pure water.   This chart is also useful in estimating heats of adsorption as discussed in Chapter VIII.

It should be noted that all the discussion in this chapter is restricted to equilibrium conditions.   Calculation of rates of sorption requires consideration of particle size, diffusion, rates of gas flow and other factors.

PROBLEMS

1. From the International Critical Tables, Vol. IV, plot a curve relating the solubility of sodium carbonate in water to temperature.   Plot solubilities as ordinates, expressed in percentage by weights of Na_2CO_3, and temperature as abscissas, expressed in degrees Centigrade, up to 60°C.   On the same axes plot the freezing-point curve of the solution, or the solubility curve of ice in sodium carbonate, locating the eutectic point of the system.

2. From the data of Figs. 10, 11, and 12 and Problem 1, tabulate in order the successive effects produced by the following changes:

(a) A solution of naphthalene in benzene is cooled from 20° to −10°C.   The solution contains 0.6 gram-mole of naphthalene per 1000 grams of benzene.

(b) A solution of sodium carbonate in water is cooled from 40° to −5°C.   The solution contains 4.5 gram-moles of Na_2CO_3 per 1000 grams of water.

(c) A mixture of aqueous sodium sulphate solution and crystals is heated from 10° to 60°C.   The original mixture contains 3.3 gram-moles of Na_2SO_4 per 1000 grams of water.

(d) Pure crystals of $Na_2SO_4 \cdot 10H_2O$ are heated from 20° to 40°C.

(e) A solution of $FeCl_3$ in water is cooled from 20° to −60°C.   The solution contains 2.5 gram-moles of $FeCl_3$ per 1000 grams of water.

(f) A solution of $FeCl_3$ is evaporated at a temperature of 34°C.   The original solution contains 5 gram-moles of $FeCl_3$ per 1000 grams of water, and it is evaporated to a concentration of 25 gram-moles of $FeCl_3$ per 1000 grams of water.

(g) An aqueous solution of $FeCl_3$ is cooled from 45° to 10°C.   The solution contains 18 gram-moles of $FeCl_3$ per 1000 grams of water.

3. In a solution of naphthalene in benzene the mole fraction of naphthalene is 0.12. Calculate the weight of this solution necessary to dissolve 100 lb of naphthalene at a temperature of 40°C.

4. An aqueous solution of sodium carbonate contains 5 grams of Na_2CO_3 per 100 cc of solution at 15°C. Calculate the pounds of anhydrous Na_2CO_3 which can be dissolved in 10 gal of this solution at 50°C.

5. A solution of sodium carbonate in water is saturated at a temperature of 10°C. Calculate the weight of $Na_2CO_3 \cdot 10H_2O$ crystals which can be dissolved in 100 lb of this solution at 30°C.

6. A solution of naphthalene in benzene contains 8.7 lb-moles of naphthalene per 1000 lb of benzene.

    (*a*) Calculate the temperature to which this solution must be cooled in order to crystallize 70% of the naphthalene.

    (*b*) Calculate the composition of the solid product if 90% of the naphthalene is crystallized.

7. The concentration of naphthalene in a solution in benzene is 1.4 lb-moles per 1000 lb of benzene. This solution is cooled to −3°C. Calculate the weight and composition of the material crystallized from 100 lb of the original solution.

8. A solution of naphthalene in benzene contains 25% naphthalene by weight. Calculate the weight of benzene which must be evaporated from 100 lb of this solution in order that 85% of the naphthalene may be crystallized by cooling to 20°C.

9. A batch of saturated Na_2CO_3 solution, weighing 1000 lb, is to be prepared at 50°C.

    (*a*) If the monohydrate ($Na_2CO_3 \cdot H_2O$) is available as the source of Na_2CO_3, how many lb of this material and how many pounds of water would be needed to form the required quantity of solution?

    (*b*) If the decahydrate ($Na_2CO_3 \cdot 10H_2O$) is available as the source of Na_2CO_3, how many pounds of this material and how many pounds of water would be required? By means of a sketch, show how the solubility chart was used in solving the problem.

10. A system containing 40% Na_2CO_3 and 60% H_2O has a total weight of 500 lb. At 50°C, 35°C and 20°C, report on:

    (*a*) The nature and composition of the phases constituting the system.

    (*b*) The weight of each phase.

By means of a sketch, indicate how the solubility chart was used in solving the problem.

11. A solution containing 35% Na_2CO_3 weighs 5000 lb.

    (*a*) To what temperature must the system be cooled in order to recover 98% of the Na_2CO_3?

    (*b*) What will be the weight of the crystals recovered, and of the residual mother liquor?

By means of a sketch, indicate how the solubility chart was used in solving the problem.

12. In Fig. 27 the temperature-composition diagram of lithium chloride is presented, showing the composition of the system as pounds of water per pound of lithium chloride. Explain the significance of each separate area, boundary line, and point of intersection.

13. A dilute solution of Na_2CO_3 and water weighs 2000 lb, and contains 5% Na_2CO_3. It is required that 95% of the Na_2CO_3 be recovered as the decahydrate. The lowest temperature that can be obtained is 5°C. Specify the process to be employed for securing the degree of recovery required. Present calculations with respect to the quantitative relations involved in the process specified.

By means of a sketch, indicate how the solubility chart was used in solving the problem.

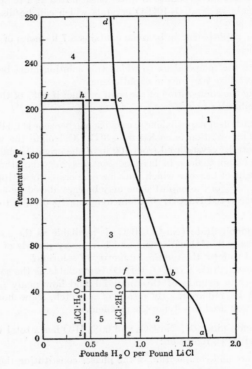

FIG. 27. Solubility of lithium chloride.

14. An aqueous solution of ferric chloride contains 12 lb-moles of $FeCl_3$ per 1000 lb of water. Calculate the yield of crystals formed by cooling 1000 lb of this solution to 28°C.

15. A solution of ferric chloride in water contains 15 gram-moles of $FeCl_3$ per 1000 grams of water.

    (a) Calculate the composition of the resulting crystals in percentage of each hydrate formed when this solution is cooled to 0°C.

    (b) Calculate the percentage of eutectic crystals present in the total crystal mass.

16. An aqueous solution contains 3.44% Na_2SO_4 and 21.0% Na_2CO_3 by weight. It is desired to cool this solution to the temperature which will produce a maximum yield of crystals of pure sodium carbonate decahydrate. Calculate the proper final temperature and the yield of crystals per 100 lb of original solution, using the data of Fig. 14.

17. It is desired to crystallize a maximum amount of pure $Na_2CO_3 \cdot 10H_2O$ from the solution of Problem 11 by evaporating water at a temperature of 25°C. Calculate the quantity of water which must be evaporated and the yield of crystals produced per 100 lb of original solution.

18. A solution contains 25 grams of Na_2SO_4 and 4.0 grams of Na_2CO_3 per 100 grams of water. From 100 lb of this solution 20 lb of water is evaporated and the residual solution cooled to 20°C. Calculate the weight and composition of the crystals formed in the process.

19. Calculate the weight of water which must be evaporated from 100 lb of the solution of Problem 18 in order to crystallize 70% of the Na_2SO_4 as the pure decahydrate at a temperature of 15°C.

20. A solution has the following initial composition:

$$Na_2SO_4 \cdot 10H_2O = 80 \text{ parts by weight}$$
$$Na_2CO_3 \cdot 10H_2O = 60 \text{ parts by weight}$$
$$\text{Solvent water} = 100 \text{ parts by weight}$$

The solution is at 50°C, and weighs 2500 lb. If this solution is cooled to 5°C:

    (a) At what temperature does crystallization start? What phase crystallizes out first?

    (b) At what temperature will a second crystalline phase start to separate? What is its composition?

    (c) Calculate the maximum weight of pure crystals that can be obtained in the first stage of crystallization, when but one solid phase is separating.

    (d) Calculate the respective weights of the two solid phases that separate during the second stage of crystallization.

    (e) With respect to the residual mother liquid at 5°C, report its total weight, and the weight of each of the three components that were present in the original solution.

By means of a sketch, show how Fig. 14 was used in solving the problem.

21. A solution has the following initial composition:

$$Na_2SO_4 \cdot 10H_2O = 40 \text{ parts by weight}$$
$$Na_2CO_3 \cdot 10H_2O = 40 \text{ parts by weight}$$
$$\text{Solvent water} = 100 \text{ parts by weight}$$

The solution originally weighs 1500 lb. If evaporation is carried out at 25°C:

    (a) How much water must be evaporated before crystallization starts?

    (b) What is the first solid to crystallize from the solution? How much of this pure crystalline material can be removed before a second solid starts to separate?

    (c) How much water must be removed by evaporation before a second crystalline phase starts to form? What is the nature of this second crystalline phase?

By means of a sketch, show how Fig. 14 was used in solving the problem.

22. A solution has the following initial composition:

$$Na_2SO_4 \cdot 10H_2O = 10 \text{ parts by weight}$$
$$Na_2CO_3 \cdot 10H_2O = 200 \text{ parts by weight}$$
$$\text{Solvent water} = 100 \text{ parts by weight}$$

The solution weighs 2500 lb. It is required that 80% of the $Na_2CO_3 \cdot 10H_2O$ be recovered in pure form. The lowest temperature that can be obtained is 15°C. How many pounds of water must be removed by evaporation if 80% recovery is obtained on cooling to 15°C?

By means of a sketch, show how Fig. 14 was used in solving the problem.

23. A solution has the following initial composition:

$$Na_2SO_4 \cdot 10H_2O = 80 \text{ parts by weight}$$
$$Na_2CO_3 \cdot 10H_2O = 10 \text{ parts by weight}$$
$$\text{Solvent water} = 100 \text{ parts by weight}$$

The solution weighs 1600 lb. Compare the maximum yield of $Na_2SO_4 \cdot 10H_2O$ obtainable by each of the following three processes:

(a) Cooling to 10°C.
(b) Evaporation at a constant temperature of 25°C.
(c) Evaporation of a limited amount of water at 25°C, followed by cooling to 10°C.

By means of a sketch, indicate how Fig. 14 was used in solving the problem.

24. The residual liquor from a crystallizing operation has the following composition:

$$Na_2SO_4 \cdot 10H_2O = 10 \text{ parts by weight}$$
$$Na_2CO_3 \cdot 10H_2O = 60 \text{ parts by weight}$$
$$\text{Solvent water} = 100 \text{ parts by weight}$$

This liquor is to be used to extract the soluble material from a powdered mass that has the following composition:

$$Na_2SO_4 = 1.25\%$$
$$Na_2CO_3 = 10.50\%$$
$$\text{Insoluble matter} = 88.25\%$$

(a) What is the minimum weight of the residual liquor required to dissolve the soluble matter present in 1000 lb of the solid mass, assuming that the leaching operation is carried out at 25°C?

(b) If the liquor obtained from the leaching operation is subsequently cooled, what is the maximum possible percentage recovery of pure Na_2CO_3?

By means of a sketch, show how Fig. 14 was used in connection with the solution of this problem.

25. From the data of International Critical Tables plot a solubility chart similar to Fig. 14 for the system $NaNO_3$—$NaCl$—H_2O. Include the solubility isotherms of 15.5°, 50°, and 100°C.

26. A mixture of $NaNO_3$ and $NaCl$ is leached with water at 100°C to form a solution which is saturated with both salts. From the chart of Problem 25, calculate the weight and composition of the crystals formed by cooling 100 lb of this solution to 15.5°C.

27. A solution of picric acid in benzene contains 30 grams of picric acid per liter. Calculate the quantity of water with which 1 gallon of this solution at 18°C must be shaken in order to reduce the picric acid concentration to 4.0 grams per liter in the benzene phase.

28. One gallon of an aqueous solution of picric acid containing 0.15 lb of picric acid is shaken with 2 gallons of benzene. Calculate the pounds of picric acid in each solute after the treatment.

29. A mixture of phenol and water contains 45% phenol by weight. Calculate the weight fractions of upper and lower layers formed by this mixture at a temperature of 45°C.

30. Calculate the weight fractions and compositions of the upper and lower layers of a mixture of 30% water, 20% isopropyl alcohol, and 50% tetrachlorethylene at a temperature of 77°F.

31. Assuming the applicability of Henry's law, calculate the percentage CO_2 by weight which may be dissolved in water at a temperature of 20°C in contact with gas in which the partial pressure of CO_2 is 450 mm of Hg.

32. Assuming the applicability of Henry's law, calculate the partial pressure of H_2S above an aqueous solution at 30°C which contains 3.0 grams of H_2S per 1000 grams of water.

33. Calculate the volume in cubic feet of NH_3 gas under a pressure of 1 atmosphere and at a temperature of 20°C which can be dissolved in 1 gal of water at the same temperature.

34. An aqueous solution of ammonia at 10°C is in equilibrium with ammonia gas having a partial pressure of 500 mm of Hg.

    (*a*) Calculate the percentage ammonia, by weight, in the solution.

    (*b*) Calculate the partial pressure of the ammonia in this solution if it were warmed to a temperature of 40°C.

35. Assuming that equilibrium quantities of adsorbate are determined only by the relative saturation of the adsorbate vapor, estimate the quantity of SO_2 adsorbed by the silica gel of Fig. 22 at a temperature of 10°C from a gas mixture in which the partial pressure of SO_2 is 150 mm of Hg.

36. Activated charcoal similar to that of Fig. 23 is to be used for the removal of benzene vapors from a mixture of gases at 20°C and a pressure of 1 atmosphere. The relative saturation of the gases with benzene is 83%.

    (*a*) Calculate the maximum weight of benzene which may be adsorbed per pound of charcoal.

    (*b*) The adsorbed benzene is to be removed by stripping with superheated steam at a temperature of 180°C. Calculate the final partial pressure to which the benzene in the steam leaving the stripper must be reduced in order to remove 90% of the adsorbed benzene.

    (*c*) If the adsorbent is so used that the benzene-bearing gases always come into equilibrium with freshly stripped charcoal of part *b* before leaving the process, calculate the loss of benzene in these treated gases, expressed as percentage of the total benzene entering the process.

37. Fifty pounds of unsized cotton cloth containing 20% total moisture are hung in a room of 4000-cu-ft capacity. The initial air is at a temperature of 100°F, at a relative humidity of 20%, and a barometric pressure of 29.92 in. of mercury. The air is kept at 100°F with no fresh air admitted and no air vented. Neglect the space occupied by the contents of room.

    (*a*) Calculate the moisture content of the cloth and the relative humidity of the air at equilibrium.

    (*b*) Calculate the equilibrium moisture content of the cloth and the corresponding relative humidity of the air if 100 lb of wet cloth instead of 50 lb are hung in the same room.

    (*c*) What is the final pressure in the room under part (*a*)?

38. Air at atmospheric pressure is to be dried at 80°F from 70% to 10% relative humidity by mixing with silica gel, initially dry, and at 80°F. What is the final equilibrium moisture content of the gel, assuming a constant temperature of 80°F?

39. If air were to be dried to a dew point of 0°F what would be the equilibrium water contents of the following desiccants assuming exit temperature of air and desiccant to be 70°F.?

> activated alumina
> silica gel
> glycerin
> sulfuric acid
> calcium chloride solution (avoid crystallization)
> lithium chloride solution (avoid crystallization)
> triethylene glycol

40. How would it be possible to dry air to a dew point 0°F with a homogeneous LiCl solution?

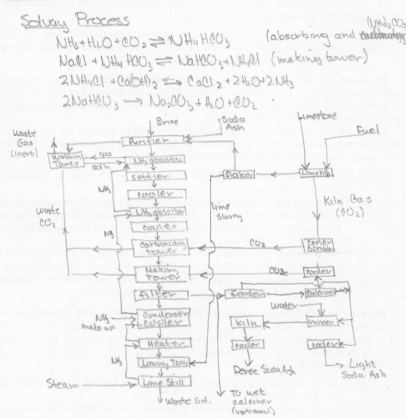

CHAPTER VI

MATERIAL BALANCES

A material balance of an industrial process is an exact accounting of all the materials which enter, leave, accumulate, or are depleted in the course of a given time interval of operation. The material balance is thus an expression of the law of conservation of mass in accounting terms. If direct measurements were made of the weight and composition of each stream entering or leaving a process during a given time interval and of the change in material inventory in the system during that time interval, no calculations would be required. Usually this is impracticable, and therefore calculations are indispensable.

The general principle of material balance calculations is to establish a number of independent equations equal to the number of unknowns of composition and mass. For example, if two streams enter a process and one stream leaves, with no change in inventory in the system during the time interval, the mass and composition of each stream completely establishes the material balance. For calculating the complete material balance the greatest number of unknowns permissible is three, selected among six possible items. Variations in solving the problem will depend upon the particular items which are unknown, whether they be of composition or mass, or of streams entering or leaving. The following principles serve as guides to direct the course of calculations.

1. If no chemical reaction is involved, nothing is gained by establishing material balances for the several chemical elements present. In such processes, material balances should be based upon the chemical compounds rather than elements, or of components of fixed composition even if not pure chemical compounds.

2. If chemical reactions occur, it becomes necessary to develop material balances based upon chemical elements, or upon radicals, compounds, or substances which are not altered, decomposed, or formed in the process.

3. For processes wherein no chemical reactions occur, use of weight units such as grams or pounds is desirable. For processes in which chemical reactions occur, it is desirable to utilize the gram-mole or pound-mole, or the gram-atom or pound-atom.

4. The number of unknown quantities to be calculated cannot exceed the number of independent material balances available; otherwise, the problem is indeterminate.

5. If the number of independent material balance equations exceeds the number of unknown weights that are to be computed, it becomes a matter of judgment to determine which of the equations should be selected to solve the problem. If all the analytical data used in setting up the equations were perfect it would be immaterial which equations would be selected for use. However, analytical data are never free from error, and a certain amount of discretion is needed to select the most nearly accurate equations for solving the problem. In general, equations based upon components forming the largest percentage of the total mass are most dependable.

6. Recognition of the maximum number of truly independent equations is important. Any material balance equation that can be derived from other equations written for the process cannot be regarded as an additional independent equation. For example, in the following Illustration 1 it would be possible to write material balance equations based upon water, HNO_3, HCl, H_2SO_4, nitrogen, sulfur, hydrogen, oxygen, chlorine, and overall weights. Of these ten equations only three are independent. If the equation based upon H_2SO_4, HNO_3 and overall weights are selected, the other seven equations can be deduced from computations alone.

7. If any two or more substances exist in fixed ratio with respect to one another in each stream where they appear, only one independent material balance equation may be written with respect to these substances. Although a balance may be written for any one substance in question, it is generally best to combine the substances appearing in constant ratio into a single group and develop a single equation for this combined group.

8. A substance which appears in but one incoming stream and one outgoing stream serves as a reference for computations and is termed a *tie-substance*. Knowledge of the percentage of a tie-substance in two streams establishes the relationship between the weights of the streams so that if one is known the other can be calculated.

Illustration 1. The waste acid from a nitrating process contains 23% HNO_3, 57% H_2SO_4, and 20% H_2O by weight. This acid is to be concentrated to contain 27% HNO_3 and 60% H_2SO_4 by the addition of concentrated sulfuric acid containing 93% H_2SO_4 and concentrated nitric acid containing 90% HNO_3. Calculate the weights of waste and concentrated acids which must be combined to obtain 1000 lb of the desired mixture.

Basis: 1000 lb of final mixture

    Let x = weight of waste acid

        y = weight of conc. H_2SO_4

        z = weight of conc. HNO_3

Overall balance:

$$x + y + z = 1000 \tag{a}$$

H_2SO_4 *balance:*

$$0.57x + 0.93y = 1000 \times 0.60 = 600 \tag{b}$$

HNO_3 *balance:*

$$0.23x + 0.90z = 1000 \times 0.27 = 270 \tag{c}$$

Equations (a), (b), and (c) may be solved simultaneously.
Eliminating z from (a) and (c):

$$\frac{270 - 0.23x}{0.90} + y + x = 1000$$

$$y + 0.744x = 700 \tag{d}$$

Eliminating y from (d) and (b):

$$0.57x + 0.93(700 - 0.744x) = 600$$
$$0.122x = 51$$
$$x = 418 \text{ lb}$$

From (b)

$$y = \frac{600 - 0.57 \times 418}{0.93} = 390 \text{ lb}$$

From (a)

$$z = 1000 - 390 - 418 = 192 \text{ lb}$$

These results may be verified by a material balance of the water in the process:

Water entering $= (418 \times 0.20) + (390 \times 0.07) + (192 \times 0.10) = 130$ lb

Since the final solution contains 13% H_2O, this result verifies the calculations.

More complicated problems may be handled by the same type of analysis. When more than two simultaneous equations are to be solved the use of determinants is recommended.

Processes Involving Chemical Reactions. Material balances of processes involving chemical reactions are of the following two general classes:

a. The compositions and weight of the various streams entering the process are known. It is required to calculate the compositions and weights of the streams leaving the process for a specified degree of completion of the reaction.

b. The compositions and weights of the entering streams are partially known. It is required to calculate the compositions and weights of all entering and leaving streams and to determine the degree of completion of the reaction.

In these calculations it is desirable to work with molal rather than ordinary weight units, particularly for components undergoing chemical transformation. The limiting reactant should be selected. The quantity of each reacting material may then be specified in terms of the percentage excess it forms of that theoretically required. The calculation is then completed on the basis of the limiting reactant which is present in a unit quantity of the reactants. The amounts of the new products formed in the reaction are determined from the degree of completion. The unconsumed reactants and inert materials pass into the product unchanged.

Illustration 2. A producer gas made from coke has the following composition by volume:

$$
\begin{array}{lr}
CO\dotfill & 28.0\% \\
CO_2\dotfill & 3.5 \\
O_2\dotfill & 0.5 \\
N_2\dotfill & \underline{68.0} \\
 & 100.0\%
\end{array}
$$

This gas is burned with such a quantity of air that the oxygen from the air is 20% in excess of the *net* oxygen required for complete combustion. If the combustion is 98% complete, calculate the weight and composition in volumetric per cent of the gaseous products formed per 100 lb of gas burned.

Discussion. The carbon monoxide is the limiting reactant, while the oxygen is the excess reactant. The amount of oxygen supplied by the air is expressed as the percentage in excess of the *net* oxygen demand, this latter term referring to the total oxygen required for complete combustion, minus that present in the fuel. Since the composition of the fuel is known on a molal basis, it is most convenient to choose 100 lb-moles of the fuel gas as the basis of calculation, and at the close of the solution to convert the results over to the basis of 100 lb of gas burned.

Basis of Calculation: 100 lb-moles of producer gas.

Constituent	Molecular Weight	Mole, %	Weight in Pounds
CO	28.0	28.0	$28.0 \times 28.0 =$ 784 lb
CO$_2$	44.0	3.5	$3.5 \times 44.0 =$ 154 lb
O$_2$	32.0	0.5	$0.5 \times 32.0 =$ 16 lb
N$_2$	28.2	68.0	$68.0 \times 28.2 =$ 1917 lb
		100.0	2872 lb

Oxygen Balance:

O$_2$ required to combine with all CO present $1/2 \times 28.0 =$..	14.0 lb-moles
O$_2$ in the producer gas	0.5 lb-mole
Net O$_2$ demand $= 14.0 - 0.5$	13.5 lb-moles
O$_2$ supplied by air $= 13.5 \times 1.20$	16.2 lb-moles
O$_2$ actually used $= 0.98 \times 28.0 \times 1/2$	13.7 lb-moles
O$_2$ in products $= 16.2 + 0.5 - 13.7$	3.0 lb-moles
or 3.0×32	96.0 lb

Carbon Balance

C in fuel gas = 28.0 + 3.5 .... 31.5 lb-atoms
C in CO of products of combustion = 0.02 × 28.0 .... 0.56 lb-atom
C in CO_2 of products of combustion = 31.5 − 0.56 .... 30.94 lb-atoms
Carbon monoxide in products .... 0.56 lb-mole
   or 0.56 × 28 .... 15.7 lb
Carbon dioxide in products......................... 30.94 lb-moles
   or 30.94 × 44 .... 1359 lb

Nitrogen Balance

N_2 in producer gas .... 68.0 lb-moles
N_2 from air = 79/21 × 16.2 .... 60.9 lb-moles
N_2 in products = 68.0 + 60.9 .... 128.9 lb-moles
   or 128.9 × 28.2 .... 3637 lb

Weight of Products of Combustion

Total weight, based on 100 lb-moles producer gas
   96 + 1359 + 16 + 3637 .... 5108 lb
Total weight, based on 100 lb producer gas
   5108 × (100/2872) .... 178 lb

Analysis of Products of Combustion

Total moles in products of combustion
   30.9 + 3.0 + 0.56 + 128.9 .... 163.4 lb-moles

	Mole or Volumetric %
CO_2 = (30.9/163.4) × 100	18.92
O_2 = (3.0/163.4) × 100	1.84
CO = (0.56/163.4) × 100	0.34
N_2 = (128.9/163.4) × 100	78.9
	100.00

Illustration 3. A solution of sodium carbonate is causticized by the addition of partly slaked commercial lime. The lime contains only calcium carbonate as an impurity and a small amount of free caustic soda in the original solution. The mass obtained from the causticization has the following analysis:

$CaCO_3$......................	13.48%
$Ca(OH)_2$......................	0.28
Na_2CO_3........................	0.61
NaOH...........................	10.36
H_2O...........................	75.27
	100.00%

The following items are desired:

(a) The weight of lime charged per 100 lb of the causticized mass, and the composition of the lime.

(b) The weight of the alkaline liquor charged per 100 lb of the causticized mass, and the composition of the alkaline liquor.

(c) The reacting material which is present in excess, and its percentage excess.

(d) The degree of completion of the reaction.

Discussion. The problem as stated cannot be solved before additional data are obtained. The needed additional information is either the analysis of the lime or the analysis of the alkaline liquor.

If the analysis of the lime were determined, the problem could be solved by the following steps: (1) By using calcium as a tie-substance the weight of lime would be determined. (2) An overall material balance would establish the weight of alkaline liquor. (3) By a carbon balance, the weight of Na_2CO_3 in the alkaline liquor would be computed. (4) By using sodium as a tie-substance the weight of NaOH in the alkaline liquor would be calculated. (5) The weight of water in the alkaline liquor would be determined by difference.

If the analysis of the alkaline liquor were determined instead, the problem would be solved according to the following procedure: (1) By using sodium as a tie-substance the total weight of the alkaline liquor would be calculated. (2) An overall material balance would establish the weight of lime. (3) By means of a carbon balance, the weight of $CaCO_3$ in the lime would be determined. (4) A calcium balance would give the weight of active CaO [free CaO plus the CaO in the $Ca(OH)_2$] in the lime. (5) The weight of free CaO and the weight of $Ca(OH)_2$ in the lime could then be computed from the available values for the weight of lime, the weight of $CaCO_3$, and the weight of active CaO.

From the foregoing, it is apparent that it is only a matter of convenience whether the lime or the alkaline liquor is analyzed. The normal choice would be to analyze the alkaline liquor, because the analysis is rapid and accurate.

An analysis of the alkaline liquor used in the process gave the following results:

NaOH........................	0.594%
Na_2CO_3........................	14.88 %
H_2O........................	84.53 %
	100.00 %

Basis of Calculation: 100 lb of the causticized mass.

Reactions for the Process:

$$CaO + H_2O \rightarrow Ca(OH)_2$$
$$Na_2CO_3 + Ca(OH)_2 \rightarrow 2NaOH + CaCO_3$$

Molecular Weights:

CaO = 56.1	Na_2CO_3 = 106.0
NaOH = 40.0	$CaCO_3$ = 100.1
$Ca(OH)_2$ = 74.1	H_2O = 18.02

Conversion into Molal Quantities

Alkaline Liquor. *Basis:* 1 lb.

	Lb	Lb-Moles	Lb-Atoms Na	Lb-Atoms C
NaOH	0.00594	0.000149	0.000149	
Na_2CO_3	0.1488	0.001404	0.002808	0.001404
H_2O	0.8453			
	1.0000		0.002957	0.001404

Causticized Mass.   Basis: 100 lb

	% or Lb	Lb-Moles	Lb-Atoms Ca	Lb-Atoms Na	Lb-Atoms C
$CaCO_3$	13.48	0.1347	0.1347		0.1347
$Ca(OH)_2$	0.28	0.00377	0.0038		
Na_2CO_3	0.61	0.00575		0.0115	0.00575
NaOH	10.36	0.2590		0.2590	
H_2O	75.27				
	100.00		0.1385	0.2705	0.1405

Sodium Balance

Object: To evaluate weight of alkaline liquor.

Na in causticized mass = ...	0.2705 lb-atom
Na in 1 lb alkaline liquor = ...	0.002957 lb-atom
Weight of alkaline liquor 0.2705/0.002957 = ...	91.50 lb

Overall Material Balance

Object: To determine the total weight of lime.

Weight of causticized mass = ...	100.00 lb
Weight of alkaline liquor = ...	91.50 lb
Weight of lime = 100.00 − 91.50 = ...	8.50 lb

Carbon Balance

Object: To evaluate the weight of $CaCO_3$ in the lime.

C in causticized mass = ...	0.1405 lb-atom
C in Na_2CO_3 = 91.50 × 0.001404 = ...	0.1285 lb-atom
C in $CaCO_3$ = 0.1405 − 0.1285 = ...	0.0120 lb-atom
Weight of $CaCO_3$ = 0.0120 × 100.1 = ...	1.20 lb

Calcium Balance

Object: To determine the active CaO [that present in the $Ca(OH)_2$ plus the free CaO] in the lime.

Ca in the causticized mass = Ca in the lime =.	0.1385 lb-atom
Ca present as $CaCO_3$ in the lime (see carbon balance) = ...	0.0120 lb-atom
Ca present in $Ca(OH)_2$ and in free CaO = 0.1385 − 0.0120 = ...	0.1265 lb-atom

Overall Balance of Constituents in the Lime

Object: To determine free CaO and $Ca(OH)_2$ in the lime

Total weight of lime charged = ...	8.50 lb
Weight of $CaCO_3$ (see carbon balance) = ...	1.20 lb
Weight of CaO + $Ca(OH)_2$ = 8.50 − 1.20 =	7.30 lb
Weight of total active CaO = 0.1265 × 56.1 =	7.10 lb
H_2O present in the $Ca(OH)_2$ = 7.30 − 7.10 =	0.20 lb
$Ca(OH)_2$ in lime = 0.20 × (74.1/18.02) = ...	0.82 lb
Weight of free CaO = 7.30 − 0.82 = ...	6.48 lb

Results

(a) Weight of lime = ... 8.50 lb

Analysis of lime

	Lb	Per Cent
CaCO₃	1.20	14.1
Ca(OH)₂	0.82	9.6
CaO	6.48	76.3
	8.50	100.0

(b) Weight of alkaline liquor .... 91.50 lb

Analysis of alkaline liquor was determined experimentally.

(c) Determination of Excess Reactant.

Total active CaO = ... 0.1265 lb-mole

Na₂CO₃ in liquor = 91.50 × 0.001404 = ... 0.1285 lb-mole

Since, according to the reaction equation, 1 mole of Na_2CO_3 requires 1 mole of active CaO, it is concluded that the Na_2CO_3 is present in excess, and that the active CaO is the limiting reactant.

Excess of Na₂CO₃ = 0.1285 − 0.1265 = ... 0.0020 lb-atom

Per cent excess = (0.0020/0.1265) × 100 = .. 1.6%

(d) Degree of Completion

Ca(OH)₂ in causticized mass = ... 0.00377 lb-mole

CaO + Ca(OH)₂ in lime charged = ... 0.1265 lb-mole

Degree of completion of the reaction =

100 − (0.00377/0.1265) × 100 = ... 97.0%

In each of the preceding three illustrations, there was but one stream of material emerging from the process. In many processes, there are two or more emergent streams. For example, there may be an evolution of gas or vapor from the reacting mass, as in the calcination of limestone, or there may be a separable residue which is removed from the major product, as when a precipitate is separated from a liquid solution. While the complexity of the problem tends to increase as the number of streams of material involved in the process increases, the general methods of solution are as indicated for the illustrations already presented. Unknown weights are evaluated through the use of material balances. Frequent use is made of tie-substances for determining unknown stream weights. The following illustrative problem is typical of the treatment applied to a process wherein there is more than one stream of emergent material.

Illustration 4. The successive reactions in the manufacture of HCl from salt and sulfuric acid may be represented by the following equations:

$$NaCl + H_2SO_4 \rightarrow NaHSO_4 + HCl$$
$$NaCl + NaHSO_4 \rightarrow Na_2SO_4 + HCl$$

In practice the salt is treated with aqueous sulfuric acid, containing 75% H_2SO_4, in slight excess of the quantity required to combine with all the salt to form Na_2SO_4. Although the first reaction proceeds readily, strong heating is required for the second. In both steps of the process HCl and water vapor are evolved from the reaction mass.

" Salt cake " prepared by such a process was found to have the following composition:

$$
\begin{array}{lr}
Na_2SO_4\dots\dots\dots\dots\dots\dots\dots\dots & 91.48\% \\
NaHSO_4\dots\dots\dots\dots\dots\dots\dots & 4.79\% \\
NaCl\dots\dots\dots\dots\dots\dots\dots\dots & 1.98\% \\
H_2O\dots\dots\dots\dots\dots\dots\dots\dots & 1.35\% \\
HCl\dots\dots\dots\dots\dots\dots\dots\dots & \underline{0.40\%} \\
& 100.00\%
\end{array}
$$

The salt used in the process is dry, and may be assumed to be 100% NaCl.

(a) Calculate the degree of completion of the first reaction and the degree of completion of the conversion to Na_2SO_4 of the salt charged.

(b) On the basis of 1000 lb of salt charged, calculate the weight of acid added, the weight of salt cake formed and the weight and composition of the gases driven off.

Discussion. The logical choice of a basis of calculation is 1000 lb of salt, since the results are wanted on that basis.

The solution of part (b) is accomplished through a series of material balances. It will be noted that sodium serves as a tie-substance between the salt cake and the salt charged, and that sulfur serves as a tie-substance between the salt cake and the aqueous acid. Accordingly, the problem may be solved by the following successive steps: (1) A sodium balance establishes the weight of salt cake. (2) A sulfur balance serves to determine the weight of aqueous acid used. (3) A chlorine balance establishes the weight of HCl driven off in the gases. (4) A water balance determines the weight of H_2O in the gases driven off.

Basis of Calculation: 1000 lb salt charged.

Molecular Weights

$$
\begin{array}{ll}
NaCl = 58.5 & HCl = 36.46 \\
H_2SO_4 = 98.1 & NaHSO_4 = 120.1 \\
Na_2SO_4 = 142.1 &
\end{array}
$$

Conversion into Molal Quantities

Salt. Basis: 1000 lb.

$$NaCl = 1000/58.5\dots\dots\dots\dots\dots\dots\dots \quad 17.08 \text{ lb-moles}$$

Sulfuric Acid. Basis: 1 lb.

$$H_2SO_4 = 0.75 \text{ lb, or } 0.75/98.1\dots\dots\dots\dots\dots \quad 0.00764 \text{ lb-mole}$$

Salt Cake. Basis: 1 lb.

	Lb	Lb-Moles	Lb-Atoms Na	Lb-Atoms S	Lb-Atoms Cl
Na_2SO_4	0.9148	$\dfrac{0.9148}{142.1} = 0.00644$	0.01288	0.00644	
$NaHSO_4$	0.0479	$\dfrac{0.0479}{120.1} = 0.000398$	0.000398	0.000398	
$NaCl$	0.0198	$\dfrac{0.0198}{58.5} = 0.000338$	0.000338		0.000338
H_2O	0.0135				
HCl	0.0040	$\dfrac{0.0040}{36.46} = 0.000110$			0.000110
	$\overline{1.0000}$		$\overline{0.01362}$	$\overline{0.00684}$	$\overline{0.000448}$

Part (a) can be worked out by focusing attention on the salt cake. Since no H_2SO_4 is present in the product, the first reaction went to completion.

Total Na present..........................	0.01362 lb-atom
Na in Na_2SO_4..............................	0.01288 lb-atom
Conversion of NaCl to Na_2SO_4 = $(0.01288/0.01362) \times 100$...............	94.5%

Sodium Balance

  Object: To determine the weight of salt cake.

Na in 1000 lb salt charged.................	17.08 lb-atoms
Na in 1 lb salt cake.......................	0.01362 lb-atom
Weight of salt cake = 17.08/0.01362	1253 lb

Sulfur Balance

  Object: To determine the weight of aqueous acid used.

S in salt cake = 1253 × 0.00684.............	8.58 lb-atoms
S in 1 lb of aqueous acid...................	0.00764 lb-atom
Weight of aqueous acid = 8.58/0.00764	1123 lb

Water Balance

  Object: To determine the weight of H_2O in gas evolved.

Water in aqueous acid = 1123 × 0.25........	281 lb
Water in salt cake = 1253 × 0.0135..........	17 lb
Water driven off = 281 − 17................	264 lb

Chlorine Balance

  Object: To determine the weight of HCl in gas evolved.

Cl in salt charged.........................	17.08 lb-atoms
Cl in salt cake = 1253 × 0.000448...........	0.56 lb-atom
Cl in HCl driven off = 17.08 − 0.56.........	16.52 lb-atoms
Weight HCl in gases = 16.52 × 36.46........	603 lb

Composition of Leaving Gases

	Lb	Per Cent
HCl	603	69.6
H_2O	264	30.4
	867	100.0

Overall Balance

  Object: To check the accuracy of the calculations.

Total weight of reactants = 1000 + 1123.............	2123 lb
Total weight of products = 1253 + 867..............	2120 lb

STEPWISE COUNTER CURRENT PROCESSING

The leaching of solids, the washing of precipitates, the extraction of oils from crushed seeds, the softening of water, and the sorption of gases are processes which are often carried out in stages by continuous or by discontinuous methods.   In a continuous counter current extraction or washing process the precipitate to be washed or solid to be leached is treated in an agitator with clear solution from tank II as shown in Fig. 28. The slurry from the agitator is transferred to tank I where upon slow agitation and settling the concentrated liquor is decanted off as the desired product and the precipitate is pumped as a slurry by a sludge pump to tank II where it is mixed with dilute solution from tank III. Again upon slow agitation and settling the more concentrated liquid is pumped into the agitator where it is mixed with the fresh solid.   The

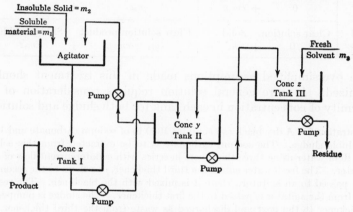

FIG. 28.   Step-wise countercurrent extraction.

sludge settling from tank II is pumped to tank III where it is treated with fresh solvent.   Again upon slow agitation and settling the clear dilute liquid is transferred to tank II and the precipitate, now nearly free from solute, is pumped out as a residual sludge.   As an approximation for establishing a material balance in this type of process it will be assumed that uniform concentration of solution is attained in each tank, that the solid pumped out of each tank as a sludge contains the same and uniform amount of solvent per unit weight of dry solid, and that the solution leaving each tank as clear liquid is of the same concentration as that retained by the solid in the sludge pumped out.

A solid containing m_2 pounds of insoluble and m_1 pounds of soluble material is fed continuously into the agitator and mixed with clear liquid from tank II, while m_3 pounds of fresh solvent are fed into tank

III. The sludge pumped from each tank contains c pounds of solvent per pound of insoluble solid. The concentration of soluble material in the solution is expressed as pounds per pound of solvent, and is designated as x in tank I, y in tank II, and z in tank III.

A material balance of the solute for each tank is as follows:

	Input		Output		
First tank	Slurry $m_1 + ym_3$	$=$	Clear solution $(m_3 - cm_2)x$	$+$	Sludge cm_2x (1)
Second tank	Clear solution m_3z	Sludge $+\ cm_2x\ =$	Clear solution m_3y	$+$	Sludge cm_2y (2)
Third tank	Clear solution 0	Sludge $+\ cm_2y\ =$	Clear solution m_3z	$+$	Sludge cm_2z (3)
Overall balance	Clear solution 0	Solid $+\ m_1\ =$	Clear solution product $(m_3 - cm_2)x$	$+$	Sludge residue cm_2z (4)

The oversimplified assumptions made in this treatment should be recognized. A more general solution requires consideration of non-uniformity of concentration in each tank for both sludge and solution.

Illustration 5. A dry black ash contains 1000 lb of sodium carbonate and 1000 lb of insoluble sludge. The sodium carbonate is to be extracted from this ash with 10,000 lb of water using three thickeners in series with countercurrent flow of sludge and water. The fresh water enters the third thickener, overflows to the second, and is then passed to an agitator, where it is mixed with the black ash. The resultant sludge from the agitator is passed to the first thickener. The sludge is pumped from one thickener to the next and discharged as waste from the third thickener. The sludge holds 3 lb of water for each pound of insoluble matter as it leaves each thickener. The concentrated sodium carbonate is drawn off and recovered from the first thickener. Calculate the weight of sodium carbonate recovered, assuming that all sodium carbonate is entirely dissolved to form a uniform solution in each agitator.

Employing the symbols given

$$m_3 = \text{fresh water supplied} = 10,000 \text{ pounds}$$
$$m_2 = \text{insoluble solid supplied} = 1,000 \text{ pounds}$$
$$m_1 = \text{sodium carbonate supplied} = 1,000 \text{ pounds}$$
$$c = \text{water per pound of insoluble solid} = 3$$

Substituting these values in Equations (1), (2), (3), and (4) for a balance of sodium carbonate, there results

First tank	$1000 + 10,000y =$	$7000x + 3000x$
Second tank	$10,000z + 3,000x =$	$10,000y + 3000y$
Third tank	$3,000y =$	$10,000z + 3000z$
Overall balance	$1,000 =$	$7,000x + 3000z$

Solving these simultaneous equations,

$$x = 0.139$$
$$y = 0.039$$
$$z = 0.009$$

The sodium carbonate recovered in the product is

$$7000x = (7000)(0.139) = 973 \quad \text{pounds}$$

The sodium carbonate lost in the final sludge residue is

$$3000z = (3000)(0.009) = 27 \text{ pounds}$$

INVENTORY CHANGES IN CONTINUOUS PROCESSES

In steady, continuous processes changes in the inventory, that is, changes in the quantity of charge, intermediates, and final products contained within the process itself are generally negligible. For the purpose of design an average material balance is selected ignoring fluctuations in inventory. However, such variations may become of great significance in interpreting experimental data obtained during a short period of operation. Problems of inventory fluctuations arise in preparing daily performance reports of continuous operations and also in analyzing the data from special test operations, especially those of short duration.

Inventory changes may result from variations in liquid or solid levels in the various operating units of the plant, from changes in the quantities of materials held in intermediate storage between successive units, and from changes in compositions at various points in the process. Less commonly inventory changes might be encountered as a result of changes in the operating pressure. Particular attention must be given to the separation of solids from solutions causing a change in liquid level with or without change in the inventory of the liquid.

Where a change in inventory occurs the material input to the plant is equal to the output plus the net accumulation of inventory or minus the net depletion in all parts of the process. A material balance may be obtained either by adding accumulation items to the output or subtracting them from the input. If the accumulation is an unchanged charge material or a finished product it is directly subtracted from the input or added to the output respectively.

If accumulation or depletion of an intermediate product takes place the problem requires conversion of the quantity of accumulation into the corresponding quantities of either charge or final product materials. In order to make this conversion, a set of yield figures must be assumed relating the intermediate product to the charge materials from which it originated or to the products to be derived from it, to whichever it is more similar.   Where calculation of yields is the object of the material balance it is necessary to assume approximate yields on which to base these corrections.   If the inventory changes are small, little error is introduced by these assumptions.   However, the calculation may be corrected by means of a second approximation based on the yields calculated from the first assumptions.

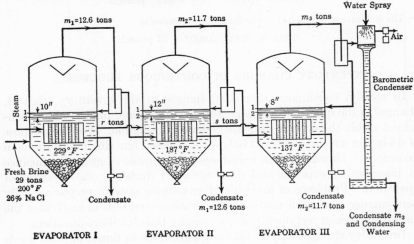

Fig. 29.   Inventory changes in triple-effect evaporation.

Illustration 6.   A triple-effect, vertical-tube, submerged-type evaporator, Fig. 29, is fed with 29 tons of salt brine solution during a given test period.   The brine is fed to the first evaporator at 200°F and contains 26% NaCl by weight.   The water evaporated from the first effect is condensed in the steam chest of the second effect and weighed as 12.6 tons; the water evaporated in the second effect is 11.7 tons. The salt crystallized during evaporation is allowed to collect in the conical hopper of each evaporator.   During the test period the liquid level was allowed to drop in each evaporator.   A drop of one inch corresponds to 12.1 cu ft of brine.

The following data were obtained:

Feed, 29 tons during period, 26% NaCl, and 200°F.

Evaporator	I	II	III
Steam temperature in steam chest	250°F	214°F	173°F
Brine temperature	229°F	187°F	137°F
Level change (inches)	10″ drop	12″ drop	8″ drop
Water evaporated	12.6 tons	11.7 tons	. . .
Solubility of NaCl,% by weight	29.0	27.9	27.0
Specific gravity of NaCl crystals	2.15		

It is desired to calculate the tons of NaCl crystallized in each evaporator and the tons of water evaporated from evaporator III.

Solution:

In each effect the brine solution is saturated, with a density which may be obtained from Fig. 1, page 18. The depletion of brine in each evaporator is not only that corresponding to the drop in liquid level but also that displaced by the salt crystallizing out and accumulating in the hopper. For each ton of salt crystallizing out $\frac{G}{2.15}$ tons of brine are displaced, where 2.15 is the specific gravity of the salt crystals and G is that of the brine solution. For example, in the first evaporator,

$$\frac{G}{2.15} = \frac{1.140}{2.150} = 0.53$$

and the total depletion of brine is

$$\frac{(10)(12.1)(62.4)(1.140)}{2000} + 0.53x \text{ tons} = 4.30 + 0.53x$$

where

$$x = \text{tons of salt crystallized in Evaporator I}$$
$$62.4 = \text{the density of water in pounds per cu ft}$$

The results are tabulated as follows:

Evaporator	I	II	III
Per cent NaCl by weight	29.0	27.9	27.0
Specific gravity of brine (Fig. 1)	1.140	1.145	1.150
Salt crystallized (tons)	x	y	z
Drop in level, inches	10	12	8
Drop in level, tons of brine	4.30	5.18	3.46
Brine displaced by salt	$0.53x$	$0.532y$	$0.535z$
Salt in total brine depletion	$1.25 + 0.154x$	$1.45 + 0.149y$	$0.935 + 0.145z$
Water in total brine depletion	$3.05 + 0.376x$	$3.73 + 0.383y$	$2.52 + 0.390z$
Water evaporated (tons)	12.6	11.7	m_3

From an overall water balance

$$29(0.74) + 3.05 + 0.376x + 3.73 + 0.383y + 2.52 + 0.390z = 12.6 + 11.7 + m_3 \quad \text{(a)}$$

or

$$m_3 = 6.45 + 0.376x + 0.383y + 0.390z \quad \text{(b)}$$

From an overall salt balance

$$29(0.26) + 1.25 + 0.154x + 1.45 + 0.149y + 0.935 + 0.145z = x + y + z \quad \text{(c)}$$

$$0.846x + 0.851y + 0.855z = 11.18 \quad \text{(d)}$$

Let

r = tons of brine leaving Evaporator I to Evaporator II

s = tons of brine leaving Evaporator II to Evaporator III

From salt and water balances for each evaporator the following are obtained:

Evaporator I

$$\text{Salt} \qquad 1.25 + 0.154x + 0.26(29) = x + 0.29r \qquad (e)$$

$$\text{Water} \qquad 3.05 + 0.376x + 0.74(29) = 12.6 + 0.71r \qquad (f)$$

$$x = 3.91$$
$$r = 18.88$$

Evaporator II

$$\text{Salt} \qquad 1.45 + 0.149y + 0.29(18.88) = y + 0.279s \qquad (g)$$

$$\text{Water} \qquad 3.73 + 0.383y + 0.71(18.88) = 11.7 + 0.721s \qquad (h)$$

$$y = 4.82$$
$$s = 10.07$$

Evaporator III

$$\text{Salt} \qquad 0.936 + 0.145 + 0.279(10.07) = z \qquad (i)$$

$$z = 4.38$$

From (b)

$$m_3 = 6.45 + 0.376x + 0.383y + 0.390z = 11.47 \qquad (j)$$

MATERIAL BALANCE

Input	Tons	Output	Tons
(a) Brine	29.00	(a) Water from I	12.60
Depletion Items		(b) Water from II	11.70
(a) Brine in I		(c) Water from III	11.47
$4.30 + 0.53x =$		Accumulation Items	
$4.30 + (0.53)(3.91) = 6.37$		(a) Salt in I, x	3.91
(b) Brine in II		(b) Salt in II, y	4.82
$5.18 + 0.532y =$		(c) Salt in III, z	4.38
$5.18 + (0.532)(4.82) = 7.75$			
(c) Brine in III			
$3.46 + 0.535z =$			
$3.46 + (0.535)(4.38) =$	5.80		
	48.92		48.88

RECYCLING OPERATIONS

The recycling of fluid streams in chemical processing is a common practice to increase yields or enrich a product. In fractionating columns part of the distillate is refluxed through the column to enrich the product. In ammonia synthesis the gas mixture leaving the converter after recovery of ammonia is recycled through the converter. In the operation of dryers part of the exit air stream is recirculated to conserve heat.

Better wetting is provided in scrubbing towers by recirculating part of the exit liquid. Recycling occurs in nearly every stage of petroleum processing.

In a recycling operation the *total* or *combined feed* is made up of a mixture of the *fresh* or *net feed* with the *recycle stock*. The *gross products* of the operation are a mixture of the *net products*, which are withdrawn from the system, and the recycle stock. In such an operation two types of material balances are of interest. In the *overall material balance* the net feed is equated against the net products. In a *once-through material balance* the total feed is equated against the gross products.

Two methods are employed for expressing the yields of a recycling operation. The net or *ultimate yields* are the percentages which the net products form of the net feed. The *once-through yields* or *yields per pass* are the percentages which the reaction products form of the total feed.

The ratio of recycle stock to fresh feed is termed the *recycle ratio* and the ratio of total or combined feed to the fresh feed is termed the combined or *total feed ratio*. It is evident that

$$\text{Total feed ratio} = \text{Recycle ratio} + 1$$

In dealing with operations in which all streams are subject to exact analysis, design problems are generally approached from data obtained in laboratory or pilot plant experiments of the once-through type, in which the conditions resulting from recycling are simulated by merely varying the composition of the total feed. However, when dealing with extremely complex mixtures such as encountered in petroleum refining, data from experiments in which recycling is actually carried out are necessary as a design basis for the establishment of ultimate yields and the characteristics of the recycle stock. In the former case the design problem is approached by first setting up a once-through material balance from which the recycle quantities and ultimate yields are derived. In the latter case both the overall and the once-through material balances must be based directly on experimental data. Illustration 7 is typical of the first class of problem while an example of the second class is shown in Chapter X, page 421.

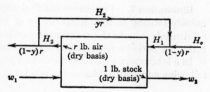

FIG. 30.   Recirculation of air in drying.

A common example of recycling is the recirculation of air in the drying of solids, shown diagrammatically in Fig. 30. On the basis of one pound of stock (dry basis) passing through the dryer, r pounds of air (dry basis) are passed through countercurrently. If y is the fraction of air (dry basis) recirculated, then $(1 - y)$ is the fraction of fresh air supplied (dry basis) and also the fraction of spent air rejected (dry basis).

Accordingly, the following material balances result:

Overall balance:

$$w_1 - w_2 = r(1 - y)(H_2 - H_0) \tag{1}$$

Once-through balance:

$$w_1 - w_2 = r(H_2 - H_1) \tag{2}$$

where

w_1 = moisture content of stock entering, pound per pound dry stock
w_2 = moisture content of stock leaving, pound per pound dry stock
H_2 = humidity of air leaving, pound per pound dry air
H_1 = humidity of air entering dryer, pound per pound dry air
H_0 = humidity of fresh air supply, pound per pound dry air

From Equations (1) and (2)

$$(H_2 - H_0)y = H_1 - H_0 \tag{3}$$

The fraction of air to be recirculated depends upon the relative costs of dryer, power, and heat, and is hence established by minimizing the total costs. With increased recirculation for the same drying capacity, a larger dryer is required with increased costs of equipment, power, and radiation offset by a reduced cost of heat requirements.

Illustration 7. Propane is catalytically dehydrogenated to produce propylene by quickly heating it to a temperature of 1000–1200°F and passing it over a granular solid catalyst. As the reaction proceeds carbon is deposited on the catalyst, necessitating its periodic reactivation by burning off the carbon with oxygen-bearing gases.

In a laboratory experiment in which pure propane is fed to the reactor, gases of the following composition leave the reactor:

Gas	Mole per cent
Propane	44.5
Propylene	21.3
Hydrogen	25.4
Ethylene	0.3
Ethane	5.3
Methane	3.2
	100.0

Based on these data it is desired to design a plant to produce 100,000 pounds per day of propylene in a mixture of 98.8% purity. The flow diagram of the proposed process is shown in Fig. 31.

Fresh propane feed, mixed with propane recycle stock is fed to a heater from which it is discharged to a catalytic reactor operating under such conditions as to produce the same conversion of propane as was obtained in the laboratory experiment. The reactor effluent gases are cooled and compressed to a suitable pressure for separation of the light gases. This separation is accomplished by absorption of the propane and propylene, together with small amounts of the lighter gases, in a cooled absorption oil which is circulated through an absorption tower. The "fat oil" from the bottom of the absorber is pumped to a stripping tower where by application of heat at the bottom of the tower the dissolved gases are distilled away from the oil which is then cooled and recirculated to the absorber.

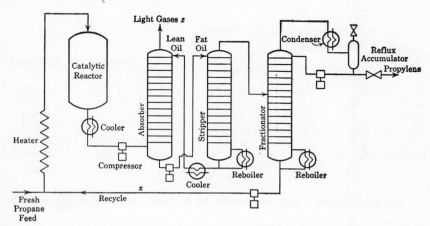

Fig. 31.   Flow chart of propane dehydrogenation plant.

The gases from the stripping tower are passed to a high pressure fractionating tower which separates them into propane recycle stock as a bottoms product and propylene and lighter gases as overhead. The following compositions are established as preliminary design bases estimated from the vaporization characteristics of the gases:

1. The light gases from the absorber are to contain 1.1% propane and 0.7% propylene by volume. Substantially all hydrogen, ethylene, and methane leaving the cooler will appear in the light gases. The ethane, however, will appear in both the light gases and the product.

2. The propane recycle stock is to contain 98% propane and 2% propylene by weight.

3. The propylene product is to contain 98.8% propylene, 0.7% ethane, and 0.5% propane by weight.

The total feed is passed over the catalyst at the rate of 4.1 pound-moles per cubic foot of catalyst per hour. The density of the catalyst is 54 pounds per cubic foot.

It may be assumed that the small amount of propylene in the feed to the reactor passes through unchanged.

Required:

(a) The amount of carbon formed on the catalyst, expressed as weight per cent of the propane fed to the catalyst chamber.

(b) The process period in minutes required in building a carbon deposit equal to 2% by weight of the catalyst.

(c) An overall material balance.

(d) A once-through material balance.

(e) Ultimate yield of recovered propylene expressed as mole percentage of propane in fresh feed.

(f) Once-through yield of propylene made in reactor expressed as mole percentage of total propane entering reactor.

(g) Recycle ratio, weight units.

(h) Total feed ratio, weight units.

Preliminary Calculations

Gas Formed by Decomposition of Propane

Basis: 1 lb-mole of gas.

	Lb-Moles	Mole Wt	Lb	Lb-Atoms C	Lb-Atoms H
Propane C_3H_8	0.445	44.09	19.62	1.335	3.560
Propylene C_3H_6	0.213	42.08	8.96	0.639	1.278
Hydrogen H_2	0.254	2.016	0.512		0.508
Ethylene C_2H_4	0.003	28.05	0.084	0.006	0.012
Ethane C_2H_6	0.053	30.07	1.594	0.106	0.318
Methane CH_4	0.032	16.04	0.513	0.032	0.128
	1.000		31.28	2.118	5.804

Gas formed from 1 mole of propane (by hydrogen balance) $= \dfrac{8.000}{5.804} = 1.378$ lb-moles.

Process Design Calculations

Basis of design: 1 hour of operation.

$$\text{Production rate of pure propylene} = \frac{100,000}{24} = 4167 \text{ lb/hr}$$

or

$$\frac{4167}{42.08} = 99.03 \text{ lb-moles/hr}$$

Evaluation of Unknown Stream Weights

Let x = lb-moles recycle

y = lb-moles fresh feed

z = lb-moles light gases

Combined feed

$$\text{Propane} = y + 0.9791x \text{ lb-moles}$$
$$\text{Propylene} = 0.02095x \text{ lb-moles}$$

Gases leaving reactor

Gases formed from the propane entering $\quad = 1.378(y + 0.9791x)$

$\qquad\qquad\qquad\qquad\qquad = 1.378y + 1.349x$ lb-moles

Propylene passing
through unchanged $= 0.02095x$ lb-moles

Total gas leaving $= 1.378y + 1.370x$

Propane leaving $= (1.378y + 1.349x)(0.445)$
                    $= 0.613y + 0.600x$ lb-moles

Propylene leaving $= 0.213(1.378y + 1.349x) + 0.02095x$
                      $= 0.2935y + 0.3083x$ lb-moles

MATERIAL BALANCES

(lb-moles)

	Leaving Reactor	Recycle	Product	Light Gases
Propane Balance	$0.613y + 0.600x =$	$0.9791x +$	$\dfrac{99.03}{0.9855}(0.00476)$	$+ 0.011z$
Propylene Balance	$0.2935y + 0.3083x =$	$0.02095x +$	99.03	$+ 0.007z$
Total	$1.378y + 1.370x =$	$x +$	$\dfrac{99.03}{0.9855}$	$+ z$

Simplifying the above three equations:

$$-0.379x + 0.613y - 0.011z = 0.478$$
$$0.2873x + 0.2935y - 0.007z = 99.03$$
$$0.370x + 1.378y - z = 100.5$$

Solving:

$$x = 210.8 \text{ lb-moles recycle}$$
$$y = 134.2 \text{ lb-moles fresh feed}$$
$$z = 162.9 \text{ lb-moles light gases}$$

Gases Leaving Reactor

Formed from propane $= (1.378)(134.2) + (1.349)(210.8) = 469.3$

Propylene from recycle $= (0.02095)(210.8)$ $= 4.42$

		Lb-Moles	Mole Wt	Lb
$C_3H_8 = (0.445)(469.3)$	$=$	208.8	44.09	9,206
$C_3H_6 = (0.213)(469.3) + 4.4$	$=$	104.4	42.08	4,393
$H_2 = (0.254)(469.3)$	$=$	119.3	2.016	240
$C_2H_4 = (0.003)(469.3)$	$=$	1.4	28.05	39.2
$C_2H_6 = (0.053)(469.3)$	$=$	24.9	30.07	749
$CH_4 = (0.032)(469.3)$	$=$	15.0	16.04	241
		473.8		14,868

$$\text{Molecular weight} = \frac{14,868}{473.8} = 31.38$$

Light Gases

	Lb-Moles	Mole Wt	Lb
C_3H_8 (0.011)(162.9)	1.792	44.09	79.0
C_3H_6 (0.007)(162.9)	1.140	42.08	48.0
H_2 (from reactor)	119.3	2.016	240.5
C_2H_4 (from reactor)	1.4	28.05	39.2
CH_4 (from reactor)	15.0	16.04	240.6
C_2H_6 Leaving reactor 24.9			
In product 0.98			
In light gases 23.9	23.9	30.07	718.7
	162.5		1,366.0

$$\text{Molecular weight} = \frac{1366}{162.5} = 8.41$$

Recycle

	Lb-Moles	Mole Wt	Lb
C_3H_8 (0.9791)(210.8)	206.4	44.09	9,100
C_3H_6 (0.02095)(210.8)	4.42	42.08	186
	210.8		9,286

$$\text{Molecular weight} = \frac{9,286}{210.8} = 44.05$$

Product

$$\text{Total weight} = \frac{100,000}{(0.988)(24)} = 4,217 \text{ lb}$$

C_3H_6 $\dfrac{100,000}{24} = 4,167 \text{ lb}$

C_3H_8 $(0.007)(4,217) = 29.5 \text{ lb}$

C_2H_6 $(0.005)(4,217) = 21.1 \text{ lb}$

Fresh Feed

(134.2)(44.09)   = 5,917 lb

Carbon Deposit

Propane to reactor	134.2 + 206.4 = 340.6 lb-moles
C in propane	= (3)(340.6)(12) = 12,262 lb
C in gases formed from propane	= (1.378)(340.6)(2.118)(12) = 11,929 lb
C deposited on catalyst	= 333 lb

Required Results:

(a) Amount of carbon deposited on catalyst, as weight per cent of the propane fed to the reactor.

$$\text{Carbon deposit} = \frac{333}{5,917 + 9,100} = 2.22\%$$

(b) Process Period

Total feed	$= 210.8 + 134.2 =$		345 lb moles/hr
Volume of catalyst	$= \dfrac{345.0}{4.1}$	$=$	84.2 cu ft
Weight of catalyst	$= (84.2)(54)$	$=$	4,547 lb
Carbon tolerance	$= (0.02)(4,547)$	$=$	90.9 lb
Rate of carbon deposition $=$		$=$	333 lb/hr
Process period	$= \dfrac{90.9}{333}(60)$	$=$	16.4 min

(c) Overall Material Balance

Input

Fresh feed   5,917 lb

Output

Light gases	1,366 lb	
Product	4,217	
Carbon deposit	333	
Total	5,916	5,916 lb

(d) Once-Through Material Balance

Input

Fresh feed	5,917 lb	
Recycle	9,286	
Total	15,203	15,203 lb

Output

Light gases	1,366 lb	
Product	4,217	
Carbon deposit	333	
Recycle	9,286	
Total	15,202	15,202 lb

(e) Ultimate Yield of Propylene

Propylene in product	99.03 lb moles
Propane feed	134.2 lb moles

Per cent yield $= \dfrac{99.03}{134.2}(100) = 73.8\%$

(f) Once-Through Yield

Propylene produced in Reactor $= 99.03 + 1.14 =$

100.2 lb moles

Propane entering reactor $= 134.2 + 206.4 = 340.6$ lb moles

Per cent yield $= \dfrac{100.2}{340.6}(100) = 29.4\%$

(g) Recycle Ratio

Weight recycle   9,286 lb

Weight fresh feed   5,917 lb

$$\text{Recycle ratio} = \frac{9,286}{5,917} = 1.57$$

(h) Total Feed Ratio

Weight recycle   9,286 lb

Weight fresh feed   5,900 lb

Total Weight   15,186 lb

$$\text{Total feed ratio} = \frac{15,186}{5,917} = 2.57$$

Accumulation of Inerts in Recycling. One limitation sometimes encountered in the recycling of fluid streams is the gradual accumulation of inerts or impurities in the recycled stock. Unless some provision is made for removing such impurities they will gradually accumulate until the process comes to a stop. This problem can be solved by bleeding off a fraction of the recycle stock. The recycle stock bled off may be passed through some special process to remove impurities and to recover the useful components, or discarded if such recovery is too costly.

Where processes take place in the medium of a special solvent the solvent may be circulated throughout the plant in a closed system. As the product is withdrawn from the solvent some impurities remain in solution and gradually accumulate. The accumulation is fixed at a certain limiting concentration by continually withdrawing a definite fraction of the recycled solvent. This type of control is maintained in the electrolytic refining of copper wherein the electrolyte is continually recirculated while a portion is bled off and replaced by fresh electrolyte. The spent electrolyte is treated separately by a special process for the recovery of residual copper, for the removal of accumulated impurities such as nickel sulfate, and the purified sulfuric acid is returned to the process. In the synthesis of ammonia from atmospheric nitrogen and hydrogen the percentage conversion of a 1:3 mixture is 25% in a single pass through the reactor. The ammonia formed is removed by cooling under high pressure and the unconverted nitrogen and hydrogen are recirculated to the reactor. The argon from the atmospheric nitrogen is allowed to accumulate to a fixed upper limit by bleeding off a fraction of the recycled gas or by leakage from the system.

Illustration 8. In the operation of a synthetic ammonia plant, shown diagrammatically in Fig. 32, a 1:3 nitrogen-hydrogen mixture is fed to the converter resulting in a 25% conversion to ammonia. The ammonia formed is separated by condensation and the unconverted gases recycled to the reactor. The initial nitrogen-hydrogen mixture contains 0.20 part of argon to 100 parts of N_2–H_2 mixture. The toleration limit of argon in the reactor is assumed to be 5 parts to 100 parts of N_2 and H_2 by volume.   Estimate the fraction of recycle which must be continually purged.

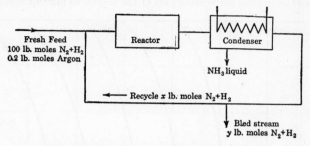

FIG. 32.   Purging of inerts in recycle stock.

Basis of Calculations: 100 pound-moles N_2-H_2 in fresh feed.

Let x = moles N_2 and H_2 recycled to reactor

   y = moles N_2 and H_2 bled off or purged

   Moles N_2 and H_2 entering reactor = $100 + x$

   Moles N_2 and H_2 leaving reactor = $0.75(100 + x)$

   Moles of NH₃ formed = $\dfrac{0.25(100 + x)}{2}$

   Moles of argon in fresh feed = 0.20

   Moles of argon in total feed = $0.05(100 + x)$

   Moles of argon per mole of N_2-H_2 mixture leaving condenser = $\dfrac{0.05}{0.75} = 0.067$

   Moles of argon bled off = $0.067y$

When a steady state of operation is attained the argon purged is equal to the argon in the fresh gas supply.

Hence

$$0.067y = 0.20$$

or

$$y = 2.98$$

From an H_2-N_2 balance around bleed point

$$0.75(100 + x) = x + y$$

Since $y = 2.98$; $x = 288$ pound-moles

   Summary:

   Fresh N_2 and H_2 = 100 pound-moles

   Recycle N_2 and H_2 = 288 pound-moles

   Purged N_2 and H_2 = 3.0 pound-moles

   Ammonia formed   = 48.5 pound-moles

   Argon   = 0.20 pound-mole

   Recycle ratio = $\dfrac{288}{100} = 2.88$

   Purge ratio = $\dfrac{3}{288} = 0.0104$

In the actual operation of a high pressure plant the unavoidable leakage generally is sufficient to keep argon in the reactor below the toleration limit so that no special provision need be applied for venting part of the recycle to prevent accumulation.

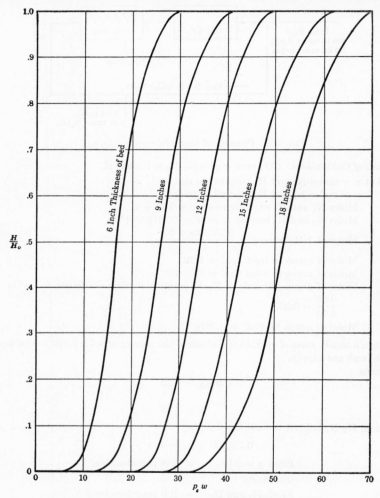

p_s = vapor pressure of water in atmosphere at temperature of bed
w = pounds of dry air passed through bed per square foot

FIG. 33. Fraction of moisture removed from air by a bed of silica gel.

By-passing Streams. *By-passing* of a fluid stream by splitting it into two parallel streams is often practiced where accurate control in concentration is desired. For example, in conditioning air with a single stationary bed of silica gel to a lower humidity a more nearly uniform humidity is obtained in the final air mixture if part of the air is dehumidified to a low moisture content and then mixed with the unconditioned air to produce the desired mixture. The fraction of the entering humidity remaining in the air after drying in beds of silica gel of various thicknesses is shown in Fig. 33 plotted against the product of $p_s w$ where

w = the total pounds of dry air passed through gel per square foot of cross-section

p_s = the vapor pressure of water in atmospheres at the temperature of the bed

$w = G\tau = u\rho\tau$

where

G = pounds of dry air per sq ft-min

τ = time in minutes

ρ = density of entering air, pounds of dry air per cu ft of mixture

u = velocity of entering air, feet per minute

It will be observed that the humidity of the air leaving the gel is by no means constant. For an extended period of time, as indicated by the abscissas, the air leaves with nearly no humidity, then it rises abruptly as more air is passed through. The bed behaves as though a wedge of water of the shape of the curves shown in Fig. 33 were advancing through the bed; at first the advancing wedge is nearly imperceptible, then suddenly the full wedge appears.

Illustration 9. 1000 cu ft of air per minute at 85°F, 70% relative humidity is to be conditioned to a relative humidity of 30% and kept at 85°F, using a stationary bed of silica gel 12 in. thick and 4 ft in diameter. To maintain the relative humidity constant at 30%, part of the air is by-passed around the dryer and mixed with the part which is dried. Constant humidity in the final air mixture is controlled automatically by a damper regulating the fraction of air to be by-passed. This is accomplished by a constant wet-bulb controller. Assuming isothermal operation,

(a) Construct a plot showing the fraction of air to be by-passed against time.

(b) Calculate the time of operation before regeneration becomes necessary.

Area of bed $= \dfrac{\pi D^2}{4} = 12.52$ sq ft

Total rate of air flow $= \dfrac{1000}{359} \times \dfrac{492}{545} \times 0.9716 \times 29 = 70.9$ lb per min (dry basis)

Rate of air flow per sq ft of bed $= \dfrac{70.9}{12.52} = 5.66$ lb per min (dry basis)

H_0 = initial humidity = 0.0182 $\qquad$ p_s = vapor pressure at 85°F = 0.0406 atm

H_1 = desired humidity = 0.0077 $\qquad$ partial pressure at 70% r.h. = 0.0284 atm

H = humidity leaving dryer $\qquad$ partial pressure at 30% r.h. = 0.0122 atm

Let x = fraction of air by-passed

From a material balance for water

$$0.0182x + H(1 - x) = 0.0077$$

$$x = \frac{0.0077 - H}{0.0182 - H} \qquad \text{(a)}$$

Let Δw = dry air passed through dryer in time interval $\Delta \tau$

Then, $\dfrac{\Delta w}{1 - x} = 5.66\Delta \tau$ = total dry air delivered at 30% relative humidity.

The calculated results are tabulated at stated intervals of $p_s w$, the abscissas of Fig. 33. The corresponding values of H/H_0 are obtained from Fig. 33 and tabulated in Column 2. From the value of $p_s = 0.0406$ atm at 85°F, the values of w are tabulated in Column 3. The dry air delivered for each interval of $p_s w$ is obtained as the difference of succeeding items in Column 3 and tabulated in Column 4 as Δw. The value of the humidity of air H leaving the gel is obtained from Column 2 with $H_0 = 0.0182$ and tabulated in Column 5. The average humidity H_a during each interval is obtained from Fig. 33 as the mean value over the interval of $p_s w$ indicated in Column 1. The average fraction x of air by-passed during the interval Δw is obtained from the equation $x = \dfrac{0.0077 - H_a}{0.0182 - H_a}$ and tabulated in Column 7. The total air delivered

1	2	3	4	5	6	7	8	9	10
$p_s w$	H/H_0 (Fig. 33)	w	Δw interval	H	H_a for interval Δw	x	$\dfrac{\Delta w}{1 - x}$	$\Delta \tau$	τ min
0	0	0	.....	0	....	.....	...	.....	0
....	.....	...	554.0	....	0	0.423	960	169.6	
22.5	0	554		0	....	.....			170
....	.....	...	61.5	....	0.00032	0.413	105	18.6	
25.0	0.035	616		0.000637	....	.....			188
....	.....	...	61.5	....	0.00128	0.379	99	17.5	
27.5	0.105	677		0.00191	....	.....			206
....	.....	...	61.5	....	0.00291	0.313	90	15.9	
30.0	0.215	739		0.00392	....	.....			222
....	.....	...	61.5	....	0.00519	0.193	76	13.4	
32.5	0.355	800		0.00646	....	.....			235
....	.....	...	24.0	....	0.00708	0.055	25	4.4	
33.5	0.423	824		0.0077	....	.....			239
							$\Sigma \Delta \tau = 239$ min		

per interval Δw is equal to $\dfrac{\Delta w}{1 - x}$ and is tabulated in Column 8. The time interval $\Delta \tau$ corresponding to the air interval Δw is obtained by dividing items in Column 8 by 5.66; these are tabulated in Column 9. The total time elapsed is obtained from a summation of items in column 9, $\tau = \Sigma \Delta \tau$, and tabulated in Column 10.

A plot of x against τ gives the desired schedule for by-passing air. For the first 170 minutes, 42% of the air is by-passed, after which time the by-pass is gradually decreased to no by-pass, according to schedule, until a total of 239 minutes has elapsed. The silica gel must then be regenerated.

PROBLEMS

1. In the manufacture of soda-ash by the LeBlanc process, sodium sulfate is heated with charcoal and calcium carbonate. The resulting " black ash " has the following composition:

Na_2CO_3...	42%
Other water-soluble material.........................	6%
Insoluble material (charcoal, CaS, etc.)..............	52%

The black ash is treated with water to extract the sodium carbonate. The solid residue from this treatment has the following composition:

Na_2CO_3...	4%
Other water-soluble salts............................	0.5%
Insoluble matter....................................	85%
Water...	10.5%

(a) Calculate the weight of residue remaining from the treatment of 1.0 ton of black ash.

(b) Calculate the weight of sodium carbonate extracted per ton of black ash.

2. A contract is drawn up for the purchase of paper containing 5% moisture at a price of 7 cents per pound. It is provided that if the moisture content varies from 5%, the price per pound shall be proportionately adjusted in order to keep the price of the bone-dry paper constant. In addition, if the moisture content exceeds 5%, the purchaser shall deduct from the price paid the manufacturer the freight charges incurred as a result of the excess moisture. If the freight rate is 90 cents per 100 lb, calculate the price to be paid for 3 tons of paper containing 8% moisture.

3. A laundry can purchase soap containing 30% of water at a price of $7.00 per 100 lb f. o. b. the factory. The same manufacturer offers a soap containing 5% of water. If the freight rate is 60 cents per 100 lb, what is the maximum price that the laundry should pay the manufacturer for the soap containing 5% water?

4. The spent acid from a nitrating process contains 43% H_2SO_4, 36% HNO_3, and 21% H_2O by weight. This acid is to be strengthened by the addition of concentrated sulfuric acid containing 91% H_2SO_4, and concentrated nitric acid containing 88% HNO_3. The strengthened mixed acid is to contain 40% H_2SO_4 and 43% HNO_3. Calculate the quantities of spent and concentrated acids which should be mixed together to yield 1000 lb of the desired mixed acid.

5. The waste acid from a nitrating process contains 21% HNO_3, 55% H_2SO_4, and 24% H_2O by weight. This acid is to be concentrated to contain 28% HNO_3 and 62% H_2SO_4 by the addition of concentrated sulfuric acid containing 93% H_2SO_4 and concentrated nitric acid containing 90% HNO_3. Calculate the weights of waste and concentrated acids which must be combined to obtain 1000 lb of the desired mixture.

6. In the manufacture of straw pulp for the production of a cheap straw-board paper, a certain amount of lime is carried into the beaters with the cooked pulp. It is proposed to neutralize this lime with commercial sulfuric acid, containing 77% H_2SO_4 by weight.

In a beater containing 2500 gal of pulp it is found that there is a lime concentration equivalent to 0.45 gram of CaO per liter.

(a) Calculate the number of pound-moles of lime present in the beater.

(b) Calculate the number of pound-moles and pounds of H_2SO_4 which must be added to the beater in order to provide an excess of 1.0% above that required to neutralize the lime.

(c) Calculate the weight of commercial acid which must be added to the beater for the conditions of b.

(d) Calculate the weight of calcium sulfate formed in the beater.

7. Phosphorus is prepared by heating in the electric furnace a thoroughly mixed mass of calcium phosphate, sand, and charcoal. It may be assumed that in a certain charge the following conditions exist: the amount of silica used is 10% in excess of that theoretically required to combine with the calcium to form the silicate; the charcoal is present in 40% excess of that required to combine, as carbon monoxide, with the oxygen which would accompany all the phosphorus as the pentoxide.

(a) Calculate the percentage composition of the original charge.

(b) Calculate the number of pounds of phosphorus obtained per 100 lb of charge, assuming that the decomposition of the phosphate ,by the silica is 90% complete and that the reduction of the liberated oxide of phosphorus, by the carbon, is 70% complete.

8. A coal containing 81% total carbon and 6% unoxidized hydrogen is burned in air.

(a) If air is used 30% in excess of that theoretically required, calculate the number of pounds of air used per pound of coal burned.

(b) Calculate the composition, by weight, of the gases leaving the furnace, assuming complete combustion.

9. In the Deacon process for the manufacture of chlorine, a dry mixture of hydrochloric acid gas and air is passed over a heated catalyst which promotes oxidation of the acid. Air is used in 30% excess of that theoretically required.

(a) Calculate the weight of air supplied per pound of acid.

(b) Calculate the composition, by weight, of the gas entering the reaction chamber.

(c) Assuming that 60% of the acid is oxidized in the process, calculate the composition, by weight, of the gases leaving the chamber.

10. In order to obtain barium in a form which may be put into solution, the natural sulfate, barytes, is fused with sodium carbonate. A quantity of barytes, containing only pure barium sulfate and infusible matter, is fused with an excess of pure, anhydrous soda ash. Upon analysis of the fusion mass it is found to contain 11.3%

barium sulfate, 27.7% sodium sulfate, and 20.35% sodium carbonate. The remainder is barium carbonate and infusible matter.

(a) Calculate the percentage completion of the conversion of the barium sulfate to the carbonate and the complete analysis of the fusion mass.

(b) Calculate the composition of the original barytes.

(c) Calculate the percentage excess in which the sodium carbonate was used above the amount theoretically required for reaction with all the barium sulfate.

11. In the manufacture of sulfuric acid by the contact process iron pyrites, FeS_2, is burned in dry air, the iron being oxidized to Fe_2O_3. The sulfur dioxide thus formed is further oxidized to the trioxide by conducting the gases mixed with air over a catalytic mass of platinum-black at a suitable temperature. It will be assumed that in the operation sufficient air is supplied to the pyrites burner that the oxygen shall be 40% in excess of that required if all the sulfur actually burned were oxidized to the trioxide. Of the pyrites charged, 15% is lost by falling through the grate with the " cinder " and is not burned.

(a) Calculate the weight of air to be used per 100 lb of pyrites charged.

(b) In the burner and a " contact shaft " connected with it, 40% of the sulfur burned is converted to the trioxide. Calculate the composition, by weight, of the gases leaving the contact shaft.

(c) By means of the platinum catalytic mass, 96% of the sulfur dioxide *remaining* in the gases leaving the contact shaft is converted to the trioxide. Calculate the total weight of SO_3 formed per 100 lb of pyrites charged.

(d) Assuming that all gases from the contact shaft are passed through the catalyzer, calculate the composition by weight of the resulting gaseous products.

(e) Calculate the overall degree of completion of the conversion of the sulfur in the pyrites charged to SO_3 in the final products.

12. In the LeBlanc soda process the first step is carried out according to the following reaction:

$$2NaCl + H_2SO_4 = NaCl + NaHSO_4 + HCl$$

The acid used has a specific gravity of 58°Baumé, containing 74.4% H_2SO_4. It is supplied in 2% excess of that theoretically required for the above reaction.

(a) Calculate the weight of acid supplied per 100 lb of salt charged.

(b) Assume that the reaction goes to completion, all the acid forming bisulfate, and that in the process 85% of the HCl formed and 20% of the water present are removed. Calculate the weights of HCl and water removed per 100 lb of salt charged.

(c) Assuming the conditions of part (b), calculate the percentage composition of the remaining salt mixture.

13. In the common process for the manufacture of nitric acid, sodium nitrate is treated with aqueous sulfuric acid containing 95% H_2SO_4 by weight. In order that the resulting " niter cake " may be fluid, it is desirable to use sufficient acid so that there will be 34% H_2SO_4 by weight in the final cake. This excess H_2SO_4 will actually be in combination with the Na_2SO_4 in the cake, forming $NaHSO_4$, although for purposes of calculation it may be considered as free acid. It may be assumed that the cake will contain 1.5% water, by weight, and that the reaction will go to completion, but that 2% of the HNO_3 formed will remain in the cake.   Assume that the sodium nitrate used is dry and pure.

(a) Calculate the weight and percentage composition of the niter cake formed per 100 lb of sodium nitrate charged.

(b) Calculate the weight of aqueous acid to be used per 100 lb of sodium nitrate.

(c) Calculate the weights of nitric acid and water vapor distilled from the niter cake, per 100 lb of $NaNO_3$ charged.

14. Pure carbon dioxide may be prepared by treating limestone with aqueous sulfuric acid.   The limestone used in such a process contained calcium carbonate and magnesium carbonate, the remainder being inert, insoluble materials.   The acid used contained 12% H_2SO_4 by weight.   The residue from the process had the following composition:

$$
\begin{aligned}
&CaSO_4\ldots\ldots\ldots\ldots\ldots\ldots\ldots\quad 8.56\% \\
&MgSO_4\ldots\ldots\ldots\ldots\ldots\ldots\ldots\quad 5.23\% \\
&H_2SO_4\ldots\ldots\ldots\ldots\ldots\ldots\ldots\quad 1.05\% \\
&Inerts\ldots\ldots\ldots\ldots\ldots\ldots\ldots\quad 0.53\% \\
&CO_2\ldots\ldots\ldots\ldots\ldots\ldots\ldots\ldots\quad 0.12\% \\
&Water\ldots\ldots\ldots\ldots\ldots\ldots\ldots\quad 84.51\%
\end{aligned}
$$

During the process the mass was warmed and carbon dioxide and water vapor removed.

(a) Calculate the analysis of the limestone used.

(b) Calculate the percentage of excess acid used.

(c) Calculate the weight and analysis of the material distilled from the reaction mass per 1000 lb of limestone treated.

15. Barium carbonate is commercially important as a basis for the manufacture of other barium compounds.   In its manufacture, barium sulfide is first prepared by heating the natural sulfate, barytes, with carbon.   The barium sulfide is extracted from this mass with water and the solution treated with sodium carbonate to precipitate the carbonate of barium.

In the operation of such a process it is found that the solution of barium sulfide formed contains also some calcium sulfide, originating from impurities in the barytes.   The solution is treated with sodium carbonate and the precipitated mass of calcium and barium carbonates is filtered off.   It is found that 16.45 lb of dry precipitate are removed from each 100 lb of filtrate collected.   The analysis of the precipitate is:

$$
\begin{aligned}
&CaCO_3\ldots\ldots\ldots\ldots\ldots\ldots\ldots\quad 9.9\% \\
&BaCO_3\ldots\ldots\ldots\ldots\ldots\ldots\ldots\quad 90.1\%
\end{aligned}
$$

The analysis of the filtrate is found to be:

$$
\begin{aligned}
&Na_2S\ldots\ldots\ldots\ldots\ldots\ldots\ldots\quad 6.85\% \\
&Na_2CO_3\ldots\ldots\ldots\ldots\ldots\ldots\ldots\quad 2.25\% \\
&H_2O\ldots\ldots\ldots\ldots\ldots\ldots\ldots\quad 90.90\%
\end{aligned}
$$

The sodium carbonate for the precipitation was added in the form of anhydrous soda ash which contained calcium carbonate as an impurity.

(a) Determine the percentage excess sodium carbonate used above that required to precipitate the BaS and CaS.

(b) Calculate the composition of the original solution of barium and calcium sulfides.   (*Note:* Barium sulfide is actually decomposed in solution, existing as the compound $OHBaSH\cdot5H_2O$.   However, in this reaction the entire calculation may be carried out and compositions expressed as though the compound in solution were BaS.)

(c) Calculate the composition of the dry soda ash used in the precipitation.

16. A mixture of 45% CaO, 20% MgO, and 35% NaOH by weight is used for treating ammonium carbonate liquor in order to liberate the ammonia.

(a) Calculate the number of pounds of CaO to which 1 lb of this mixture is equivalent.

(b) Calculate the composition of the mixture, expressing the quantity of each component as the percentage it forms of the total reacting value of the whole.

(c) Calculate the number of pounds of ammonia theoretically liberated by 1 lb of this mixture.

17. A water is found to contain the following metals, expressed in milligrams per liter: Ca, 32; Mg, 8.4; Fe (ferrous), 0.5.

(a) Calculate the " total hardness " of the water, expressed in milligrams of equivalent $CaCO_3$ per liter, the calcium of which would have the same reacting value as the total reacting value of the metals actually present.

(b) Assuming that these metals are all combined as bicarbonates, calculate the cost of the lime required to soften 1000 gal of the water. Commercial lime, containing 95% CaO, costs $8.50 per ton.

18. A glass for the manufacture of chemical ware is composed of the silicates and borates of several basic metals. Its composition is as follows:

SiO_2............	66.2%	ZnO............	10.3%
B_2O_3............	8.2%	MgO............	6.0%
Al_2O_3..........	1.1%	Na_2O..........	8.2%

Determine whether the acid or the basic constituents are in excess in this glass, and the percentage excess reacting value above that theoretically required for a neutral glass. (Assume that Al_2O_3 acts as a base and B_2O_3 as an acid, HBO_2.)

19. In Illustration 6, assume that the level is kept constant in each effect with the same evaporations as noted, namely, 12.6, 11.70, and 11.76 tons from the three evaporators. How much salt would be crystallized in each effect?

20. In Illustration 6, assume that the salt was removed from the hoppers, that the drops in liquid levels were the same, and that the water evaporated was 11.1 tons in the first effect and 10.2 tons in the second effect. Calculate the salt crystallized in each evaporator and the water evaporated from the third effect.

21. A mixture of benzene, toluene, and xylenes is separated by continuous fractional distillation in two towers, the first of which produces benzene as an overhead product and a bottoms product of toluene and xylenes, which is charged to the second tower. This tower produces toluene as overhead and xylenes as bottoms. A flow diagram of the operation is shown in Fig. 34. During a twelve-hour period of operation the following products are removed from the towers:

	#1 Tower Net overhead	#2 Tower Net overhead	#2 Tower Bottoms
Gallons at 60°F	9,240	6,390	10,800
Composition % by volume			
Benzene	98.5	1.0	....
Toluene	1.5	98.5	0.7
Xylenes	....	0.5	99.3

During the period the liquid levels in the accumulator sections of the plant varied, as shown by level indicators measuring the heights of the levels above arbitrary fixed points. The liquid volume, corrected to 60°F of the accumulator sections per inch of height and the initial and final levels were as follows:

	Corrected liquid vol. gal/in.	Initial level, in.	Final level in.
#1 Tower Reflux Accumulator	24	36	58
#1 Tower Bottom Accumulator	46	60	45
#2 Tower Reflux Accumulator	35	80	65
#2 Tower Bottom Accumulator	18	52	54

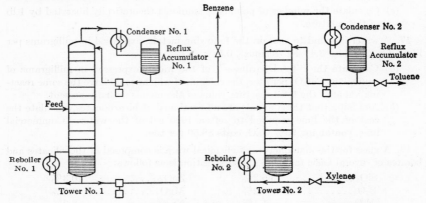

Fig. 34.   Inventory changes in three-component distillation.

Calculate:

(a) The volume and composition of the feed during the twelve hour operating period.

(b) The rate of flow from the bottom of Tower #1.

(c) The rate of production of distillate and bottoms from each tower assuming the liquid levels had been maintained constant for the same rate of feed.

22. Stock containing 1.562 pounds of water per pound of dry stock is to be dried to 0.099 pound. For each pound of stock (dry basis) 52.5 pounds of dry air pass through the dryer, leaving at a humidity of 0.0525. The fresh air is supplied at a humidity of 0.0152.

(a) Calculate the fraction of air recirculated.

(b) If the size of dryer required is inversely proportional to the average wet-bulb depression of the air in the dryer, calculate the relative size of dryers required with and without recirculation when operated at a constant temperature of 140°F.

23. It is required to condition 1000 cu ft of air per minute from 70°F and 80% relative humidity to a constant value of 10% relative humidity by means of a stationary bed of silica gel. Part of the stream may be by-passed and a damper regulator automatically controlled to maintain a constant wet-bulb temperature in the final air mixture. The air velocity through the bed shall not exceed 100 ft per minute (total cross section basis) and the time cycle, before regeneration is necessary, shall be 3 hours. Assume isothermal operation. Calculate the diameter and thickness of bed required.

CHAPTER VII

THERMOPHYSICS

In Chapter II, the general concepts of energy, temperature, and heat were introduced under the broad classification of potential and kinetic energies. Both these forms were subclassified into external forms determined by the position and motion of a mass of matter relative to the earth or other masses of matter and into internal forms determined by the inherent composition, structure, and state of matter itself, independent of its external position or motion as a whole.

Internal Energy. The internal energy of a substance is defined as the total quantity of energy which it possesses by virtue of the presence, relative positions, and movements of its component molecules, atoms, and subatomic units. A part of this energy is contributed by the translational motion of the separate molecules and is particularly significant in gases where translational motion is nearly unrestricted in contrast to the situation in liquids and solids. Internal energy also includes the rotational motion of molecules and of groups of atoms which are free to rotate within the molecules. It includes the energy of vibration between the atoms of a molecule and the motion of electrons within the atoms. These kinetic portions of the total internal energy are determined by the temperature of the substance and by its molecular structure. The remainder of the internal energy is present as potential energy resulting from the attractive and repulsive forces acting between molecules, atoms, electrons, and nuclei. This portion of the internal energy is determined by molecular and atomic structures and by the proximity of the molecules and atoms to one another. At the absolute zero of temperature all translational energy disappears but a great reservoir of potential energy and a small amount of vibrational energy remains.

The total internal energy of a substance is unknown but the amount relative to some selected temperature and state can be accurately determined. The crystalline state or hypothetical gaseous state at absolute zero temperature are commonly used as references for scientific studies, whereas engineering calculations are based upon a variety of reference conditions arbitrarily selected.

Energy in Transition; Heat and Work. In reviewing the several forms of energy previously referred to, it will be noted that some are capable of storage, unchanged in form. Thus, the potential energy of an elevated weight or the kinetic energy of a rotating flywheel are stored as such until by some transformation they are converted, in part, at least, to other forms. The internal energy of matter may be considered as in storage awaiting transformation.

Heat represents energy in transition under the influence of a temperature difference. When heat flows from a hot metal bar to a cold one the internal energy stored in the cold bar is increased at the expense of that of the hot bar and the amount of heat energy in transition may be expressed in terms of the change in internal energy of the source or of the receiver. Under the influence of a temperature gradient heat flows also by the bodily convection and mixing of hot and cold fluids and by the emission of radiant energy from a hotter to a colder body without the aid of any tangible intermediary.

It is inexact to speak of the storage of heat. The energy stored within a body is internal energy and when heat flows into the body it becomes internal energy and is stored as such. Zemansky[1] writes as follows: "The phrase, 'the heat in a body' has absolutely no meaning. Perhaps an analogy will clinch the matter. Consider a fresh water lake. During a shower, a certain amount of rain enters the lake. After the rain has stopped, there is no rain in the lake. There is water in the lake. Rain is a word used to denote water that is entering the lake from the air above. Once it is in the lake, it is no longer rain."

Another form of energy in transition of paramount interest is *work*, which is defined as the energy which is transferred by the action of a mechanical force moving under restraint through a tangible distance. It is evident that work cannot be stored as such but is solely the manifestation of the transformation of one form of energy to another. Thus, when a winch driven by a gasoline engine is used to lift a weight, the internal energy of the gasoline is transformed in part to the potential energy of the elevated weight and the work done is the energy transferred from one state to the other.

Energy Units. The fundamental energy unit is derived from the definition and concept of work. Thus, in the cgs system the basic energy unit is the erg, which is the amount of work done by a force of one dyne acting under restraint through a distance of one centimeter. Since this unit is inconveniently small, the joule, equal to 10^7 ergs, is more commonly used. In English units there is some confusion in terminology resulting from the fact that the pound is used both as a unit of mass and as a unit of force. By definition, a force of one pound is the standard gravitational force exerted on a mass of one pound at sea

[1] *Heat and Thermodynamics*, McGraw-Hill Book Co. (1937). With permission.

level. This pound force will accelerate a mass of one pound at a rate of 32.17 feet per second per second. An alternate unit of force is the *poundal*, which is defined as the force which will accelerate a mass of one pound at a rate of one foot per second per second. Thus, a force of one pound is equal to 32.17 poundals. The *foot-pound* is the work done by a force of one pound acting through a distance of one foot and the *foot-poundal* is the work done by a force of one poundal acting through a distance of one foot.

Since heat is a form of energy it may also be expressed in ergs, joules, or foot-pounds. However, in problems dealing with the production, generation, and transfer of heat it is customary to use special units of energy called heat units. These units are expressed in various terms depending upon which temperature scale and system of weights are employed. The smallest heat unit is the *calorie*, which is the amount of heat required to increase the temperature of one gram of water from $15°$ to $16°C$. One calorie is equivalent to 4.1833×10^7 ergs. It will be recognized that the calorie is an arbitrary unit expressed in terms of the thermal capacity of water and is not based upon the cgs system of measurements. A calorie has also been defined as one-hundredth of the amount of heat required to increase the temperature of one gram of water from $0°$ to $100°C$. This latter unit, called the mean calorie, is no longer in use. The mean calorie is 0.017 per cent larger than the $15°$ calorie. The calorie is also rigorously defined directly in electrical energy units as $\frac{1}{860}$ international watt-hour.

In industrial calculations it is always convenient to use heat units larger than the calorie, such as the *kilocalorie, British thermal unit*, or *Centigrade heat unit*. The kilocalorie is equal to 1000 calories, or it is the amount of heat absorbed in increasing the temperature of one kilogram of water from $15°$ to $16°C$. When measurements are made in English units, employing pounds and the Fahrenheit temperature scale, the British thermal unit or Btu is used. This unit is the quantity of heat absorbed in increasing the temperature of one pound of water from $60°$ to $61°F$. One Btu is mechanically equivalent to 778 foot-pounds of energy. Industries of the United States have often been willing to adopt the Centigrade temperature scale but have refused to abandon the English system of weights. This condition has given rise to a hybrid heat unit known as the Centigrade heat unit, or Chu, which is the amount of heat absorbed in increasing the temperature of one pound of water from $15°$ to $16°C$. The numerical values of heats of formation and reaction, as explained later, are the same when expressed in either kilocalorie per kilogram or Centigrade heat units per pound. For these reasons the Centigrade heat unit has received some favorable acceptance.

In the Appendix are factors for the conversion from one system of units to another.

Energy Balances. In accordance with the principle of conservation of energy, also called the first law of thermodynamics, and referred to in Chapter II, page 28, energy is indestructible and the total amount of energy entering any system must be exactly equal to that leaving plus any accumulation of energy within the system. A mathematical or numerical expression of this principle is termed an *energy balance*, which in conjunction with a material balance is of primary importance in problems of process design and operation.

In establishing a general energy balance for any process, it is convenient to use as a basis a unit time of operation, for example, one hour for a continuous operation and one cycle for a batch or intermittent operation. It is necessary to distinguish between a *flow process*, which is one in which streams of materials continually enter and leave the system, and a *non-flow process*, which is intermittent in character and in which no continuous streams of material enter or leave the system during the course of operation. An ideal flow process is also characterized by steady states of flow, temperatures, and compositions at any point in the process in contrast to the changing conditions of composition in a batch or non-flow process.

In an energy balance the inputs are equated to the outputs plus the accumulation of energy inventory within the system over the unit period of time in a flow process, or for a given cycle of operation for the non-flow process. The separate forms of energy are conveniently classified as follows, neglecting electrostatic and magnetic energies, which are ordinarily small:

a. Internal energy, designated by the symbol U per unit mass or mU for mass m.

b. The energy added in forcing a stream of materials into the system under the restraint of pressure. This flow work is equal to mpV, where p is pressure of the system and V is the volume per unit mass. A similar flow-work term is involved in forcing a stream of materials from the system. These terms appear only in the energy balance of a flow process.

c. The external potential energies of all materials entering and leaving the system. External potential energy is expressed relative to an arbitrarily selected datum plane and is equal to mZ, where Z is the height of its center of gravity above the datum plane.

d. The kinetic energies of all streams entering or leaving the system. The kinetic energy of a single stream is equal to $\frac{1}{2}mu^2$, where u is its average linear velocity, and energy is expressed in ergs, joules, or foot-poundals, depending upon the units of m and u. Expressed in foot-pounds the kinetic energy of a stream is $\dfrac{mu^2}{2g_c}$, where g_c is the standard gravitational constant and equal to 32.17 feet per second per second.

e. The surface energies of all materials entering and leaving the system. Surface energy per unit mass is designated as E_σ and is generally negligible except where large surface areas are involved, as in the formation of sprays or emulsions.

f. The net energy *added* to the system as heat, and designated as q. This net heat input represents the difference between the sum of all heat flowing into the system from all sources and the sum of all heat flowing out of the system from all sources.

g. The net energy removed as *work done by the system*, designated as w. This net work includes all forms of work *done by* the system, such as mechanical and electrical work minus all such work added to the system from all sources.

h. The net change in energy content within the system during the course of the operation and designated by the symbol ΔE. In a steady-state flow process with constant inventory, the energy change term ΔE is zero. In a general process ΔE represents the change in energy content of the system as a result of any change in inventory of the system, of change in temperature, of change in composition, of change in potential energy of elevation, or of change in kinetic energy from stirring, for example. In a non-flow process, the energy terms of the entering and leaving streams disappear and the term ΔE becomes of major importance.

The general energy equation may be set up, designating input items by the subscript *1* and output items by the subscript *2*. All input energy items are balanced against the output and change items, the summation symbol $\sum$ indicating the sum of each form of energy in all streams entering or leaving. Thus,

$$\sum m_1 U_1 + \sum m_1 p_1 V_1 + \sum \frac{m_1 u_1^2}{2g_c} + \sum m_1 Z_1 + \sum m_1 E_{\sigma 1} + q$$

$$= \sum m_2 U_2 + \sum m_2 p_2 V_2 + \sum \frac{m_2 u_2^2}{2g_c} + \sum m_2 Z_2$$

$$+ \sum m_2 E_{\sigma 2} + w + \Delta E \tag{1}$$

Simplifications of this general equation result in most specific cases. In a steady flow process, without fluctuations in temperature, composition, and inventory, the term ΔE becomes zero. In a non-flow process, where no stream of materials enters or leaves the system during the course of operation, the flow work, kinetic energy, and potential energy terms of the streams do not appear.   For such a non-flow process where surface energy is negligible the general equation reduces to

$$q = w + \Delta E \tag{2}$$

where

$$\Delta E = \sum\left(m_2'U_2' + m_2'Z_2' + m_2'\frac{u_2'^2}{2g_c} - m_1'U_1' - m_1'Z_1' - \frac{m_1'u_1'^2}{2g_c}\right) \tag{3}$$

where the prime values all refer to the properties of the inventory, with subscript *1* referring to initial conditions and subscript *2* to final conditions.   The kinetic energy term in a non-flow process results from agitation or stirring.

For a *non-flow process at constant volume*, no work and no potential energy changes are involved, and if the kinetic and surface energy terms are absent or negligible, Equation (1) combined with (13) becomes

$$q = \sum m_2'U_2 - \sum m_1'U_1 \tag{4}$$

For a *non-flow process at constant pressure*, where changes in kinetic, potential, and surface energy terms are negligible

$$q = \sum m_2'U_2 - \sum m_1'U_1 + w \tag{5}$$

Where work is performed only as a result of expansion against the constant pressure p

$$w = \sum m_2'pV_2 - \sum m_1'pV_1 \tag{6}$$

Combining Equations (5) and (6),

$$q = \sum m_2'(U_2 + pV_2) - \sum m_1'(U_1 + pV_1) \tag{7}$$

Where only a single phase is present Equation (7) becomes

$$q = m'(U_2 + pV_2) - m'(U_1 + pV_1) = m'\Delta(U + pV) \tag{8}$$

In many flow processes of the type encountered in chemical engineering practice where large internal energy changes are involved and where the energy associated with work and changes in potential kinetic and surface energies are relatively small the general energy Equation (1) reduces to

$$\sum m_1(U_1 + p_1V_1) + q = \sum m_2(U_2 + p_2V_2) \tag{9}$$

Equation (9) may be satisfactorily applied to the great majority of transformation processes where the primary objective is manufacture and not the production of power.   However, this equation should always be recognized as an approximation and the significance of the neglected terms verified when in doubt.

Enthalpy. In the energy equations for both flow and non-flow processes it will be seen that the term $(U + pV)$ repeatedly occurs. It is convenient to designate this term by the name *enthalpy*[2] and by the symbol H, thus

$$H = U + pV \tag{10}$$

In a flow system the term pV represents flow energy but in a non-flow system it merely represents the product of pressure and volume, having the units of energy but not representing energy.

In a *non-flow* process proceeding at *constant pressure* and without generation of electrical energy the heat added is seen from Equation (7) to be equal to the increase in enthalpy of the system, or

$$q = \sum m_2 H_2 - \sum m_1 H_1$$

In a flow system where the kinetic energy and potential energy terms are negligible the heat added is equal to the gain in enthalpy plus the work done, including both electrical and mechanical work, or

$$q = \sum m_2 H_2 - \sum m_1 H_1 + w \tag{11}$$

Where the work done is negligible in relative magnitude, the heat added is equal to the gain in enthalpy, or

$$q = \sum m_2 H_2 - \sum m_1 H_1$$

To summarize: For most industrial flow processes, such as the operation of boilers, blast furnaces, chemical reactors, or distillation equipment, the kinetic energy, potential energy, and work terms are negligible or cancel out and the heat added is equal to the increase in enthalpy. Similarly, in non-flow processes at constant pressure where work other than that of expansion is negligible, the heat added is equal to the increase in enthalpy. However, where work, kinetic, or potential energies are not negligible, or in non-flow processes at constant volume, the increase in enthalpy is *not* equal to the heat added. The distinction between increases of enthalpy and additions of heat must be kept constantly in mind.

Like internal energy, absolute values of enthalpy for any substance are unknown but accurate values can be determined relative to some arbitrarily selected reference state. The temperature, form of aggregation, and pressure of the reference state must be definitely specified. The values so reported are *relative enthalpies*. For a gas the reference state of zero enthalpy is usually taken at 32°F and one atmosphere pressure, and for steam as liquid water at 32°F under its own vapor

---

[2] This word was coined by Kammerling Onnes (1909), who purposely placed the accent on the second syllable to distinguish it in speech from the commonly associated term, entropy.

pressure at 32°F. For example, the enthalpy of steam at 200°F and an absolute pressure of 10 pounds per square inch represents the increase in enthalpy when liquid water at 32°F and its own vapor pressure is vaporized and heated to a temperature of 200°F while its pressure is increased to 10 pounds per square inch.

Heat Balance. " Heat balance'' is a loose term referring to a special form of energy balance which has come into general use in all thermal processes where changes in kinetic energy, potential energy, and work done are negligible. For such processes, so-called heat or enthalpy balances are applicable to flow processes at any pressure and to non-flow processes at constant pressure.

From consideration of the general energy equation it is evident that neither heat nor enthalpy input and output items balance in the general case. Although the term " heat balance " has become entrenched in engineering literature, its use is undesirable because of the misleading implications of the name. Accordingly, all such balances will be referred to by the proper term of " energy balance," even where the kinetic, potential, and work items are neglected.

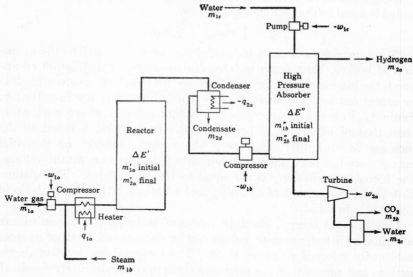

FIG. 35. Energy balance of a process for the production of hydrogen and carbon dioxide from water gas.

Illustration 1. Application of the General Energy Balance. The principle of the general energy balance of Equation (1) may be illustrated by application to the recovery of hydrogen and carbon dioxide from water gas (Fig. 35). Dust-free water gas is compressed, mixed with steam, heated, and passed to a reactor where in contact with a catalyst the CO is converted to CO_2. The products from the reactor are cooled, with resultant condensation of water vapor, compressed further, and passed into an absorber where CO_2 is dissolved in water at high pressure. Hydrogen gas is delivered at high pressure in a nearly pure state. The high pressure carbon dioxide solution generates power in a turbine and the CO_2 gas and water are thereby released at atmospheric pressure, and separated.

The following symbols are used to designate the various streams. All mass and energy units correspond to the period of time selected.

m_{1a} = mass of water gas entering reactor

m_{1b} = mass of steam entering reactor

m_{1c} = mass of water entering absorber

m_{2a} = mass of hydrogen stream leaving absorber

m_{2b} = mass of CO_2 stream leaving turbine

m_{2c} = mass of water leaving turbine

m_{2d} = mass of water leaving condenser

m'_{1a} = initial inventory in reactor

m'_{2a} = final inventory in reactor

m''_{1b} = initial inventory in absorber

m''_{2b} = final inventory in absorber

$-w_{1a}$ = net work delivered to water gas in compression

$-w_{1b}$ = net work delivered to absorber gases in compression

$-w_{1c}$ = net work delivered to water in pumping to absorber

$+w_{2a}$ = work delivered to turbine by solution leaving absorber

q_{1a} = heat added in heating gas mixture entering reactor

$-q_{2a}$ = heat removed from gas stream leaving reactor by condenser

$\sum -q_r$ = heat lost by radiation from all parts of plant

$\Delta E'$ = change in energy content of mass in reactor over period of run

$\Delta E''$ = change in energy content of mass in absorber over period of run

The markings to U, p, V, u, Z all correspond to those used above.

Applying these conditions to the general equation (1)

$$\sum m_1 U_1 = m_{1a} U_{1a} + m_{1b} U_{1b} + m_{1c} U_{1c} \tag{a}$$

$$\sum m_2 U_2 = m_{2a} U_{2a} + m_{2b} U_{2b} + m_{2c} U_{2c} + m_{2d} U_{2d} \tag{b}$$

$$\sum m_1 p_1 V_1 = m_{1a} p_{1a} V_{1a} + m_{1b} p_{1b} V_{1b} + m_{1c} p_{1c} V_{1c} \tag{c}$$

$$\sum m_2 p_2 V_2 = m_{2a} p_{2a} V_{2a} + m_{2b} p_{2b} V_{2b} + m_{2c} p_{2c} V_{2c} + m_{2d} p_{2d} V_{2d} \tag{d}$$

$$\sum \frac{m_1 u_1^2}{2g_c} = \frac{m_{1a} u_{1a}^2}{2g_c} + \frac{m_{1b} u_{1b}^2}{2g_c} + \frac{m_{1c} u_{1c}^2}{2g_c} \tag{e}$$

$$\sum \frac{m_2 u_2^2}{2g_c} = \frac{m_{2a} u_{2a}^2}{2g_c} + \frac{m_{2b} u_{2b}^2}{2g_c} + \frac{m_{2c} u_{2c}^2}{2g_c} + \frac{m_{2d} u_{2d}^2}{2g_c} \tag{f}$$

$$\sum m_1 Z_1 = m_{1a} Z_{1a} + m_{1b} Z_{1b} + m_{1c} Z_{1c} \tag{g}$$

$$\sum m_2 Z_2 = m_{2a} Z_{2a} + m_{2b} Z_{2b} + m_{2c} Z_{2c} + m_{2d} Z_{2d} \tag{h}$$

$$q = q_{1a} + q_{2a} + q_r \tag{i}$$

$$w = w_{2a} + w_{1a} + w_{1b} + w_{1c} \tag{j}$$

$$\Delta E = \Delta E' + \Delta E'' \tag{k}$$

$$\Delta E' = m'_{2a} \left(U'_{2a} + p'_{2a} V'_{2a} + \frac{u'^2_{2a}}{2g_c} + Z'_{2a} \right) - m'_{1a} \left(U'_{1a} + p'_{1a} V'_{1a} + \frac{u'^2_{1a}}{2g_c} + Z'_{1a} \right) \quad (1)$$

$$\Delta E'' = m''_{2b} \left(U''_{2b} + p''_{2b} V''_{2b} + \frac{u''^2_{2b}}{2g_c} + Z''_{2b} \right) - m''_{1b} \left(U''_{1a} + p''_{1b} V''_{1b} + \frac{u''^2_{1b}}{2g_c} + Z''_{1b} \right) \quad (m)$$

In general the inventory of both mass and energy remains constant, the kinetic energy terms are negligible, the potential energy terms cancel, and the equation reduces to

$$\Sigma m_1 H_1 + q = \Sigma m_2 H_2 + w$$

It should be noted that the work terms refer to the net work energy added to the system or supplied to the turbine and as a result of mechanical inefficiencies do not correspond to the work required to drive the pumps and compressors or to that generated by the turbine. The heat developed as a result of these inefficiencies has been neglected. In chemical processing the work terms are usually negligible in the total energy balance although they may be of major importance in cost.

HEAT CAPACITY OF GASES

In a restricted sense *heat capacity* is defined as the amount of heat required to increase the temperature of a body by one degree. *Specific heat* is the ratio of the heat capacity of a body to the heat capacity of an equal weight of water. Specific heat is a property, characteristic of a substance and independent of any system of units, but dependent on the temperatures of both the substance and the reference water. Water at 15°C is usually chosen as the reference.

The heat capacity of any quantity of a substance is expressed mathematically as

$$C = \frac{dq}{dT} \quad (12)$$

where

C = heat capacity
dq = heat added to produce a temperature change, dT

If a system is heated in such a manner that its volume remains constant, $dq = dU$. The heat capacity under these conditions of constant volume is expressed by

$$C_v = \left(\frac{\partial U}{\partial T} \right)_v \quad (13)$$

Thus, the heat capacity at constant volume, C_v, is equal to the change of internal energy with temperature.

If a system is heated in such a manner that the total pressure remains constant and the volume is permitted to change, heat will be absorbed both to increase the internal energy and to supply the heat equivalent of, the external work which is done by the system in expanding. Thus, Equation (12) becomes

$$C_p = \left(\frac{\partial U}{\partial T}\right)_p + \left(\frac{p\partial V}{\partial T}\right)_p = \left(\frac{\partial H}{\partial T}\right)_p \tag{14}$$

where C_p is the heat capacity under constant pressure. For an ideal gas,

$$\left(\frac{\partial V}{\partial T}\right)_p = \frac{nR}{p} \tag{15}$$

Then, the heat capacity at constant pressure, c_p, of one mole of an ideal gas is represented by

$$c_p = \left(\frac{\partial U}{\partial T}\right)_p + R = c_v + R \tag{16}$$

or

$$c_p = c_v + R \tag{17}$$

where c_p and c_v are the molal heat capacities at constant pressure and volume, respectively.

The heat capacity of a substance which expands with rise in temperature is always greater when heated under a constant pressure than when heated under constant volume by the heat equivalent of the external work done in expansion. For an ideal gas the molal heat capacities under the two different conditions differ by the magnitude of the constant, R. The numerical value of R is 1.99 calories per gram-mole per degree Kelvin, or 1.99 Btu per pound-mole per degree Rankine.

For an ideal monatomic gas, such as helium, at a low pressure, it may be assumed that, as a result of the simple molecular structure, the only form of internal energy is the translational kinetic energy of the molecules. The translational kinetic energy per mole of gas may be obtained from Equation 3, Chapter II, page 30, and is equal to the internal energy U in this particular case. Thus,

$$U = \tfrac{1}{2}mnu^2 = \tfrac{3}{2}RT \tag{18}$$

Then from Equation (13), for a monatomic gas,

$$c_v = \tfrac{3}{2}R \tag{19}$$

Since R is approximately 2.0 calories per gram-mole per degree Kelvin, the molal heat capacity of a monatomic gas at constant volume is equal to 3.0 calories per gram-mole per degree Kelvin. The molal heat capacity at constant pressure, from Equation (16), will be 3.0 + 2.0 or 5.0

calories.   For monatomic gases the ratio of heat capacities, γ, is

$$\gamma = \frac{c_p}{c_v} = \frac{5}{3} = 1.67 \qquad (20)$$

For all gases, other than monatomic gases, the molal heat capacity at constant volume is greater than 3.0.   For a multatomic gas an increase in enthalpy is used not only in imparting additional translational kinetic energy, as evidenced by an increase in temperature and an increasing velocity of translation, but also to impart increased energies of rotation and vibration of the molecular and submolecular units.

TABLE IV

TRUE MOLAL HEAT CAPACITIES OF GASES

Constant Pressure (p = 0 atm abs) in calories per gram-mole per °C*

t°C	H_2	D_2	N_2	O_2	CO	NO	H_2O	CO_2	SO_2	Air	CH_4	C_2H_4	C_2H_2
0	6.86	6.97	6.96	6.99	6.96	7.16	7.98	8.61	9.31	6.94	8.24	10.02	10.13
100	6.96	6.98	6.98	7.13	7.00	7.15	8.10	9.69	10.17	6.99	9.40	12.42	11.76
200	6.99	7.01	7.05	7.37	7.09	7.25	8.32	10.47	10.94	7.10	10.70	14.74	12.98
300	7.01	7.06	7.16	7.61	7.23	7.42	8.56	11.23	11.53	7.24	12.15	16.73	13.71
400	7.03	7.16	7.31	7.84	7.40	7.62	8.84	11.79	12.03	7.40	13.40	18.42	14.39
500	7.06	7.27	7.47	8.02	7.57	7.79	9.12	12.25	12.38	7.57	14.60	19.90	15.01
600	7.12	7.39	7.63	8.18	7.75	7.95	9.41	12.63	12.65	7.72	15.65	21.17	15.55
700	7.20	7.52	7.78	8.31	7.90	8.09	9.72	12.94	12.86	7.87	16.60	22.30	16.04
800	7.28	7.67	7.91	8.41	8.03	8.22	10.02	13.20	13.02	7.99	17.40	23.30	16.49
900	7.38	7.81	8.03	8.50	8.15	8.32	10.30	13.41	13.15	8.10	18.23	24.17	16.90
1000	7.49	7.94	8.14	8.60	8.24	8.41	10.58	13.60	13.25	8.21	18.93	24.94	17.26
1100	7.59	8.06	8.24	8.66	8.33	8.48	10.84	13.74	13.34	8.29			
1200	7.69	8.16	8.32	8.73	8.41	8.55	11.08	13.87	13.41	8.37			
1300	7.80	8.26	8.38	8.79	8.47	8.61	11.31	13.98	13.46	8.43			
1400	7.89	8.34	8.44	8.85	8.53	8.65	11.52	14.07	13.51	8.49			
1500	7.98	8.43	8.49	8.90	8.57	8.69	11.71	14.15	13.56	8.54			
1600	8.08	8.51	8.54	8.96	8.62	8.73	11.88	14.22	13.59	8.59			
1700	8.16	8.58	8.59	9.01	8.66	8.77	12.04	14.28	13.62	8.64			
1800	8.24	8.64	8.63	9.08	8.69	8.80	12.19	14.33	13.65	8.68			
1900	8.32	8.71	8.66	9.14	8.72	8.82	12.33	14.38	13.67	8.72			
2000	8.38	8.76	8.70	9.19	8.75	8.85	12.45	14.42	13.69	8.77			
2100	8.45	8.82	8.73	9.24	8.78	8.87	12.57	14.46	13.70	8.80			
2200	8.51	8.87	8.76	9.29	8.81	8.89	12.68	14.49	13.72	8.83			
2300	8.57	8.91	8.78	9.34	8.83	8.91	12.78	14.52	13.73	8.86			
2400	8.62	8.96	8.80	9.39	8.85	8.93	12.87	14.54	13.74	8.89			
2500	8.68	9.00	8.83	9.43	8.87	8.95	12.95	14.57	13.76	8.92			
2600	8.73	9.04	8.84	9.47	8.89	8.96	13.02	14.59	13.77	8.94			
2700	8.78	9.08	8.86	9.51	8.90	8.98	13.08	14.61	13.77	8.96			
2800	8.83	9.11	8.88	9.55	8.91	8.99	13.14	14.63	13.78	8.98			
2900	8.88	9.14	8.89	9.59	8.92	9.01	13.18	14.64	13.79	9.00			
3000	8.93	9.16	8.90	9.62	8.93	9.02	13.23	14.66	13.79	9.02			
Mol. wt.	2.02	4.03	28.03	32.00	28.00	30.01	18.02	44.00	64.07	28.964	16.03	28.04	26.03

* E. Justi and H. Lüder, *Forsch. Gebiete Ingenieurw.*, **6**, 210–1 (1935).

In Table IV are given values of the true molal heat capacities of a few common gases at zero pressure and at temperature intervals of 100°C from 0° to 3000°C. These same values are plotted in Fig. 36 against degrees Fahrenheit. These data have been calculated from spectroscopic data by Justi and Lüder and may be safely used at all pressures up to one atmosphere. At high pressures corrections can be applied as discussed in Chapter XII.

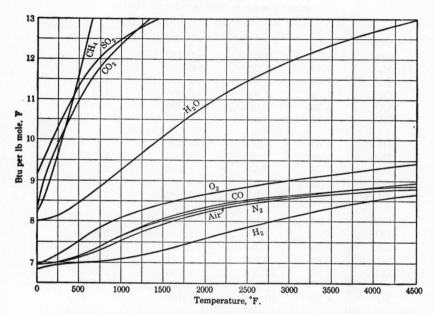

Fig. 36.   True molal heat capacities of gases at constant pressure.

(Reproduced in " C.P.P. Charts ")

Empirical Equations for Molal Heat Capacities. Rigorous heat capacity equations such as the Planck-Einstein equation based upon spectroscopic data and statistical mechanics are too complicated for engineering use. Many empirical equations have been proposed to represent the experimental data, the most useful form of which is of the quadratic type,

$$c_p = a + bT + cT^2 \tag{21}$$

The constants a, b, and c were established by W. C. M. Bryant from the best data available up to 1933, and are recorded in Table V for a number of common gases. The constants for oxygen, nitric oxide, and air have been recalculated from the more recent data of Justi and Lüder. The values for sulfur trioxide were calculated from vibration frequencies by the method described in Chapter XVI. The maximum deviations from experimental results over the range 0 to 1800°C are given. Data for hydrocarbon gases are presented in Chapter IX, pages 335 and 336.

TABLE V

EMPIRICAL EQUATIONS FOR MOLAL HEAT CAPACITIES OF GASES*

$c_p = a + bT + cT^2$ where T is in degrees K

calories per gram-mole per degree Kelvin

Gas	a	$b \times 10^3$	$c \times 10^6$	Maximum Deviation 0 to 1800°C
†Air	6.27	2.09	−0.459	1.0%
H_2	6.88	0.066	+0.279	1.6
N_2	6.30	1.819	−0.345	1.2
†O_2	6.13	2.99	−0.806	1.5
CO	6.25	2.091	−0.459	1.2
†NO	6.17	2.60	−0.66	<2.0
HCl	6.64	0.959	−0.057	2.0
H_2O	6.89	3.283	−0.343	1.0
H_2S	6.48	5.558	−1.204	<2.0
SO_2	8.12	6.825	−2.103	2.0
HCN	7.01	6.600	−1.642	<3.0
CO_2	6.85	8.533	−2.475	<3.0
COS	8.32	7.224	−2.146	2.0
CS_2	9.76	6.102	−1.894	2.0
NH_3	5.92	8.963	−1.764	<1.0
C_2H_2	8.28	10.501	−2.644	2.0
‡SO_3	8.20	10.236	−3.156	

* Bryant, *Ind. Eng. Chem.* 822 **25,** (1933). With permission.
† Recalculated from 1935 data of Justi and Lüder.
‡ Estimated.

Effect of Pressure on the Heat Capacity of Gases. The heat capacities of gases are not greatly affected by small variations in pressure at low pressures. However, at low temperatures, heat capacities increase rapidly with increase in pressure and become infinite at the critical point. The effects of high pressures on heat capacities are discussed in detail in Chapter XII. As the temperature is increased above the critical, the effect of pressure upon heat capacity gradually diminishes.

Special Units for Heat Capacity of Gases. Industrial gases are usually measured by volume, in cubic feet or cubic meters. From the equations given above for the molal heat capacities of gas, values may be readily calculated for any other units of weight or volume, such as kilocalories per kilogram, kilocalories per cubic meter, Btu per pound, and Btu per cubic foot.

Whenever heat capacities are based on unit volumes, the basic quantity of gas involved is that contained in a unit volume measured at standard conditions of temperature and pressure and not at the existing conditions. The heat capacity per unit volume refers to the heat capacity of a definite and constant mass of gas, regardless of its temperature and pressure. For example, the heat capacity per cubic meter of oxygen at 1000°C signifies the heat capacity at 1000°C of the weight of gas contained in 1 cubic meter at standard conditions, or of 1.44 kilograms of oxygen. It does not signifiy the heat capacity of the oxygen contained in 1 cubic meter of gas at the given temperature and pressure. This distinction must always be kept in mind in speaking of the heat-capacities of unit volumes of gases at various temperatures.

In the fuel gas industries an unusual unit of gas quantity is employed as the standard. This unit is the quantity of gas contained in 1000 cubic feet, measured at a pressure of 30 inches of mercury, a temperature of 60°F, and saturated with water vapor. This volume of gas corresponds to 2.597 pound-moles of dry gas containing 0.046 pound-mole of water vapor, or 2.643 pound-moles of the mixture. The heat capacity of the mass of gas equivalent to 1000 cubic feet as measured in the gas industry at 30 inches of mercury, saturated, and at 60°F can be obtained by multiplying the molal heat capacity of that gas, expressed in Btu per pound-mole per degree Fahrenheit, by the factor 2.643.

Mean Heat Capacities of Gases. The heat capacity equations given in Table V represent the values of heat capacities at any temperature, $T°K$. In heating a gas from one temperature to another it is desirable to know the mean or average heat capacity over that temperature range. The total heat required in heating the gas can then be readily calculated by simply multiplying the number of moles of gas by the mean molal heat capacity and by the temperature rise. This method is easier than employing direct integration of the heat capacity formulas for each problem. For gases of the air group where the temperature coefficient is nearly constant, and over short temperature ranges for other gases, accurate results may be obtained by simply employing the heat capacity at the average temperature as the mean heat capacity. Even for a gas such as carbon dioxide, whose temperature coefficient of heat capacity changes markedly with temperature, the heat capacity at 500°C is only 0.6 per cent higher than the correct mean value over the temperature range from 0° to 1000°C.

The mean heat capacity over any given temperature range may be calculated by integrating the general equation for true values over the desired temperature range as follows:

$$\text{Mean } c_p = \frac{\int_{T_1}^{T_2} c_p \, dT}{T_2 - T_1} = \frac{\int_{T_1}^{T_2} (a + bT + cT^2) \, dT}{T_2 - T_1}$$

$$= a + \frac{b}{2}(T_2 + T_1) + \frac{c}{3}(T_2^2 + T_2 T_1 + T_1^2) \tag{22}$$

where T_2 is the higher and T_1 the lower temperature.

TABLE VI

MEAN MOLAL HEAT CAPACITIES OF GASES BETWEEN 18°C AND t°C

Constant Pressure ($p = 0$ atm abs) in calories per gram-mole per °C*

t°C	H$_2$	D$_2$	N$_2$	O$_2$	CO	NO	H$_2$O	CO$_2$	SO$_2$	Air	CH$_4$	C$_2$H$_4$	C$_2$H$_2$
18	6.86	6.97	6.96	7.00	6.96	7.16	7.99	8.70	9.38	6.94	8.30	10.20	10.25
100	6.92	6.97	6.97	7.06	6.97	7.16	8.04	9.25	9.81	6.96	8.73	11.45	11.12
200	6.95	6.98	7.00	7.16	7.00	7.17	8.13	9.73	10.22	7.01	9.48	12.64	11.79
300	6.97	6.99	7.04	7.28	7.06	7.22	8.23	10.14	10.59	7.06	10.20	13.73	12.36
400	6.98	7.02	7.09	7.40	7.12	7.31	8.35	10.48	10.91	7.14	10.88	14.76	12.81
500	6.99	7.06	7.15	7.51	7.20	7.39	8.49	10.83	11.18	7.21	11.53	15.69	13.21
600	7.02	7.10	7.21	7.61	7.28	7.47	8.62	11.11	11.42	7.28	12.15	16.52	13.58
700	7.04	7.15	7.28	7.70	7.35	7.55	8.76	11.35	11.62	7.35	12.74	17.26	13.90
800	7.07	7.21	7.36	7.79	7.44	7.63	8.91	11.57	11.79	7.43	13.28	18.07	14.22
900	7.10	7.28	7.43	7.87	7.51	7.71	9.06	11.76	11.95	7.50	13.79	18.62	14.49
1000	7.13	7.33	7.50	7.94	7.58	7.77	9.20	11.94	12.08	7.57	14.27	19.24	14.75
1100	7.16	7.41	7.57	8.00	7.65	7.84	9.34	12.11	12.20	7.63			
1200	7.21	7.46	7.63	8.06	7.71	7.90	9.47	12.25	12.29	7.69			
1300	7.25	7.52	7.68	8.13	7.77	7.95	9.60	12.37	12.39	7.74			
1400	7.29	7.57	7.74	8.18	7.82	8.00	9.74	12.50	12.46	7.79			
1500	7.33	7.63	7.79	8.22	7.85	8.04	9.86	12.61	12.53	7.85			
1600	7.37	7.67	7.83	8.25	7.91	8.09	9.98	12.71	12.60	7.89			
1700	7.41	7.73	7.87	8.27	7.95	8.13	10.11	12.80	12.65	7.93			
1800	7.46	7.78	7.92	8.34	7.99	8.16	10.22	12.88	12.71	7.97			
1900	7.50	7.83	7.95	8.39	8.03	8.20	10.32	12.96	12.74	8.00			
2000	7.54	7.88	7.99	8.43	8.06	8.23	10.43	13.03	12.80	8.04			
2100	7.58	7.93	8.02	8.46	8.10	8.27	10.54	13.10	12.84	8.07			
2200	7.63	7.96	8.06	8.49	8.13	8.30	10.63	13.17	12.88	8.09			
2300	7.67	8.00	8.09	8.53	8.16	8.32	10.73	13.23	12.92	8.13			
2400	7.71	8.04	8.11	8.57	8.19	8.35	10.81	13.27	12.96	8.15			
2500	7.75	8.08	8.15	8.60	8.22	8.37	10.89	13.33	12.98	8.19			
2600	7.79	8.11	8.18	8.64	8.24	8.39	10.98	13.37	13.01	8.21			
2700	7.82	8.14	8.20	8.66	8.26	8.41	11.05	13.42	13.03	8.24			
2800	7.86	8.18	8.23	8.69	8.28	8.43	11.13	13.46	13.06	8.26			
2900	7.90	8.21	8.25	8.73	8.31	8.45	11.20	13.51	13.09	8.28			
3000	7.93	8.24	8.27	8.77	8.33	8.46	11.25	13.55	13.12	8.30			
Mol. wt.	2.02	4.03	28.03	32.00	28.00	30.01	18.02	44.00	64.07	28.96	16.03	28.04	26.03

* E. Justi and H. Lüder, *Forsch. Gebiete Ingenieurw.*, **6**, 211 (1935).

The available data for the mean heat capacity of gases over the temperature range from 18°C to t°C are tabulated in Table VI for temperature intervals of 100°C from 0°C to 3000°C from the data of Justi and Lüder.[3] These data are also plotted in Fig. 37 for the temperature range 65° to t°F. This particular range has been selected because the lower temperature corresponds to the reference temperature of thermochemical data.

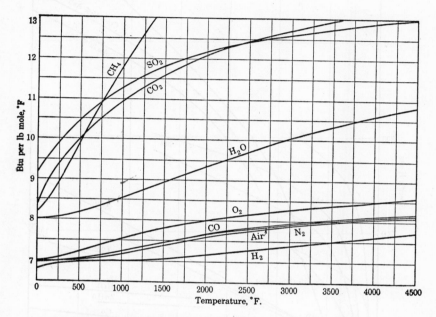

FIG. 37.   Mean molal heat capacities of gases at constant pressure
(mean value 65° to t°F).

(Reproduced in " C.P.P. Charts ")

HEAT CAPACITIES OF SOLIDS

According to the law of Petit and Dulong the atomic heat capacities of the crystalline solid elements are constant and equal to 6.2 calories per gram-atom. This rule applies satisfactorily to all elements having atomic weights above 40 when applied to constant volume conditions at room temperatures. From kinetic theory, Boltzmann showed that the atomic heat capacity of the elements at constant volume reaches a maximum value of $3R = 5.97$ calories per degree. The atomic heat

[3] *Forsch. Gebiete Ingenieurw.*, **6**, 211 (1935).

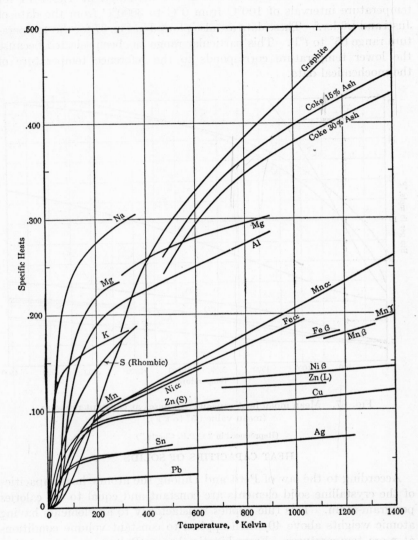

Fig. 38.   Specific heats of common elements and cokes.

capacities of elements such as carbon, hydrogen, boron, silicon, oxygen, fluorine, phosphorus, and sulfur are much lower than the rule indicates. At increasing temperatures, however, the atomic heat capacities of these elements also approach the value 6.2. The atomic heat capacities of all elements decrease greatly with decrease in temperature, approaching a value of zero at absolute zero temperature when in the crystalline state.

Kopp's Rule. The heat capacity of a solid compound is approximately equal to the sum of the heat capacities of the constituent elements. This generalization was first shown by Kopp to be approximately correct, provided the following atomic heat capacities are assigned to the elements: C, 1.8; H, 2.3; B, 2.7; Si, 3.8; O, 4.0; F, 5.0;

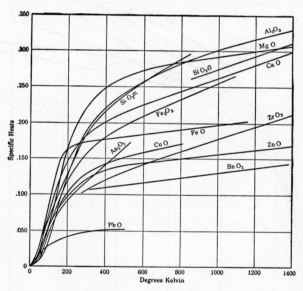

FIG. 39. Specific heats of some common oxides.

P, 5.4; S, 5.4; all others, 6.2. This rule should be used only where experimental values are lacking. Since the heat capacities of solids increase with temperature it is obvious that the above empirical rule is inexact. In general, the heat capacities of compounds are higher in the liquid than in the solid state. At the melting point the two heat capacities are nearly the same.

The heat capacity of a heterogeneous mixture is a simple additive property, the total heat capacity being equal to the sum of the heat capacities of the component parts. When true solutions are formed, this simple additive property no longer exists.

The specific heats of various elements and oxides are presented graphically in Figs. 38 and 39. In Fig. 40 are specific heats of a few calcium compounds. The heat capacities of many other common solids are

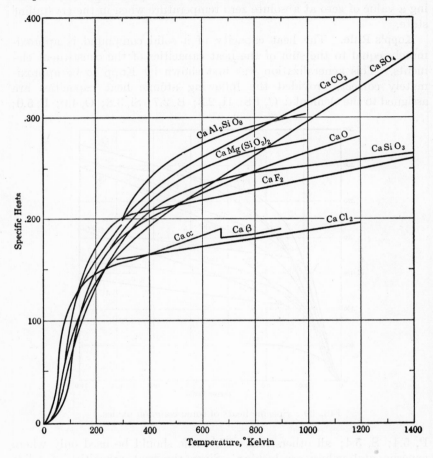

FIG. 40. Specific heats of some calcium compounds.

tabulated in Table VII. It will be noted that no general prediction can be made as to the quantitative effect of temperature on these heat capacities. Transition points, TP, indicating changes in crystalline structure, and melting points, MP, correspond to abrupt changes in the heat capacity relationships.

TABLE VII

HEAT CAPACITIES OF SOLID SUBSTANCES

Data from the International Critical Tables unless otherwise indicated

C_p = calories per gram per °C or = Btu per lb per °F
= kilocalories per kilogram per °C

Inorganic Compounds

Substance	Formula	t°C	C_p
Aluminum sulfate	$Al_2(SO_4)_3$	50	0.184
Aluminum sulfate	$Al_2(SO_4)_3 \cdot 17H_2O$	34	0.353
Ammonium chloride	NH_4Cl	0	0.357
Antimony trisulfide (stibnite)	Sb_2S_3	0	0.0830
		100	0.0884
Arsenious oxide	As_2O_3	0	0.117
		40	0.122
Barium carbonate	$BaCO_3$	0	0.100
		100	0.110
		400	0.123
		800	0.130
Barium chloride	$BaCl_2$	0	0.0853
		100	0.0945
Barium sulfate	$BaSO_4$	0	0.1112
		1000	0.1448
Cadmium sulfide	CdS	0	0.0881
		50	0.0924
Cadmium sulfate	$CdSO_4 \cdot 8H_2O$	0	0.1950
Calcium carbonate	$CaCO_3$	0	0.182
		200	0.230
		400	0.270
Calcium chloride	$CaCl_2$	61	0.164
Calcium chloride	$CaCl_2 \cdot 6H_2O$	0	0.321
Calcium fluoride	CaF_2	0	0.204
		40	0.212
		80	0.216
Calcium sulfate	$CaSO_4$	0	0.1691
		400	0.2275
Calcium sulfate	$CaSO_4 \cdot 2H_2O$	0	0.2650
		50	0.198
Chromium oxide	Cr_2O_3	0	0.168
		50	0.188
Copper sulfate	$CuSO_4$	0	0.148
Copper sulfate	$CuSO_4 \cdot H_2O$	0	0.1717
Copper sulfate	$CuSO_4 \cdot 3H_2O$	9	0.2280
Copper sulfate	$CuSO_4 \cdot 5H_2O$	0	0.2560
Ferrous carbonate	$FeCO_3$	54	0.193
Ferrous sulfate	$FeSO_4$	45	0.167
Lead carbonate	$PbCO_3$	32	0.080
Lead chloride	$PbCl_2$	0	0.0649
		200	0.0704
		400	0.0800
Lead nitrate	$Pb(NO_3)_2$	45	0.1150
Lead sulfate	$PbSO_4$	45	0.0838
Magnesium chloride	$MgCl_2$	48	0.193
Magnesium sulfate	$MgSO_4$	61	0.222
Magnesium sulfate	$MgSO_4 \cdot H_2O$	9	0.239
Magnesium sulfate	$MgSO_4 \cdot 6H_2O$	9	0.349
Magnesium sulfate	$MgSO_4 \cdot 7H_2O$	12	0.361
Manganese dioxide	MnO_2	0	0.152

TABLE VII (*Continued*)

Substance	Formula	$t°C$	C_p
Manganic oxide	Mn_2O_3	58	0.162
Manganous oxide	MnO	58	0.158
Mercuric chloride	$HgCl_2$	0	0.0640
Mercuric sulfide	HgS	0	0.0506
Mercurous chloride	HgCl	0	0.0499
Nickel sulfide	NiS	0	0.116
		100	0.128
		200	0.138
Potassium chlorate	$KClO_3$	0	0.1910
		200	0.2960
Potassium chloride	KCl	0	0.1625
		200	0.1725
		400	0.1790
Potassium chromate	K_2CrO_4	46	0.1864
Potassium dichromate	$K_2Cr_2O_7$	0	0.178
		400	0.236
Potassium sulfate	K_2SO_4	0	0.1760
Potassium nitrate	KNO_3	0	0.2140
		200	0.267
		300	0.292
Silver chloride	AgCl	0	0.0848
		200	0.0974
		500	0.101
Silver nitrate	$AgNO_3$	50	0.146
Sodium borate	$Na_2B_4O_7$	45	0.234
Sodium borate (borax)	$Na_2B_4O_7 \cdot 10H_2O$	35	0.385
Sodium carbonate	Na_2CO_3	45	0.256
Sodium chloride	NaCl	0	0.204
		100	0.217
		400	0.229
		600	0.236
Sodium nitrate	$NaNO_3$	0	0.2478
		100	0.294
		250	0.358
Sodium sulfate	Na_2SO_4	0	0.202
		100	0.220
Water	H_2O	−40	0.435
(Ice)		0	0.492

Organic Compounds

Substance	Formula	$t°C$	C_p
Cyanamide	CH_2N_2	20	0.548
Oxalic acid	$C_2H_2O_4 \cdot 2H_2O$	0	0.338
		100	0.416
Tartaric acid	$C_4H_6O_6 \cdot H_2O$	0	0.308
Picric acid	$C_6H_3N_3O_7$	0	0.240
		100	0.297
Nitrobenzene	$C_6H_5NO_2$	20	0.349
Benzene	C_6H_6	0	0.376
Benzoic acid	$C_7H_6O_2$	20	0.287
Naphthalene	$C_{10}H_8$	0	0.281
		100	0.392
Anthracene	$C_{14}H_{10}$	50	0.308
Palmitic acid	$C_{16}H_{32}O_2$	0	0.382
Stearic acid	$C_{18}H_{36}O_2$	15	0.399

TABLE VII (*Continued*)

SPECIFIC HEATS OF MISCELLANEOUS MATERIALS*

Substance	Specific Heat Cals/g °C	Temperature Range °C
Alumina	0.2	100
	0.274	1500
Alundum	0.186	100
Asbestos	0.25	
Asphalt	0.22	
Bakelite	0.3 to 0.4	
Brickwork	Approx. 0.2	
Carbon	0.168	26–76
	0.314	40–892
	0.387	56–1450
Carbon (gas retort)	0.204	
Cellulose	0.32	
Cement	0.186	
Charcoal (wood)	0.242	
Chrome brick	0.17	
Clay	0.224	
Coal	0.26 to 0.37	
Coal tars	0.35	40
	0.45	200
Coke	0.265	21–400
	0.359	21–800
	0.403	21–1300
Concrete	0.156	70–312
	0.219	72–1472
Cryolite	0.253	16–55
Diamond	0.147	
Fire clay brick	0.198	100
	0.298	1500
Fluorspar	0.21	30
Glass (crown)	0.16 to 0.20	
(flint)	0.117	
(pyrex)	0.20	
(silicate)	0.188 to 0.204	0–100
	0.24 to 0.26	0–700
(wool)	0.157	
Graphite	0.165	26–76
	0.390	56–1450
Granite	0.20	20–100
Gypsum	0.259	16–46
Limestone	0.217	
Litharge	0.055	
Marble	0.21	18
Magnesia	0.234	100
	0.188	1500
Magnesite-brick	0.222	100
	0.195	1500
Pyrites (copper)	0.131	19–50
(iron)	0.136	15–98
Quartz	0.17	0
	0.28	350
Sand	0.191	
Silica	0.316	
Steel	0.12	

* From *Chemical Engineers' Handbook*, John H. Perry, McGraw-Hill Book Company, Inc. (1941), with permission.

HEAT CAPACITIES OF LIQUIDS AND SOLUTIONS

Few generalizations can be stated regarding the heat capacities of liquids. The heat capacities of most liquids increase with an increase in temperature.

The heat capacity of most substances is greater for the liquid state than for either the solid or the gas. Where experimental data are lacking Kopp's rule (page 219) may be applied by assigning the following values of atomic heat capacities to the atoms of the liquid, according to Wenner[4]: C = 2.8, H = 4.3, B = 4.7, Si = 5.8, O = 6.0, F = 7.0, P = 7.4, S = 7.4, and to most other elements a value of 8.

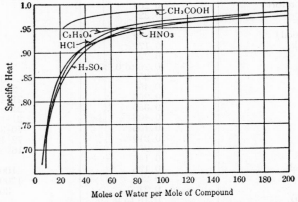

FIG. 41. Specific heats of aqueous solutions of acids at 20°C.

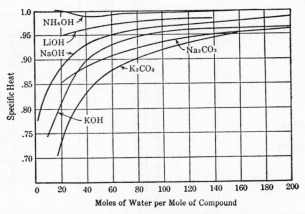

FIG. 42. Specific heats of aqueous solutions of bases at 20°C.

[4] Wenner *Thermochemical Calculations*, McGraw-Hill Book Co. (1941), with permission.

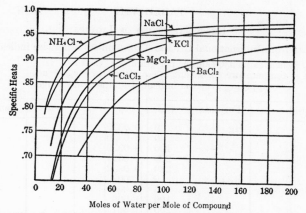

FIG. 43. Specific heats of aqueous solutions of chlorides at 20°C.

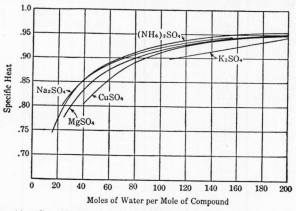

FIG. 44. Specific heats of aqueous solutions of sulfates at 20°C.

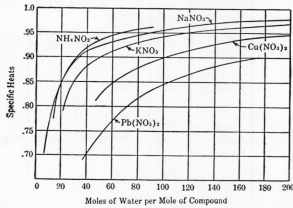

FIG. 45. Specific heats of aqueous solutions of nitrates at 20°C.

TABLE VIII

Heat Capacities of Liquids

Data from International Critical Tables unless otherwise indicated.

c_p = heat capacity, calories per gram per °C at t°C.
a = temperature coefficient in equation: $c_p = c_{p_0} + at$,
applying over the indicated temperature range

Liquid	Formula	t°C	c_p	c_{p_0}	a	Temp. Range
Mercury	Hg	0	0.0335			
		60	0.0330			
		100	0.0329			
		200	0.0329			
		280	0.0332			
Water *	H_2O	0	1.008			
		15	1.000			
		100	1.006			
		200	1.061			
		300	1.155			
Sulfur dioxide	SO_2	−20	0.3130			
				0.318	0.00028	11° to 140°
Sulfuric acid	H_2SO_4			0.339	0.00038	10° to 45°
Ammonia	NH_3	−40	1.051			
		0	1.098			
		60	1.215			
		100	1.479			
Silicon tetrachloride	$SiCl_4$	12 to 50	0.200			
Sodium nitrate	$NaNO_3$	350	0.430			
Carbon tetrachloride	CCl_4	20	0.201	0.198	0.000031	0° to 70°
Carbon disulfide	CS_2			0.235	0.000246	−100° to +150°
Chloroform	$CHCl_3$	15	0.226	0.221	0.000330	−30° to +60°
Formic acid	CH_2O_2	0	0.496		0.000709	40° to 140°
Methyl alcohol	CH_4O	0	0.566			
		40	0.616			
Acetic acid	$C_2H_4O_2$			0.468	0.000929	0° to 80°
Ethyl alcohol	C_2H_6O	−50	0.473			
		0	0.535			
		25	0.580			
		50	0.652			
		100	0.824			
		150	1.053			
Glycol	$C_2H_6O_2$	0	0.544	0.544	0.001194	−20° to +200°
Allyl alcohol	C_3H_6O	0	0.3860			
		21 to 96	0.665			
Acetone	C_3H_6O			0.506	0.000764	−30° to +60°
Propane	C_3H_8	0	0.576	0.576	0.001505	−30° to +20°
Propyl alcohol	C_3H_8O	−50	0.456			
		0	0.525			
		+50	0.654			
Glycerol	$C_3H_8O_3$	−50	0.485			
		0	0.540			
		+50	0.598			
		+100	0.668			
Ethyl acetate	$C_4H_8O_2$	20	0.478			
n-Butane	C_4H_{10}	0	0.550	0.550	0.00191	−15° to +20°
Ether	$C_4H_{10}O$	0	0.529			
		30	0.548			
		120	0.802			
Isopentane	C_5H_{12}	0	0.512			
		8	0.526			

* Handbook of Chemistry and Physics (1939), with permission.

TABLE VIII.   (Continued)

Liquid	Formula	$t°C$	c_p	c_{p_0}	a	Temp. Range
Nitrobenzene	$C_6H_5NO_2$	10	0.358			
		50	0.329			
		120	0.393			
Benzene	C_6H_6	5	0.389			
		20	0.406			
		60	0.444			
		90	0.473			
Aniline	C_6H_7N	0	0.478			
		50	0.521			
		100	0.547			
n-Hexane	C_6H_{14}	20 to 100	0.600			
Toluene	C_7H_8	0	0.386			
		50	0.421			
		100	0.470			
n-Heptane	C_7H_{16}	0 to 50	0.507			
		30	0.518	0.476	0.00142	30° to 80°
Decane (BP 172°)	$C_{10}H_{22}$	0 to 50	0.502			
n-Hexadecane	$C_{16}H_{34}$	0 to 50	0.496			
Stearic acid	$C_{18}H_{36}O_2$	75 to 137	0.550			
Diphenyl*	$C_{12}H_{10}$			0.300	0.00120	80° to 300°

* Forrest, Brugmann, and Cummings, *Ind. Eng. Chem.* **23**, 340 (1931).

A more nearly accurate method of estimating heat capacities of the liquid state from data on the gaseous state is discussed in Chapter XII. Since, as previously mentioned, heat capacities of gases can be calculated by generalized empirical relationships based on spectroscopic data, these methods make it possible to predict the heat capacity of a substance at any conditions of the liquid or gaseous state from a minimum of experimental data.

Water has a higher specific heat than any other substance, with the exception of liquid ammonia and a few organic compounds. The heat capacity of water is a minimum at 30°C. The specific heats of aqueous solutions in general decrease with increasing concentration of solute. In dilute solutions the heat capacity of aqueous solutions is nearly equal to the heat capacity of the water present. The heat capacities of some common aqueous solutions of acids, bases, and salts are shown graphically in Figs. 41, 42, 43, 44, and 45. The heat capacities of the solutions all correspond to a temperature of about 20°C. In Table VIII are values of the specific heats of some common liquids. Data for petroleum fractions are presented in Chapter IX, page 334.

LATENT HEAT

Heat of Fusion. The fusion of a crystalline solid at its melting point to form a liquid at the same temperature is accompanied by an increase in enthalpy or an absorption of heat. Since the volume changes and hence the external work in fusion are small, this heat of fusion is largely utilized in increasing the internal energy through rearrangement of the atoms. Attempts have been made to establish general relationships between latent heats of fusion and other more easily measured properties. None of these generalizations are accurate. For most elements[5] the ratio of $\dfrac{\lambda_f}{T_f}$ varies from 2 to 3, for most inorganic compounds from 5 to 7, and for most organic compounds from 9 to 11, where

λ_f = heat of fusion, calories per gram formula weight
T_f = melting point, °K

There are a few marked exceptions to these rules.

TABLE IX

HEATS OF FUSION*

λ_f = heat of fusion, calories per gram-atom or mole' or Chu per pound-atom or mole. To convert to Btu per pound-mole multiply by 1.8.
t_f = melting point, °C.
T_f = melting point, °K.

Elements

	λ_f	t_f	λ_f/T_f
Ag.	2800	961	2.3
Al.	2340	657	2.5
Cu.	2650	1083	2.0
Fe*.	3660	1535	1.5
Na.	629	98	1.7
Ni.	4280	1450	2.5
Pb.	1160	327	1.9
S (rhombic)	300	115	0.8
Sn.	1600	232	3.2
Zn.	1660	419	2.4

Compounds

	λ_f	t_f
H_2O.	1,435	0.0
Sb_2S_3.	5,950	540
CO_2.	2,000	− 56.2
$CaCl_2$.	6,040	774
NaOH.	1,600	318
NaCl.	7,210	804
Carbon tetrachloride.	640	− 24
Methyl alcohol.	525	− 97
Acetic acid.	2,690	16.6
Ethyl alcohol.	1,150	−114
Benzene.	2,370	5.4
Aniline.	1,950	− 7.0
Naphthalene.	4,550	80
Diphenyl.	4,020	71
Stearic acid.	13,500	64

* See also Fig. 89, page 414.

[5] Wenner, *ibid.*

Experimentally determined values of heats of fusion are given in Table IX.

Heat of Transition.   Many crystalline substances exhibit transformation or transition points at which temperature changes in crystalline structure take place.   The equilibrium temperature of transformation is nearly constant although the actual temperature of transformation is frequently a function of the rate at which the substance is heated or cooled prior to the transformation.   The transition usually takes place at a slightly higher temperature when the substance is being heated than when it is being cooled.

Crystalline transformations are accompanied by either absorption or evolution of heat.   The transformation of the phase which is stable at low temperatures into the phase stable at high temperatures requires an absorption of heat.

Data for heats of transition of a few solids are recorded in Table X.

TABLE X

HEATS OF TRANSITION

λ_t = heat absorbed in transition, calories per gram-atom or Chu per pound-atom or mole.   To convert to Btu per pound-mole multiply by 1.8.
t_t = temperature of transformation, °C.

Data from International Critical Tables

Transition	λ_t	t_t
Sulfur:		
rhombic → monoclinic..............	7.0	114–151°C
Iron (electrolytic) (see also Fig. 90, page 419):		
$\alpha \rightarrow \beta$...........................	363	770
$\beta \rightarrow \gamma$...........................	313	910
$\gamma \rightarrow \delta$...........................	106	1400
Manganese:		
$\alpha \rightarrow \beta$...........................	1325	1070–1130
Nickel:		
$\alpha \rightarrow \beta$...........................	78	320–330
Tin:		
white → gray...................	530	0

HEAT OF VAPORIZATION

The heat required to vaporize a substance consists of the energy absorbed in overcoming the intermolecular forces of attraction in the liquid and the work performed by the vapors in expanding against an external pressure.   The external work performed by one mole in vaporizing under a constant pressure is equal to the product of the pressure and the increase in volume. Thus,

$$\lambda_w = p\,(v_g - v_l) \tag{23}$$

where

λ_w = external work of vaporization
p = pressure
v_g = molal volume of saturated vapor
v_l = molal volume of saturated liquid

At ordinary pressures the volume of the liquid may be neglected. If the vapor follows the ideal gas law the molal external work of vaporization is equal to RT or $2T$ calories per gram-mole, where T is the temperature of vaporization in degrees Kelvin. The latent heat of vaporization is much larger than the latent heat of fusion because the forces of molecular attraction must be overcome to a greater extent in vaporization than in fusion.

As pointed out in Chapter III, page 59, an exact relationship between heat of vaporization and vapor pressure is expressed by the Clapeyron equation, derived from the second law of thermodynamics. This equation permits accurate calculation of latent heats of vaporization at any temperature if data on the relationship between vapor pressure and temperature and on the molal volumes of the liquid and vapor are available. The results are rigorous at all temperatures and pressures. However, the necessary data for use of the rigorous equation are available for only a few substances.

Trouton's Rule. According to a rule proposed by Trouton, the ratio of the molal heat of vaporization, λ, of a substance at its normal boiling point to the absolute temperature of the boiling point T_s is a constant.

Thus
$$\frac{\lambda}{T_s} = K \tag{24}$$

where K is termed Trouton's constant or, better, Trouton's ratio. For many substances this ratio is equal to approximately 21 where the latent heat is expressed in calories per gram-mole and the temperature in degrees Kelvin. However, this ratio is by no means a constant. For polar liquids Trouton's rule breaks down completely, values of the ratio being much greater than 21. In nonpolar liquids the variation is smaller and the ratio increases as the normal boiling point increases.

Kistyakowsky Equation. A thermodynamic equation was proposed by Kistyakowsky for the calculation of Trouton's ratio at the normal boiling point of a nonpolar liquid:

$$\frac{\lambda}{T_s} = 8.75 + 4.571 \log_{10} T_s \tag{25}$$

where

λ = heat of vaporization in calories per gram-mole at $T_s{}^\circ$K
T_s = normal boiling point in degrees Kelvin.

This equation is in excellent agreement with experimental results for a wide variety of nonpolar liquids but is inapplicable to polar liquids.

Heats of Vaporization from Empirical Vapor-Pressure Equations.[6] Although, as mentioned above, the data necessary for the rigorous calculation of heats of vaporization from the Clapeyron equation are seldom available, satisfactory approximations at low pressures may be obtained by combining this equation with the empirical vapor-pressure equation developed by Calingaert and Davis, page 67. This method is particularly useful for estimating the heats of vaporization of polar compounds at their normal boiling points, for which compounds the Kistyakowsky equation is not applicable.

As a fair approximation the difference between the molal volumes of the vapor and liquid of any material at its normal boiling point is represented by

$$(v_g - v_l) = 0.95RT_b/p_b \tag{26}$$

where

T_b = normal boiling point
p_b = pressure at normal boiling point = 1.0 atmosphere

The factor 0.95 serves to correct for deviations from the ideal gas law and for the volume occupied by the liquid. By substituting Equation (26) in Equation (III–1), page 59, a modified Clausius-Clapeyron equation is obtained, applicable at the normal boiling point:

$$\frac{dp}{dT} = \frac{\lambda p_b}{0.95RT_b{}^2} \tag{27}$$

Another expression for dp/dT may be obtained by differentiating Equation (III–7), page 67. Thus,

$$\frac{dp}{dT} = \frac{pB}{(T - 43)^2} \tag{28}$$

Combining Equations (28) and (27) for the normal boiling point,

$$\lambda_b = 0.95RB \left(\frac{T_b}{T_b - 43}\right)^2 \tag{29}$$

The constant B of the Calingaert-Davis equation is readily evaluated from vapor pressure values at any two temperatures. The normal boiling point and the critical point may be used to establish this constant.

[6] K. M. Watson, *Ind. Eng. Chem.*, **35**, 398 (1943), with permission.

However, with accurate vapor pressure data, better results are obtained by selecting two values relatively close to the normal boiling point. Thus,

$$B = \frac{\ln p_2/p_1}{\left(\dfrac{1}{T_1 - 43} - \dfrac{1}{T_2 - 43}\right)} \tag{30}$$

Equation (29) may be used with fair accuracy for estimating heats of vaporization at low pressures other than one atmosphere by substituting the proper boiling points for T_b. However, it is recommended that its application be limited to estimating values at the normal boiling point. Values at other temperatures may then be obtained by the empirical method described later.

Illustration 2. Ethyl alcohol has a normal boiling point at 78.3°C and a vapor pressure of 15.61 atmospheres at 170°C. Estimate the molal heat of vaporization at its normal boiling point.

From Equation (30),

$$B = \frac{\ln \dfrac{15.61}{1}}{\left(\dfrac{1}{351.3 - 43} - \dfrac{1}{443 - 43}\right)} = 3720$$

From Equation (29)

$$\lambda = (0.95)(1.99)(3720)\left(\frac{351.3}{308.3}\right)^2 = 9140 \text{ cal per gr-mole}$$

More reliable heats of vaporization may be obtained by differentiating Equation (III–16) to obtain dp/dT for substitution in Equation (27). Further improvement results from use of the actual compressibility factor of the vapor instead of the average value of 0.95.

Low Pressure Heats of Vaporization from Reference Substance Vapor Pressure Plots.

From Equation (5) of Chapter III, page 65, it is evident that the slope of a vapor pressure line on an equal temperature reference substance plot is equal to the ratio of the heat of vaporization of the compound in question to that of the reference substance at the same temperature. Thus, the heat of vaporization at any temperature is obtained by multiplying that of the reference substance at the same temperature by the constant value of the slope of the vapor pressure line. These slopes are indicated on Fig. 7.

A similar but less convenient relationship between the slopes of equal pressure reference substance plots may be developed by applying the Clausius-Clapeyron equation to both the compound in question and the reference substance at the same pressure.   Thus,

$$\lambda = \lambda' \left(\frac{T}{T'}\right)^2 \frac{dT'}{dT} \qquad (31)$$

where

λ = heat of vaporization at temperature T and vapor pressure p

λ' = heat of vaporization of reference substance at pressure p and temperature T'

$\dfrac{dT'}{dT}$ = slope of Dühring line

As discussed in Chapter III, the satisfactory linearity of reference substance plots up to the critical point appears to result from a fortuitous compensation of errors in the basic assumptions involved.   Even though such a plot may be a straight line the actual ratio of the latent heats may vary over a wide range and become meaningless as the critical point of either substance is approached.   For this reason great care must be exercised in deriving heats of vaporization from reference substance plots at elevated pressures.   It is recommended that their use be restricted to the low pressure range and that the methods of the following sections be used for all pressures substantially above atmospheric.

Empirical Relationship Between Heat of Vaporization and Temperature.   The heat of vaporization of a substance diminishes as its temperature and pressure are increased.   At the critical temperature, as pointed out in Chapter III, the kinetic energies of translation of the molecules become sufficiently great to overcome the potential energies of the attractive forces which hold them together and molecules pass from the liquid to the vapor state without additional energization.   At the critical point there is no distinction between the liquid and vapor states, either in enthalpy or other physical properties and the heat of vaporization becomes zero.

It was pointed out by Watson[7] that if values of Trouton's ratio are plotted against reduced temperature, curves of the same shape are obtained for all substances both polar and nonpolar.   These curves may be superimposed by multiplying the ordinates of each by the proper constant factor.   It was found that the following empirical equation satisfactorily represents these curves over the entire range of available data:

$$\frac{\lambda}{\lambda_1} = \left(\frac{1 - T_r}{1 - T_{r_1}}\right)^{0.38} \qquad (32)$$

[7] *Ind. Eng. Chem.*, **23**, 360 (1931); **35**, 398 (1943).

TABLE XI

HEATS OF VAPORIZATION AND CRITICAL CONSTANTS*

λ = heat of vaporization at $t°C$ calories/gr-mole or Chu/lb-mole
t = temperature
t_c = critical temperature in °C
p_c = critical pressure in atm

Substance	λ	$t°C$	t_c	p_c
Air			−140.7	37.2
Ammonia —	5,581	−33.4	132.4	111.5
Argon	1,590	−185.8	−122	48.0
Bromine	7,420	58.0	302	
Carbon dioxide	6,030	−78.4	31.1	73.0
Carbon disulfide			273.0	76.0
Carbon monoxide	1,444	−191.5	−139.0	35.0
Carbon oxysulfide	4,423	−50.2	105.0	61.0
Carbon tetrachloride	7,280	77	283.1	45.0
Chlorine	4,878	−34.1	144.0	76.1
Dichlorodifluoromethane	4,888	−29.8	111.5	39.56
(Freon 12)				
Dichloromonofluoromethane (F-21)		8.9	178.5	51.0
Helium	22	−268.4	−267.9	2.26
Hydrogen	216	−252.7	−239.9	12.8
Hydrogen bromide	4,210	−66.7	90.0	84.0
Hydrogen chloride	3,860	−85.0	51.4	81.6
Hydrogen cyanide	6,027	25.7	183.5	50.0
Hydrogen fluoride	7,460	33.3	230.2	
Hydrogen iodide			151.0	82.0
Hydrogen sulfide	4,463	−60.3	100.4	88.9
Mercury	13,980	361	>1550	>200
Nitric oxide	3,307	−151.7	−94.0	65.0
Nitrogen	1,336	−195.8	−147.1	33.5
Nitrous oxide	3,950	−88.5	36.5	71.7
Oxygen	1,629	−183.0	−118.8	49.7
Phosgene			182.0	56.0
Silicon tetrafluoride	6,130	−94.8	−1.5	50.0
Sulfur	20,200	444.6	1,040	
Sulfur dioxide	5,960	−5.0	157.2	77.7
Sulfur trioxide	10,190	44.8	218.3	83.6
Trichloromonofluoromethane (F-11)		23.6	198.0	43.2
Water	9,729	100.0	374.0	217.7

* *Chemical Engineers' Handbook*, John H. Perry, McGraw-Hill Book Company, Inc. (1941), with permission.

This equation permits estimation of the heat of vaporization of any substance at any temperature if its critical temperature and the heat of vaporization at some one temperature are known. In combination with the methods described in the previous sections for estimating heats of vaporization at low pressures it is possible to predict data at all conditions up to the critical with errors generally less than 5%.

Illustration 3. The latent heat of vaporization of ethyl alcohol is experimentally found to be 204 calories per gram at its normal boiling point of 78°C. Its critical temperature is 243°C.   Estimate the heat of vaporization at a temperature of 180°C.

$$T_b/T_c = 351/516 = 0.680 = T_{r1}$$
$$T/T_c = 453/516 = 0.880 = T_{r2}$$

From Equation (32)

$$\lambda_2 = 204 \left[\frac{1 - 0.880}{1 - 0.680}\right]^{0.38} = 140 \text{ cal per gram}$$

Reduced Reference Substance Plot.   D. H. Gordon[7] has pointed out that the limitations of the conventional reference substance plots for the estimation of heats of vaporization may be greatly minimized by basing the reference substance plot on equal *reduced* conditions.   The most convenient form is a logarithmic plot of vapor pressures of the substance in question and of the reference substance at equal reduced temperatures. It is found that this method of plotting yields good approximations to linear curves for a wide variety of substances, both polar and nonpolar over wide ranges of conditions up to the critical.

Since $T = T_r T_c$, the Clausius-Clapeyron equation (page 60) may be written in terms of reduced temperatures:

$$d \ln p = \frac{\lambda dT_r}{RT_r^2 T_c} \qquad (33)$$

Applying Equation (33) to a substance and a reference substance at the same reduced temperature,

$$\frac{d \ln p}{d \ln p'} = \frac{\lambda}{\lambda'} \frac{T_c'}{T_c} = s_r \qquad (34)$$

or

$$\lambda = s_r \frac{T_c}{T_c'} \lambda' \qquad (35)$$

where the primed quantities designate the reference substance.   The group of terms $s_r \dfrac{T_c}{T_c'}$ is a constant for any one pair of substances.

The term s_r is the slope of the line resulting when the logarithm of the vapor pressure of the given substance is plotted against the logarithm of the vapor pressure of a reference substance at the same reduced temperature.

Values of $s_r \dfrac{T_c}{T_c'}$ are given in Table XIII for various refrigerants with water as the reference substance.   In Table XII are given the heats of vaporization of water in Btu per pound-mole at various values of reduced temperature, T_r, and the corresponding vapor pressures.

[7] Univ. of Wis. Ph.D. Thesis (1942).

TABLE XII

MOLAL HEATS OF VAPORIZATION AND VAPOR PRESSURES OF WATER

$T_c = 1165.4°R$ λ = Btu per lb-mole p = pounds per sq in. abs

T_r	λ	p
0.423	19,370	0.092
0.44	19,170	0.198
0.46	18,940	0.446
0.48	18,700	0.934
0.50	18,460	1.824
0.52	18,210	3.365
0.54	17,960	5.896
0.56	17,770	9.871
0.58	17,440	15.871
0.60	17,170	24.613
0.62	16,880	36.959
0.64	16,590	54.006
0.66	16,280	76.785
0.68	15,950	106.65
0.70	15,600	145.08
0.72	15,230	193.68
0.74	14,840	254.1
0.76	14,420	328.3
0.78	13,980	418.4
0.80	13,490	526.1
0.82	12,970	654.2
0.84	12,400	804.5
0.86	11,770	980.3
0.88	11,070	1184.7
0.90	10,290	1419.3
0.91	9,684	1548.7
0.92	9,401	1688.4
0.93	8,907	1836.0
0.94	8,365	1995.6
0.95	7,774	2164.3
0.96	7,103	2346.1
0.97	6,336	2538.6
0.98	5,393	2746.8
0.99	4,143	2967.5
1.00	0	3206.2

TABLE XIII

Heat of Vaporization Factors and Critical Temperatures of Refrigerants*

Refrigerant	Formula	s_r	Critical Temp. °F	$\dfrac{s_r T_c}{T_c'}$ †
Ammonia	NH_3	0.933	270.4	0.584
Benzene	C_6H_6	0.923	551.3	0.800
Butyl alcohol	C_4H_9OH	1.327	548.6	1.149
Carbon dioxide	CO_2	0.872	88.43	0.410
Carbon disulfide	CS_2	0.839	523.0	0.707
Carbon tetrachloride	CCl_4	0.869	541.6	0.746
Chlorine	Cl_2	0.776	295.0	0 502
Chlorobenzene	C_6H_5Cl	0.933	678.6	0.912
Chloroform	$CHCl_3$	0.916	505.0	0.749
Ethane	C_2H_6	0.810	89.8	0.382
Ethyl alcohol	C_2H_5OH	1.280	469.6	1.022
Ethyl chloride	C_2H_5Cl	0.885	369.0	0.644
Ethyl ether	$(C_2H_5)_2O$	0.952	380.8	0.687
Freon 11	CCl_3F	0.897	388.4	0.652
Freon 12	CCl_2F_2	0.869	232.7	0.516
Freon 21	$CHCl_2F$	0.891	353.3	0.621
Freon 113	$C_2Cl_3F_3$	0.962	417.4	0.724
Isobutane	C_4H_{10}	0.879	273.0	0.551
Methane	CH_4	0.716	−116.5	0.211
Methyl alcohol	CH_3OH	1.179	464.0	0.936
Methyl chloride	CH_3Cl	0.845	289.4	0.543
n-Butane	C_4H_{10}	0.897	307.0	0.591
Nitrogen dioxide	NO_2	0.851	97.7	0.417
n-Propyl alcohol	nC_3H_7OH	1.303	506.8	1.083
Propane	C_3H_8	0.851	206.2	0.486
Sulfur dioxide	SO_2	0.949	315.0	0.631
Toluene	C_7H_8	0.956	609.1	0.877

* D. H. Gordon, Univ. of Wis., Ph.D. Thesis (1942).
† Reference substance, water.

Illustration 4. Estimate the heat of vaporization of Freon 12 (difluorodichloromethane) at 200°F.

From Table XIII the critical temperature of Freon 12 is 232.7°F and the value of $s_r \dfrac{T_c}{T_c'}$ is 0.516. The molecular weight is 121.

$$T_r = \frac{660}{692.7} = 0.951$$

At $T_r = 0.951$ the molal heat of vaporization of water from Table XII is 7707 Btu per pound-mole.

From Equation (35),

$$\lambda = s_r \frac{T_c}{T_c'} \lambda' = (0.516)(7707) = 3977 \text{ Btu per pound-mole}$$

or $\dfrac{3977}{121}$ = 32.8 Btu per pound

This is in agreement with published data.

Gordon has studied the application of this method to much of the available data on heats of vaporization and found good agreement, with errors generally less than 5%. The reliability of the method is apparently of the same order as the empirical equation discussed in the previous section. It is particularly convenient to use where a number of heat of vaporization values at different conditions are required for a single substance.

Another good method for the prediction of heats of vaporization at all conditions has been developed by Meissner.[8] This method is applicable to all substances and has approximately the same accuracy as the methods herein described.

EVALUATION OF ENTHALPY

As pointed out on page 207, the absolute enthalpy or energy content of matter is unknown. However, the enthalpy of a given substance relative to some reference state can be calculated from its thermophysical properties. This state can be taken arbitrarily as a temperature of 0°C (32°F), atmospheric pressure, and the state of aggregation normally existent at this temperature and pressure. The reference state for steam is usually taken as the liquid state, under its own vapor pressure, at 0°C.

The enthalpy of a substance is calculated as the change in enthalpy in passing from the reference state to the existing conditions. As previously pointed out, at constant pressure the increase in enthalpy is equal to the heat absorbed. Ordinarily at moderate pressures the effect of pressure on the enthalpy of liquids and solids may be neglected except when conditions are close to the critical point. This subject is discussed in Chapter XII.

Illustration 5. Calculate the enthalpy of 1 lb of steam at a temperature of 350°F and a pressure of 50 lb per sq in., referred to the liquid state at 32°F.

Solution. From the vapor-pressure data of water (Table I) it is found that the saturation temperature under an absolute pressure of 50 lb per sq in. is 281°F. The steam is therefore superheated 69°F above its saturation temperature. The enthalpy will be the heat absorbed in heating 1 lb of liquid water from 32°F to 281°F, vaporizing it to form saturated steam at this temperature, and heating the steam at constant pressure to a temperature of 350°F. The total enthalpy is the sum of the enthalpy of the liquid, the latent heat of vaporization, and the superheat of the vapors. The effect of pressure on the enthalpy of the liquid water is neglected.

[8] H. P. Meissner, *Ind. Eng. Chem.*, **33**, 1440 (1941), with permission.

The mean specific heat of water between 32°F and 281°F is 1.006. The mean heat capacity of water vapor between 281°F and 350°F at a pressure of 50 lb per sq in. is 9.2 Btu per lb-mole per °F. The latent heat of vaporization of water at 281°F is 926.0 Btu per lb.

Enthalpy of liquid water at 281°F =
    (281 − 32) 1.006 = ...   250.3 Btu per lb
Heat of vaporization at 281°F.........................   926.0 Btu per lb
Superheat of vapor = (350 − 281) $\dfrac{9.20}{18}$ = ...   35.2 Btu per lb
                                                         _____
    Enthalpy...................................   1211.5 Btu per lb

Extensive steam tables have been compiled giving the enthalpies and other properties of steam under widely varying conditions, for both saturated and superheated vapors. In calculating these tables it is necessary to take into account the variation of the heat capacity with pressure, as discussed in Chapter XII. Tables and charts of enthalpies have been worked out for a number of substances for which frequent thermal calculations are made in engineering practice.

Calculations of enthalpy often include several changes of state. For example, in calculating the enthalpy of zinc vapor at 1000°C and atmospheric pressure, relative to the solid at standard conditions, it is necessary to include the sensible enthalpy of the solid metal at the melting point, the latent heat of fusion, the heat absorbed in heating the liquid from the melting point to the normal boiling point, the latent heat of vaporization, and the heat absorbed in heating the zinc vapor from the boiling point up to 1000°C at constant pressure.

Illustration 6. Calculate the enthalpy of zinc vapor at 1000°C and atmospheric pressure, relative to the solid at 0°C. Zinc melts at 419°C and boils under atmospheric pressure at 907°C.

The mean heat capacities of the solid and liquid may be estimated from Fig. 38, page 218.

Mean specific heat of solid, 0°C to 419°C = ...   0.105
Mean specific heat of liquid, 419°C to 907°C = ...   0.109

From Table IX, page 228, the heat of fusion is 1660 calories per gram-atom. The heat of vaporization at the normal boiling point may be estimated from Equation (25)

$$\lambda/1180 = 8.75 + 4.571 \log 1180 = 22.80$$
$$\lambda = 26{,}900 \text{ calories per gram-mole}$$

Since zinc vapor is monatomic its molal heat capacity at constant pressure is constant and equal to 4.97 calories per gram-mole

Heat absorbed by solid = 0.105(419 − 0) = ...   44 calories per gram
Heat of fusion = $\dfrac{1660}{65.4}$ = ...   25 calories per gram

Heat absorbed by liquid =
$$0.109(907 - 419) = \ldots \ldots \ldots \ldots \ldots \ldots \ldots \quad \text{53 calories per gram}$$

$$\text{Heat of vaporization} = \frac{26,900}{65.4} = \ldots \ldots \ldots \ldots \ldots \quad \text{412 calories per gram}$$

$$\text{Heat absorbed by vapor} = \frac{4.97}{65.4}(1000 - 907) = \ldots \quad \text{7 calories per gram}$$

$$\text{Total enthalpy} \ldots \ldots \ldots \ldots \ldots \ldots \ldots \ldots \quad \overline{\text{541}} \text{ calories per gram}$$

Frequently it is difficult to determine experimentally the individual heats of transition involved in heating a substance. Under such conditions the enthalpy is measured directly and tabulated as such for various temperatures. For example, the enthalpy of steel at various temperatures is determined by cooling in a calorimeter from these initial temperatures. This determination includes all heats of transition undergone in the cooling process.

ENTHALPY OF HUMID AIR

The properties of humid air are conveniently expressed on the basis of the weight of humid air which contains either 1 pound or 1 pound-mole of moisture-free air. The enthalpy of the quantity of humid air containing a unit quantity of moisture-free air is the sum of the sensible enthalpy of the dry air and that of the water vapor which is associated with it. The reference states ordinarily chosen are air and liquid water at 0°C. The water vapor in the air may be considered as derived from liquid water at 0°C by the following series of processes:

1. The liquid water is heated to the dew point of the humid air.
2. The water is vaporized at the dew point temperature to form saturated vapor.
3. The water vapor is superheated to the dry-bulb temperature of the air.

The enthalpy of the water is the sum of the heat absorbed by the liquid, the heat of vaporization at the dew point, and the superheat absorbed by the vapor.

Illustration 7. Calculate the enthalpy, per pound of dry air, of air at a pressure of 1 atmosphere, a temperature of 100°F, and with a percentage humidity of 50%.

Solution. From the humidity chart, Fig. 8, page 100, it is seen that air under these conditions contains 0.0345 pound-mole of water per mole of dry air or $\dfrac{0.0345(18)}{29}$ = 0.0215 pound of water per pound of dry air. This corresponds to a dew point of 79°F.

From Fig. 36, the mean molal heat capacity of water vapor between 79°F and 100°F is 8.02 and that of air between 32°F and 100°F is 6.95.

The heat of vaporization at 79°F may be estimated from Fig. 8 as 18,840 Btu per pound-mole or 1046 Btu per pound.

Sensible enthalpy of air $= (100 - 32)\dfrac{6.95}{29.0} = \dots\dots\dots$ 16.3 Btu

Sensible enthalpy of liquid water $= (79 - 32)0.0215 = \dots$ 1.0 Btu

Latent heat of water $= 1046 \times 0.0215 = \dots\dots\dots\dots\dots$ 22.5 Btu

Superheat of water vapor $= (100 - 79) \times 0.0215 \times$

$\dfrac{8.02}{18} = \dots\dots\dots\dots\dots\dots\dots\dots\dots\dots\dots\dots\dots\dots\dots\dots\dots\dots$ 0.2 Btu

Total enthalpy$\dots\dots\dots\dots\dots\dots\dots\dots\dots\dots\dots\dots$ 40.0 Btu per lb
of dry air

Humid Heat Capacity of Air. It has been pointed out that when dealing with humid air it is convenient to use 1 pound or 1 pound-mole of dry air as the basis of calculations, regardless of the humidity of the air. In problems dealing with the heating or cooling of air where no change in moisture content takes place the total change in enthalpy is equal to the sum of the change in the sensible enthalpy of the dry air and the change in sensible enthalpy of the water vapor. For example, in heating 1 pound of dry air associated with H pounds of water vapor from t_1 to t_2 degrees Fahrenheit, the total heat, q, required is given by the equation,

$$q = C_{pa}(t_2 - t_1) + H (C_{pw}) (t_2 - t_1) \qquad (36)$$

where

C_{pa} = the mean specific heat of air at constant pressure

C_{pw} = the mean specific heat of water vapor at constant pressure

Instead of considering the air and water vapor separately it is convenient to employ a heat capacity term which combines the two.

Thus, $q = S (t_2 - t_1)$ (37)

where

S = heat capacity of one pound of dry air and of the water associated with it, expressed in Btu per pound of dry air per degree F

Combining (36) and (37)

$$S = C_{pa} + HC_{pw} \qquad (38)$$

The combined heat capacity, S, is termed the *humid heat capacity* of the air. Over the low temperature range from 30° to 180°F the mean heat capacity of dry air is 0.240 Btu per pound and that of water vapor is 0.446 Btu per pound, from Fig. 36. Accordingly the humid heat capacity of air when expressed in Btu per pound of air per degree Fahrenheit is given by the equation,

$$S = 0.240 + 0.446H \qquad (39)$$

Adiabatic Humidification. In the discussion of the humidity chart, Chapter IV, page 99, it was pointed out that a line of constant wet-bulb temperature also represents the relationship between dry-bulb temperature and humidity in the adiabatic vaporization of water into air. This relationship may be derived from the thermophysical data of the system and is used in locating the wet-bulb temperature lines on the humidity chart. The same procedure may be used to establish wet-bulb temperatures in any other system of liquid and gas in which it is known that the wet-bulb temperature does not change appreciably during adiabatic vaporization. The derivation is as follows:

When air is cooled by the adiabatic vaporization of water into it, sensible heat is derived from the humid air to supply the heat necessary in vaporizing the water at the wet-bulb temperature and in heating the evolved vapor to the existing dry-bulb temperature. Since the total enthalpy of the system remains constant, the heat lost by the humid air must equal that gained by the water in vaporization and superheating. This equality may be expressed mathematically for the evaporation of dH moles of water into humid air containing 1 mole of dry air. Thus,

$$dH[\lambda + c_{pw}(t - t_w)] = -S \, dt \qquad (40)$$

where,

H = molal humidity
λ = molal heat of vaporization at temperature t_w
c_{pw} = mean molal heat capacity of water vapor
t = dry-bulb temperature
t_w = temperature of adiabatic evaporation
S = mean molal humid heat capacity of air

Assuming that the wet-bulb temperature remains constant, as humidification proceeds the final dry-bulb temperature reached by the entire weight of air will be the wet-bulb temperature t_w, corresponding to saturation and a humidity H_w. In the temperature range from 32° to 200°F the molal heat capacities of air and water vapor may be taken from Fig. 36 as constant at 6.95 and 8.04, respectively. Then, from Equation (38), $S = 6.95 + 8.04H$. Substituting these values in Equation (40), and rearranging,

$$\frac{dH}{6.95 + 8.04H} = -\frac{dt}{\lambda + 8.04(t - t_w)} \qquad (41)$$

Integrating between the limits H, t and H_w, t_w,

$$\frac{1}{8.04} \ln \frac{6.95 + 8.04H}{6.95 + 8.04H_w} = \frac{1}{8.04} \ln \frac{\lambda}{\lambda + 8.04(t - t_w)}$$

or $6.95\lambda + 8.04\lambda H + 8.04(t - t_w)(6.95 + 8.04H) = 6.95\lambda + 8.04\lambda H_w$

or $$t = \frac{(H_w - H)\lambda}{6.95 + 8.04H} + t_w \qquad (42)$$

The temperature, t_w, of adiabatic evaporation corresponds to the experimental value of wet-bulb temperature provided evaporation from the wet-bulb thermometer proceeds adiabatically, that is, with no gain or loss of heat by radiation, and also provided the actual vapor-pressure equilibrium is established at the liquid-air interface. The first condition is realized where the air is passed rapidly over the wet-bulb thermometer such that radiation errors become negligible; the second condition is true where the rate of evaporation by diffusion keeps pace with the rate of heat transfer by convection. This latter condition is realized without appreciable error at temperatures below 150°F. *The lines of adiabatic evaporation are therefore commonly referred to as wet-bulb temperature lines and will be so designated.*

The adiabatic cooling or wet-bulb temperature lines of Fig. 8 were constructed from Equation (42). Corresponding to a selected value of t_w, values of dry-bulb temperature were calculated to correspond to various humidities, thus establishing a complete curve. The wet-bulb temperature lines of Fig. 8 which apply to gases of appreciable carbon dioxide content were constructed from a similar equation in which the effect of the carbon dioxide on the humid heat capacity of the gas was considered. The molal heat capacity of carbon dioxide may be assumed to be 9.3 (Fig. 36). Then,

$$S = 6.95 (1 - x) + 9.3x + 8.04H \qquad (43)$$

where

$$x = \text{mole fraction of } CO_2 \text{ in the dry gas}$$

With this modification Equation (42) becomes

$$t = \frac{(H_w - H)\,\lambda}{6.95(1 - x) + 9.3x + 8.04H} + t_w \qquad (44)$$

This equation permits calculations of adiabatic cooling or wet-bulb temperature lines to apply to combustion gases or other mixtures containing appreciable amounts of carbon dioxide.

Enthalpy of Humid Air. The ordinary psychrometric chart is limited to direct use at an atmospheric pressure of 29.92 inches Hg. For other pressures different sets of wet-bulb temperature and percentage (or relative) humidity lines are required. W. Goodman[9] has designed a psychrometric chart which covers a range of atmospheric pressures from 22 to 32 inches Hg as shown in Fig. 46. In this chart lines of constant enthalpy have been constructed instead of the usual wet-bulb temperature lines, where enthalpy of the humid air is expressed on the basis of one pound of dry air. The horizontal lines in this chart represent absolute humidities; the diagonal lines, constant enthalpies; the nearly vertical lines, dry-bulb temperatures; and the curved lines, humidities at saturation for various constant atmospheric pressures. The tempera-

[9] *Air Conditioning Analysis*, Macmillan (1943), with permission.

ture lines are given a slight slope to allow for the increase in heat capacity of air and water vapor with temperature and thus avoid curvature in the constant enthalpy lines.

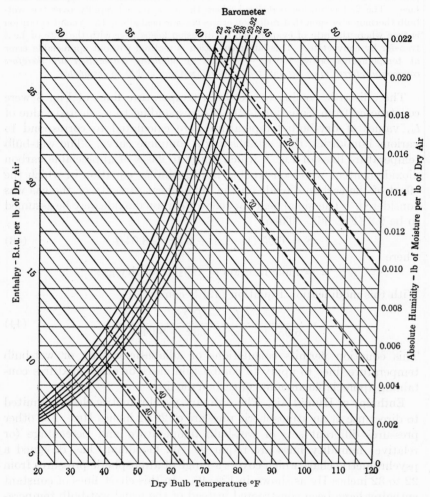

Fig. 46. Enthalpy chart for water vapor-air system. From W. Goodman, *Air Conditioning Analysis*, Macmillan (1943), with permission.

Reference State: Dry Air at 0°F; Liquid Water at 32°F
(Reproduced in " C.P.P. Charts ")

In adiabatic evaporation the wet-bulb temperature is constant and the system including the humid air and the water to be evaporated is hence at constant enthalpy. At low temperatures the enthalpy of the liquid water to be evaporated is negligible so that the enthalpy of the humid air is nearly equal to that of the system and hence the slope of the constant enthalpy lines is nearly equal to that of the constant wet-bulb lines. The constant enthalpy lines instead of wet-bulb temperature lines have the advantage that the humidity lines are nearly independent of atmospheric pressure and may be used directly in establishing heat requirements in air-conditioning problems. In Fig. 46 wet-bulb temperature lines have been constructed for 70°F and for 40°F. It will be observed that the wet-bulb temperature lines have nearly the same slope as the enthalpy lines, and become more nearly the same as the temperature is lowered. It will also be observed that the location of a given wet-bulb temperature line depends upon the atmospheric pressure. The slopes of these particular lines were obtained from Equation (42); other lines can be drawn in similarly or the slopes may be estimated from the existing 40° and 70°F wet-bulb temperature lines.

PROBLEMS

1. (a) From the data of Table V, page 214, calculate the mean heat capacity of one of the following gases: oxygen, hydrogen, water vapor, sulfur dioxide, ammonia:

   (1) In kilocals per kilogram per degree Centigrade from 0° to t°C.

   (2) In Chu per pound per degree Centigrade from 0° to t°C.

   (3) In kilocals per cubic meter per degree Centigrade from 0° to t°C.

   (4) In kilocals per kilogram per degree Centigrade from 1000° to 2000°C.

   (5) In Btu per pound-mole per degree Fahrenheit from 32° to t°F.

   (6) In Btu per cubic foot per degree Fahrenheit from 32° to t°F.

   (7) In Btu per pound per degree Fahrenheit from 1000° to 2000°F.

   (b) Calculate the heat capacity of the assigned gas in kilocals per kilogram per degree Centigrade at 1500°C.

2. From the experimental data for the molal heat capacities of oxygen at various temperatures derive the constants in the following types of empirical equations over the temperature range from 0°C to 1500°C:

$$c_p = a + bT + cT^2$$

$$c_p = a + \frac{b}{T} + \frac{c}{T^2}$$

$$c_p = a + \frac{b}{\sqrt{T}} + \frac{c}{T}$$

3. Calculate the amount of heat given off when 1 cu m of air (standard conditions) cools from 500° to −100°C at a constant pressure of 1 atmosphere, assuming the heat capacity formula of Table V, page 214, to be valid over this temperature range.

4. Calculate the number of kcals required to heat, from 200° to 1200°C, 1 cu m (standard conditions) of a gas having the following composition by volume:

CO_2............................... 20%
N_2............................... 77%
O_2............................... 2%
H_2............................... 1%

5. Calculate the number of Btu required to heat 1 lb each of the following liquids from a temperature 32° to 100°F:

(a) Acetone.
(b) Carbon tetrachloride.
(c) Ether.
(d) Propyl alcohol.

6. Calculate the number of calories required to heat 1000 grams of each of the following aqueous solutions from 0° to 100°C.

(a) 5% NaCl by weight.
(b) 20% NaCl by weight.
(c) 20% H_2SO_4 by weight.
(d) 20% KOH by weight.
(e) 20% NH_4OH by weight.
(f) 20% $Pb(NO_3)_2$ by weight.

7. From Fig. 38 determine the heat required to raise 1 pound of graphite from 32°F to 1450°F. Show how the graph is utilized to determine the quantity of heat required.

8. Calculate the specific heat at 20°C of MgO, SiO_2 (quartz), CaO, CuO, PbO from Kopp's rule and compare with the experimental values.

9. Calculate the heat equivalent in Btu of the external work of vaporization of 1 lb of water at a temperature of 80°F, assuming that water vapor follows the ideal gas law.

10. Calculate the heat of vaporization in Btu per pound of diethyl ether ($C_2H_5OC_2H_5$) at its normal boiling temperature by the following methods:

(a) From the equation of Kistyakowsky.
(b) From Equation (29).
(c) From Equation (35).

11. Obtaining the necessary boiling-point and critical data from Fig. 4, estimate the heat of vaporization, in Btu per pound of n-butane (C_4H_{10}) at a pressure of 200 lb per sq in.

(a) By Equations (25) and (32).
(b) By Equations (29) and (32).
(c) By Equation (35) and Tables XII and XIII.

12. Using Fig. 4 for the necessary boiling-point and critical data, estimate the heat of vaporization, in Btu per pound of carbon disulfide (CS_2) when under a pressure of 77.5 lb/in.² abs using the three methods of Problem 11.

13. Cyclohexane (C_6H_{12}) has a normal boiling point of 80.8°C and at this temperature the density of the liquid is 0.719 grams per cc. Estimate:

(a) The critical temperature.
(b) The boiling point at 10 atmospheres pressure.
(c) The heat of vaporization at 10 atmospheres pressure, expressed as Btu per pound.

14. Utilizing the thermal data for diphenyl, $(C_6H_5 \cdot C_6H_5)$ tabulated below, estimate the following:

(a) Critical temperature.
(b) Boiling point at 25 lb per sq in.
(c) Heat of vaporization at 25 lb per sq in., as Btu per lb.
(d) Enthalpy of 1 lb of saturated diphenyl vapor at 25 lb/sq in. relative to solid diphenyl at 32°F.

Data for diphenyl

Normal boiling point.	255°C
Density of the liquid at the normal boiling point.	0.75 gram/cc
Melting point.	71°C
Specific heat of solid diphenyl.	0.385 Btu per lb per deg. F
Heat of fusion.	46.9 Btu per lb
Specific heat of liquid diphenyl. . . . $c_p = 0.300 + 0.00120t$°C	

15. Calculate the heat of vaporization, in calories per gram, of water at a temperature of 100°C by means of the Clapeyron equation. At this temperature $dp/dt = 27.17$, where p is the vapor pressure in millimeters of mercury and t is the temperature in degrees Centigrade.

16. The vapor pressure of zinc in the range from 600° to 985°C is given by the equation

$$\log p = -\frac{6160}{T} + 8.10$$

where

$$p = \text{vapor pressure, millimeters of mercury}$$
$$T = \text{temperature, degrees K}$$

Estimate the heat of vaporization of zinc at 907°C, the normal boiling point. Compare this result with that calculated from the equation of Kistyakowsky.

17. From the International Critical Tables obtain the following data:

(a) The heat capacity, in calories per gram per degree C, of

    (*1*) Liquid o-nitroaniline $(C_6H_2N_2O_2)$ at 100°C.
    (*2*) Liquid $SiCl_4$ at 25°C.
    (*3*) A solution containing 50 mole per cent ether $(C_4H_{10}O)$ in benzene (C_6H_6) at a temperature of 20°C.
    (*4*) Solid FeS_2 at 100°C.

(b) The heat of fusion of

    (*1*) $BaCl_2$.
    (*2*) Benzoic acid $(C_7H_6O_2)$.
    (*3*) Stearic acid $(C_{18}H_{36}O_2)$.

(c) The heat of vaporization of

    (*1*) Nitrogen at −202°C.
    (*2*) $SiCl_4$ at 57°C.
    (*3*) n-octyl alcohol $(C_8H_{18}O)$ at 196°C.

18. Calculate the enthalpy in kcals per kilogram referred to the solid at 0°C, of molten copper at a temperature of 1200°C.

19. Obtaining the latent heat data from the steam tables, calculate the enthalpy in Btu per pound relative to the liquid at 32°F, of steam at a temperature of 500°F superheated 200°F above its saturation point.

20. Calculate the enthalpy in Btu per pound relative to 32°F, of pure molten iron at a temperature of 2850°F. In heating iron from 32°F to its melting point it undergoes three transformations, from α to β, from β to γ, and from γ to δ forms.

21. Using the latent heat data calculated in Problem 14, calculate the enthalpy in Btu per pound relative to the solid at 32°F, of saturated diphenyl vapors under a pressure of 40 lb per sq in.

22. Calculate the enthalpy in Btu per pound of dry air relative to air and liquid water at 32°F, of humid air at a temperature of 150°F, a pressure of 1 atmosphere, and a percentage humidity of 40%.

23. Humid air at a pressure of 1 atmosphere has a dry-bulb temperature of 180°F and a wet-bulb temperature of 120°F. This air is cooled to a dry-bulb temperature of 115°F. Calculate the heat evolved, in Btu per pound of dry air.

24. Hot gases are passing through a chimney at a rate of 1200 cu ft per minute, measured at the existing conditions of 600°C and a pressure of 740 mm of Hg. The gases have the following composition by volume on the dry basis:

$$CO_2 \dots\dots\dots\dots\dots\dots\dots\dots\dots\dots\dots\dots\dots\dots 12\%$$
$$N_2 \dots\dots\dots\dots\dots\dots\dots\dots\dots\dots\dots\dots\dots\dots 80\%$$
$$O_2 \dots\dots\dots\dots\dots\dots\dots\dots\dots\dots\dots\dots\dots\dots 8\%$$

The dew point of the gases is 50°C and they contain 20 grams of carbon soot per cubic meter measured at the chimney conditions. Calculate the enthalpy of the material passing through the chimney per minute, in Btu relative to gases, solid carbon, and liquid water at 18°C.

Chemical Caustic

$$Na_2CO_3 + Ca(OH)_2 \rightleftharpoons 2NaOH + CaCO_3$$

Batch process

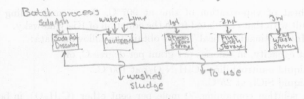

↓ washed sludge   ↓ To use

Electrolytic Caustic & Chlorine

Diaphragm cells, mercury cathode cells, etc

$$2NaCl + 2H_2O \xrightarrow{} NaOH + Cl_2 + H_2$$

Purification of Caustic

Concentration to ~50%, salt crystallizes out.
To get salt free add Na_2SO_4 to get NaOH-NaCl, Na_2SO_4 out.

Sodium Hypochlorite

$$NaOH + Cl_2 \rightarrow NaOCl + NaCl + H_2O$$

Bleaching Powder

$$Ca(OH)_2 + Cl_2 \rightarrow Ca(OH)(Cl) \cdot H_2O$$

CHAPTER VIII

THERMOCHEMISTRY

All chemical reactions are accompanied either by an absorption or evolution of energy, which usually manifests itself as heat. That branch of science which deals with the changes of energy in chemical reactions is called thermochemistry.

As discussed in Chapter VII the internal energy of a given substance is dependent upon its temperature, pressure, and state of aggregation and is independent of the means by which this state was brought about. Likewise the change in internal energy, ΔU, of a system which results from any physical change or chemical reaction depends only on the initial and final state of the system. The total change in internal energy will be the same whether or not energy is absorbed or evolved in the form of heat, radiant energy, electrical energy, mechanical work, or other forms.

For a flow reaction proceeding with negligible changes in kinetic energy, potential energy and with no electrical work and no mechanical work beyond that required for flow, the heat added is equal to the increase in enthalpy of the system,

$$q = \Delta H \qquad (1)$$

For a non-flow reaction proceeding at constant pressure the heat added is also equal to the gain in enthalpy,

$$q = \Delta H \qquad (2)$$

For a non-flow reaction proceeding at constant volume the heat added is equal to the gain in internal energy of the system,

$$q = \Delta U \qquad (3)$$

Standard Heat of Reaction. The heat of a chemical reaction is the heat absorbed in the course of the reaction, or, in a more general sense it is equal to the change in enthalpy of the system for the reaction proceeding at constant pressure. This heat of reaction is dependent not only on the chemical nature of each reacting material and product but also on their physical states. For purposes of organizing thermochemical data it is necessary to define a standard heat of reaction which may be recorded as a characteristic property of the reaction and from which heats of reaction under other conditions may be calculated. The *standard heat of reaction* is defined as the *change in enthalpy resulting from the procedure of the reaction under a pressure of 1 atmosphere, starting and ending with all materials at a constant temperature of 18°C.* A temperature of 25°C is also frequently used as a standard reference temperature.

For example, 1 gram-atom (65.38 grams) of zinc may be allowed to react with 2073 grams of 1.0 molal aqueous hydrochloric acid containing 2.0 gram-moles of HCl. The reaction may be carried out in a calorimeter under atmospheric pressure with all reactants at an initial temperature of 18°C. During the course of the reaction the system will become heated, hydrogen gas will be evolved, and a 0.5 molal solution of zinc chloride will be formed. When the reaction is completed the resultant solution and the hydrogen gas may be cooled to 18°C. If no evaporation of water takes place it will be found that 35,900 calories will be evolved by the system. The net result of the reaction is the conversion of 2 moles of hydrochloric acid in aqueous solution into 1 mole of zinc chloride in aqueous solution and 1 mole of hydrogen gas at atmospheric pressure, all at a temperature of 18°C. The measured amount of heat absorbed represents the standard molal heat of reaction for this particular reaction, proceeding under atmospheric pressure in an aqueous solution of the specified concentration.

Exactly the same net result is obtained by allowing the above reaction to proceed in an electrolytic cell in which one electrode is zinc and the other platinum. An electric motor might be connected to the cell and be permitted to do work as the reaction proceeds. In this case the amount of heat evolved will be less than 35,900 calories by the heat equivalent of the electrical energy produced by the cell. However, the heat of reaction is the same and equal to the algebraic sum of the amounts of energy absorbed as heat and as electrical energy.

Conventions and Symbols. As pointed out in the preceding section, the heat of reaction accompanying a chemical change is dependent on the physical state of each reactant and product as well as on its chemical nature. For this reason, in order to define a heat of reaction it is necessary to specify completely the nature and state of each material involved. The following system of conventions and symbols, to be used in conjunction with the conventional chemical equation, is adopted for this purpose.

The formula of a substance appearing in an equation designates not only the nature of the substance but also the quantity which is involved in the reaction. Thus, H_2SO_4 indicates 1 mole of sulfuric acid, and $1\frac{1}{2}$ N_2 indicates $1\frac{1}{2}$ moles of nitrogen. All equations are written with the reactants on the left and the products on the right side. The value of the heat of reaction accompanying an equation is the heat of reaction resulting from the procedure of the reaction from the left to the right of the equation as written. If the reaction proceeds in the reverse direction the heat of reaction is of opposite sign.

Unless otherwise specified it is assumed that each reactant or product is in its normal state of aggregation at a temperature of 18°C and a pressure of 1 atmosphere.

The state of aggregation of a substance is indicated by a letter in parentheses following its chemical formula. Thus (g) indicates the gaseous state, (l) the liquid, and (s) the solid.

Additional information may accompany these letters in parentheses. Thus, S (rhombic) and C (diamond) indicate sulfur in the rhombic state and carbon as diamond, respectively, while S (monoclinic) and C (graphite) indicate monoclinic sulfur and solid graphitic carbon. In the case of a gas the pressure may be specified. Thus, CO_2 $(g, 2\ atm)$ indicates gaseous carbon dioxide under a pressure of two atmospheres.

The concentration of a substance in aqueous solution is indicated by its molality (m), by the number of moles of solvent (n_1) per mole of solute, or by the mole fraction of the solute (N_2). Thus, $(m = 0.1)$ following a chemical formula indicates that the substance is in aqueous solution with a molality of 0.1. The symbol $(n_1 = 200)$ indicates an aqueous solution with 200 moles of water per mole of solute. The symbol $(N_2 = 0.55)$ indicates an aqueous solution in which the mole fraction of the solute is 0.55. If the aqueous solution is highly dilute, such that additional dilution produces no thermal effect, the symbol (aq) follows the formula of the solute.

The concentration of a substance in nonaqueous solution is indicated by the above symbols, accompanied in the parentheses by the formula of the solvent. Thus, $(C_2H_6O, N_2 = 0.55)$ indicates that a substance is in alcoholic solution with a mole fraction of 0.55.

Ionic reactions are indicated in the usual manner, for example, H^+ and Ca^{++} for the positive hydrogen and calcium ions and Cl^- and SO_4^{--} for the negative chloride and sulfate ions, respectively.

As previously pointed out, positive values of q represent an absorption of heat by the system under consideration, that is, an increase in enthalpy of the system, $q = \Delta H$. Where the initial and final temperatures of the system are the same, a subscript may be used to designate this temperature, thus ΔH_{18} is the heat of reaction, or change in enthalpy, at 18°C and 1 atmosphere pressure.

When heat is evolved in a reaction, corresponding to a decrease in enthalpy, the reaction is termed *exothermic;* when heat is absorbed the reaction is said to be *endothermic.*

With the aid of the above symbols the states of a chemical reaction are indicated by the following:

$$Zn(s) + 2HCl(m = 1.0) = ZnCl_2(m = 0.5) + H_2(g, 1.0\ atm)$$
$$\Delta H_{18} = -35,900\ \text{calories per gram-mole}$$

This equation designates the changes occurring in the reaction described in the preceding section.

Heat of Formation. The *heat of formation* of a chemical compound is a special case of the standard heat of a chemical reaction wherein the reactants are the necessary elements and the compound in question is the only product formed. Heats of formation are always expressed with reference to a standard state. The molal heat of formation of a compound represents, unless otherwise stated, the heat of reaction, ΔH_f, when 1 mole of the compound is formed from the elements in a reaction beginning and ending at 18°C and at a pressure of 1 atmosphere with the reacting elements originally in the states of aggregation which are stable at these conditions of temperature and pressure. The heat of formation of a compound is positive when its formation from the elements is accompanied by an increase in enthalpy. A compound whose heat of formation is negative is termed an *exothermic compound*. If the heat of formation is positive it is called an *endothermic compound*.

For example, the molal heat of formation of liquid water is $-68,320$ calories per gram-mole. This means that when 2.016 grams of hydrogen gas combine with 16 grams of oxygen at a temperature of 18°C and a pressure of 1 atmosphere to form 18.016 grams of liquid water at the same temperature, the heat given off to the surroundings is 68,320 calories and the enthalpy of the system is decreased by 68,320 calories. It is obvious that this reaction will not proceed at a constant temperature but during its progress will be at a very high temperature, and the product formed will be temporarily in the vapor state. However, upon cooling to 18°C this sensible and latent heat appearing temporarily in the system itself is evolved and included in the heat of formation. If water vapor were the final product at 18°C the heat of formation would be numerically less by an amount equal to the heat of vaporization of water at 18°C. The heat of vaporization of water at 18°C is 10,565 calories per gram-mole. Therefore the heat of formation of water vapor at 18°C is $-68,320 + 10,565 = -57,755$ calories per gram-mole.

The basic thermochemical data of inorganic compounds are always presented in terms of standard heats of formation. In the International Critical Tables, Vol. V, page 169, are extensive tables giving the heats of formation of a great variety of inorganic compounds, both in pure states and in solutions of varying concentrations.

TABLE XIV

HEATS OF FORMATION AND SOLUTION

Reference Conditions: 18°C and 1 atmosphere pressure

Δ_{Hf} = heat of formation

Δ_{Hs} = heat of solution

in kilocalories per gram-mole

Multiply values by 1000 to obtain kilocalories per kilogram-mole and by 1800 to obtain Btu per pound-mole.

Data taken from Bichowski and Rossini, *Thermochemistry of Chemical Substances*, Reinhold Publishing Corp. (1936), with permission.

Abbreviations

c = crystalline state	dil = dilute solution
s = solid	∞ = infinite dilution
l = liquid	ppt = precipitated solid
g = gas	amorph = amorphous state

Compound	Formula	State	Δ_{Hf} Heat of Formation	Moles of Water	Δ_{Hs} Heat of Solution
Acetic acid	CH₃COOH	l	−117.4	200	−0.345
Aluminum chloride	AlCl₃	c	−166.8	600	−77.9
Aluminum hydroxide	Al(OH)₃	ppt	−304.9		
Aluminum nitride	AlN	c	−80.0		
Aluminum oxide	Al₂O₃	c	−380.0		
Aluminum silicate	Al₂O₃·SiO₂	sillimanite	−623.7		
Aluminum disilicate	Al₂O₃·2SiO₂	amorph	−979.0		
Aluminum disilicate	Al₂O₃·2SiO₂·2H₂O	amorph	−944.0		
Trialuminum disilicate	3Al₂O₃·2SiO₂	c	−1804.0		
Aluminum sulfate	Al₂(SO₄)₃	c	−770.0	∞	−1.26
Ammonia	NH₃	g	−11.0	200	−8.35
Ammonia	NH₃	l	−16.07	200	−3.28
Ammonium carbonate	(NH₄)₂CO₃	dil	−223.4		
Ammonium bicarbonate	NH₄HCO₃	c	−202.8	400	+6.81
Ammonium chloride	NH₄Cl	c	−74.95	∞	+3.818
Ammonium hydroxide	NH₄OH		−87.67		
Ammonium nitrate	NH₄NO₃	c	−87.13	∞	+6.47
Ammonium oxalate	(NH₄)₂C₂O₄	c	−266.6	400	+7.90
Ammonium sulfate	(NH₄)₂SO₄	c	−281.46	400	+2.38
Ammonium acid sulfate	NH₄HSO₄	c	−244.64	800	−0.56
Antimony trioxide	Sb₂O₃	c	−165.4	∞	−0.6
Antimony pentoxide	Sb₂O₅	c	−230.0	∞	+2.0
Antimony sulfide	Sb₂S₃	c	−35.7		
Arsenic acid	H₃AsO₄	c	−214.9	∞	+0.4
Arsenic trioxide	As₂O₃	octahedral	−154.0	∞	+7.5
Arsenic pentoxide	As₂O₅	c	−217.9	∞	−6.0
Barium acetate	Ba(C₂H₃O₂)₂	c	−357.6	∞	−5.75
Barium carbonate	BaCO₃	ppt	−290.8		
Barium carbonate	BaCO₃	c	−290.7		
Barium chlorate	Ba(ClO₃)₂	c	−176.6	∞	+6.74
Barium chloride	BaCl₂	c	−205.28	∞	−2.45
Barium chloride	BaCl₂·2H₂O	c	−349.0	∞	+4.80
Barium hydroxide	Ba(OH)₂	c	−225.9	400	−11.78
Barium oxide	BaO	c	−133.0	∞	−36.1
Barium peroxide	BaO₂	c	−152.4		
Barium silicate	BaSiO₃	glass	−363.0		
Barium sulfate	BaSO₄	ppt	−349.4	∞	+5.24
Barium sulfide	BaS	c	−111.2	∞	−7.16

TABLE XIV — *Continued*

Compound	Formula	State	Δ_{Hf} Heat of Formation	Moles of Water	Δ_{Hs} Heat of Solution
Bismuth oxide	Bi$_2$O$_3$	c	−137.1		
Boric acid	H$_3$BO$_3$	c	−251.6	∞	+5.4
Boron oxide	B$_2$O$_3$	c	−279.9	∞	−7.3
Bromine chloride	BrCl	g	+3.07		
Cadmium chloride	CdCl$_2$	c	−93.0	400	−3.06
Cadmium oxide	CdO	c	−65.2		
Cadmium sulfate	CdSO$_4$	c	−222.22	400	−10.69
Cadmium sulfide	CdS	c	−34.6		
Calcium acetate	Ca(C$_2$H$_3$O$_2$)$_2$		−357.6	∞	−7.64
Calcium aluminate	CaO·Al$_2$O$_3$	glass	−620.0		
Calcium aluminate	2CaO·Al$_2$O$_3$	glass	−857.0		
Calcium aluminate	3CaO·Al$_2$O$_3$	glass	−1098.0		
Calcium aluminum silicate	3CaO·Al$_2$O$_3$·2SiO$_3$	c	−1292.0		
Calcium aluminum silicate	CaO·Al$_2$O$_3$·6SiO$_2$	c	−3287.0		
Calcium carbide	CaC$_2$	c	−14.1		
Calcium carbonate	CaCO$_3$	ppt	−287.8		
Calcium carbonate	CaCO$_3$	calcite	−289.3		
Calcium chloride	CaCl$_2$	c	−190.6	∞	−18.517
Calcium chloride	CaCl$_2$·6H$_2$O	c	−623.45	∞	+9.05
Calcium fluoride	CaF$_2$	ppt	−290.2		
Calcium hydroxide	Ca(OH)$_2$	c	−236.0	∞	−3.06
Calcium nitrate	Ca(NO$_3$)$_2$	c	−224.04	400	−3.94
Calcium oxalate	CaC$_2$O$_4$	ppt	−329.7		
Calcium oxide	CaO	c	−151.7	∞	−19.0
Calcium phosphate	Ca$_3$(PO$_4$)$_2$	c	−983.0		
Calcium silicate	CaO·SiO$_2$	glass	−377.9		
Calcium silicate	2CaO·SiO$_2$	glass	−432.0		
Calcium sulfate	CaSO$_4$	c	−340.4	∞	−5.14
Calcium sulfide	CaS	c	−113.4	∞	−6.4
Carbon β-graphite	C	c	0		
Diamond	C	c	+.22		
Gas carbon	C	amorph	+1.72		
Charcoal (H$_2$ free)	C	amorph	+2.22		
Coke	C	amorph	+2.6		
Sugar carbon	C	amorph	+2.389		
Carbon monoxide	CO	g	−26.620	sat.	−2.76
Carbon dioxide	CO$_2$	g	−94.03	sat.	−4.76
Carbon disulfide	CS$_2$	g	+22.44		
Carbon disulfide	CS$_2$	l	+15.84		
Carbon tetrachloride	CCl$_4$	g	−25.7		
Carbon tetrachloride	CCl$_4$	l	−33.6		
Chloric acid	HClO$_3$	dil	−20.8		
Chromium chloride(ic)	CrCl$_3$	c	−143.0	∞	−40.0
Chromium chloride(ous)	CrCl$_2$	c	−103.1	∞	−18.6
Chromium oxide	Cr$_2$O$_3$	c	−273.0		
Chromium oxide	Cr$_2$O$_3$	amorph	−266.2		
Chromium trioxide	CrO$_3$	c	−139.3	80	−2.5
Cobalt oxide	CoO	c	−57.5		
Cobalt oxide	Co$_3$O$_4$	c	−196.5		
Cobalt sulfide	CoS	ppt	−22.3		
Copper acetate	Cu(C$_2$H$_3$O$_2$)$_2$	c	−216.4	∞	−2.4
Copper carbonate	CuCO$_3$	ppt	−141.4		
Copper chloride	CuCl$_2$	c	−53.4	800	−11.2
Copper chloride	CuCl	c	−34.3		
Copper oxide	CuO	c	−38.5		
Copper oxide	Cu$_2$O	c	−42.5		
Copper sulfate	CuSO$_4$	c	−184.7	800	−15.94

TABLE XIV — *Continued*

Compound	Formula	State	ΔH_f Heat of Formation	Moles of Water	ΔH_s Heat of Solution
Copper sulfide	CuS	c	−11.6		
Copper sulfide	Cu₂S	c	−18.97		
Copper nitrate	Cu(NO₃)₂	c	−73.1	200	−10.3
Cyanogen	C₂N₂	g	+71.4	∞	−7.0
Hydrobromic acid	HBr	g	−8.65	∞	−20.02
Hydrochloric acid	HCl	g	−22.06	∞	−17.627
Hydrocyanic acid	HCN	g	+30.90	100	−6.8
Hydrofluoric acid	HF	l	−71.0	600	−4.7
Hydriodic acid	HI	g	+5.91	∞	−19.28
Hydrogen oxide	H₂O	g	−57.801		
Hydrogen oxide	H₂O	l	−68.320		
Hydrogen peroxide	H₂O₂	l	−45.2	200	−0.46
Hydrogen sulfide	H₂S	g	−5.3	∞	−4.6
Iron acetate	Fe(C₂H₃O₂)₂	dil	−355.8		
Iron carbonate	FeCO₃	c	−172.6		
Iron chloride	FeCl₃	c	−96.4	1000	−31.7
Iron hydroxide	Fe(OH)₃	ppt	−197.3		
Iron oxide	Fe₂O₃	c	−198.5		
Iron oxide	FeO	c	−64.3		
Iron oxide	Fe₃O₄	magnetite	−266.90		
Iron silicate	FeO·SiO₂	c	−273.50		
Iron sulfate	Fe₂(SO₄)₃	dil	−653.20		
Iron sulfate	FeSO₄	c	−221.30	400	−14.68
Iron sulfide	FeS	c	−23.1		
Iron sulfide	FeS₂	pyrites	−35.5		
Lead acetate	Pb(C₂H₃O₂)₂	c	−233.4	400	−1.40
Lead carbonate	PbCO₃	c	−167.8		
Lead chloride	PbCl₂	c	−85.71	∞	+3.41
Lead nitrate	Pb(NO₃)₂	c	−106.89	400	+7.61
Lead oxide	PbO	c	−52.46		
Lead peroxide	PbO₂	c	−65.0		
Lead suboxide	Pb₂O	c	−51.3		
Lead sesquioxide	Pb₃O₄	c	−172.4		
Lead sulfate	PbSO₄	c	−218.5		
Lead sulfide	PbS	ppt	−22.3		
Lithium chloride	LiCl	c	−97.65	∞	−8.665
Lithium hydroxide	LiOH	c	−116.55	∞	−4.738
Magnesium carbonate	MgCO₃	c	−267.8		
Magnesium chloride	MgCl₂	c	−153.3	800	−35.92
Magnesium hydroxide	Mg(OH)₂	ppt	−218.7		
Magnesium oxide	MgO	amorph (?)	−146.1		
Magnesium silicate	MgSiO₃	c	−347.5		
Magnesium sulfate	MgSO₄	c	−304.95	400	−20.30
Manganese carbonate	MnCO₃	ppt	−207.50		
Manganese chloride	MnCl₂	c	−112.70	400	−16.0
Manganese oxide	MnO	c	−96.5		
Manganese oxide	Mn₃O₄	c	−345.0		
Manganese dioxide	MnO₂	c	−123.0		
Manganese dioxide	MnO₂	amorph	−115.5		
Manganese silicate	MnO·SiO₂	c	−301.3		
Manganese silicate	MnO·SiO₂	glass	−292.8		
Manganese sulfate	MnSO₄	c	−251.2	400	−13.8
Manganese sulfide	MnS	ppt	−47.0		
Manganese sulfide	MnS	c	−59.7		
Mercury acetate	Hg(C₂H₃O₂)₂	c	−197.1	∞	+4.0
Mercury chloride	HgCl₂	c	−53.4	∞	+3.3
Mercury chloride	Hg₂Cl₂	ppt	−63.15		

TABLE XIV — *Continued*

Compound	Formula	State	ΔH_f Heat of Formation	Moles of Water	ΔH_s Heat of Solution
Mercury nitrate	Hg(NO₃)₂	*dil*	−56.6		
Mercury nitrate	Hg₂(NO₃)₂·2H₂O	*c*	−206.5		
Mercury oxide	HgO	*c*	−21.6		
Mercury oxide	Hg₂O	*c*	−21.6		
Mercury sulfate	HgSO₄	*c*	−166.6		
Mercury sulfate	Hg₂SO₄	*c*	−176.5		
Mercury sulfide	HgS	*amorph*	−11.0		
Mercury thiocyanate	Hg(CNS)₂	*c*	+52.4		
Molybdenum dioxide	MoO₂	*c*	−130.0		
Molybdenum trioxide	MoO₃	*c*	−176.5		
Nickel cyanide	Ni(CN)₂	*c*	+232.4		
Nickel hydroxide	Ni(OH)₃	*ppt*	−163.2		
Nickel hydroxide	Ni(OH)₂	*ppt*	−129.8		
Nickel oxide	NiO	*c*	−58.4		
Nickel sulfide	NiS	*ppt*	−20.4		
Nickel sulfate	NiSO₄	*dil*	−231.1		
Nitrogen oxide	NO	*g*	+21.6		
Nitrogen oxide	N₂O	*g*	+19.65		
Nitrogen oxide	N₂O	*l*	+18.72		
Nitrogen pentoxide	N₂O₅	*g*	+0.6	∞	−29.8
Nitrogen pentoxide	N₂O₅	*c*	−13.1	400	−18.61
Nitrogen tetroxide	N₂O₄	*g*	+3.06		
Nitrogen trioxide	N₂O₃	*g*	+20.0		
Nitric acid	HNO₃	*l*	−41.66	∞	−7.53
Oxalic acid	H₂C₂O₄·2H₂O	*c*	−339.8	300	+8.58
Oxalic acid	H₂C₂O₄	*c*	−197.2	300	+2.3
Perchloric acid	HClO₄	*l*	−19.35	500	−20.3
Permanganic acid	HMnO₄	*dil*	−122.3		
Phosphoric acid (meta)	HPO₃	*c*	−224.8	∞	−9.8
Phosphoric acid (ortho)	H₃PO₄	*l*	−300.85	400	−5.31
Phosphoric acid (pyro)	H₄P₂O₇	*l*	−529.4	∞	−10.2
Phosphorous acid (hypo)	H₃PO₂	*l*	−139.0	450	−2.41
Phosphorous acid (ortho)	H₃PO₃	*l*	−225.86	∞	−2.94
Phosphorous trichloride	PCl₃	*g*	−70.0		
Phosphorous pentoxide	P₂O₅	*c*	−360.0		
Platinum chloride	PtCl₄	*c*	−62.6	∞	−19.4
Platinum chloride	PtCl	*c*	−17.0		
Potassium acetate	KC₂H₃O₂	*c*	−174.23	400	−3.404
Potassium carbonate	K₂CO₃	*c*	−274.24	400	−6.63
Potassium chlorate	KClO₃	*c*	−91.33	∞	+10.31
Potassium chloride	KCl	*c*	−104.361	∞	+4.404
Potassium chromate	K₂CrO₄	*c*	−333.4	2185	+4.87
Potassium cyanide	KCN	*c*	−28.3	200	+3.0
Potassium dichromate	K₂Cr₂O₇	*c*	−488.5	1600	+17.81
Potassium fluoride	KF	*c*	−134.51	400	−3.86
Potassium oxide	K₂O	*c*	−86.2	300	−75.00
Potassium sulfate	K₂SO₄	*c*	−342.66	∞	+6.32
Potassium sulfide	K₂S	*c*	−121.5	∞	+10.96
Potassium sulfite	K₂SO₃	*c*	−267.7	∞	−1.8
Potassium thiosulfate	K₂S₂O₃·H₂O	*c*	−270.5	∞	+4.5
Potassium hydroxide	KOH	*c*	−102.02	∞	−12.91
Potassium nitrate	KNO₃	*c*	−118.093	∞	+8.633
Potassium permanganate	KMnO₄	*c*	−192.9	400	+10.4
Selenium oxide	SeO₂	*c*	−56.36	2000	+0.93
Silicon carbide	SiC	*c*	−27.8		
Silicon tetrachloride	SiCl₄	*l*	−150.1		
Silicon tetrachloride	SiCl₄	*g*	−142.5		

TABLE XIV — *Continued*

Compound	Formula	State	ΔH_f Heat of Formation	Moles of Water	ΔH_s Heat of Solution
Silicon dioxide	SiO₂	quartz	−203.34		
Silver bromide	AgBr	c	−23.81		
Silver chloride	AgCl	c	−30.30	∞	+14.0
Silver nitrate	AgNO₃	c	−29.4	200	+5.44
Silver sulfide	Ag₂S	c	−5.5		
Sodium acetate	NaC₂H₃O₂	c	−171.12	400	−3.935
Sodium arsenate	Na₃AsO₄	c	−366.0	500	−15.65
Sodium tetraborate	Na₂B₄O₇	c	−742.6	900	−10.0
Sodium borax	Na₂B₄O₇·10H₂O	c	−1453.1	1600	+25.85
Sodium bromide	NaBr	c	−86.73	200	+0.601
Sodium carbonate	Na₂CO₃	c	−269.67	400	−5.63
Sodium carbonate	Na₂CO₃·10H₂O	c	−975.26	400	+16.39
Sodium bicarbonate	NaHCO₃	c	−226.2	1800	+4.1
Sodium chlorate	NaClO₃	c	−83.6	∞	+5.37
Sodium chloride	NaCl	c	−98.33	∞	+1.164
Sodium cyanide	NaCN	c	−22.73	200	+0.37
Sodium fluoride	NaF	c	−135.95	∞	+0.271
Sodium hydroxide	NaOH	c	−101.96	∞	−10.179
Sodium iodide	NaI	c	−69.28	200	−1.562
Sodium nitrate	NaNO₃	c	−111.72	∞	+5.051
Sodium oxalate	Na₂C₂O₄	c	−314.3	600	+4.12
Sodium oxide	Na₂O	c	−99.45	∞	−56.4
Sodium triphosphate	Na₃PO₄	c	−457.0	800	−14.0
Sodium diphosphate	Na₂HPO₄	c	−414.85	800	−5.41
Sodium monophosphate	NaH₂PO₄	dil	−364.64		
Sodium phosphite	Na₂HPO₃	c	−335.33	800	−9.30
Sodium selenate	Na₂SeO₄	c	−254.0	∞	−7.3
Sodium selenide	Na₂Se	c	−59.1	440	−18.8
Sodium sulfate	Na₂SO₄	c	−330.48	∞	−0.28
Sodium sulfate	Na₂SO₄·10H₂O	c	−1033.20	400	+18.9
Sodium bisulfate	NaHSO₄	c	−269.04	800	−1.74
Sodium sulfide	Na₂S	c	−89.8	400	−15.16
Sodium sulfide	Na₂S·4½H₂O	c	−417.54	1000	+5.0
Sodium sulfite	Na₂SO₃	c	−261.2	800	−2.7
Sodium bisulfite	NaHSO₃	dil	−206.5		
Sodium silicate	Na₂SiO₃	glass	−374.0		
Sodium silicofluoride	Na₂SiF₆	c	−669.0	660	+10.0
Sulfur dioxide	SO₂	g	−70.92	2000	−8.56
Sulfur trioxide	SO₃	g	−93.90	∞	−53.58
Sulfuric acid	H₂SO₄	l	−193.75	∞	−22.05
Tellurium oxide	TeO₂	c	−77.58	∞	+0.9
Tin chloride(ic)	SnCl₄	l	−127.4	200	−29.9
Tin chloride(ous)	SnCl₂	c	−81.1	200	−0.4
Tin oxide(ic)	SnO₂	c	−138.1		
Tin oxide(ous)	SnO	c	−67.7		
Titanium oxide	TiO₂	amorph	−214.1		
Titanium oxide	TiO₂	c	−218.0		
Tungsten oxide	WO₂	c	−130.5		
Vanadium oxide	V₂O₅	c	−437.0		
Zinc acetate	Zn(C₂H₃O₂)₂	c	−260.2	400	−9.8
Zinc carbonate	ZnCO₃	ppt	−193.1		
Zinc chloride	ZnCl₂	c	−99.55	400	−15.72
Zinc hydroxide	Zn(OH)₂	amorph	−153.5		
Zinc oxide	ZnO	c	−83.5		
Zinc sulfate	ZnSO₄	c	−233.4	400	−18.55
Zinc sulfide	ZnS	c	−44.0		
Zirconium oxide	ZrO₂	glass	−253.0		

In Table XIV are listed a few selected values of heats of formation. It will be observed in these tables that the heat of formation is made synonymous with increase in enthalpy. This results in a negative sign for all exothermic compounds. This practice is in agreement with the nomenclature of the American Standards Association which is here adopted. The opposite sign for heats of formation are to be found in some tables and in previous work of the authors.

When a compound is hydrated, its heat of formation in Table XIV includes the heat of formation of the water forming the hydrate. For example, the heat of formation of solid $CaCl_2 \cdot 6H_2O$ is given as $-623,450$. This represents the heat of reaction accompanying the formation at 18°C of 1 mole of $CaCl_2 \cdot 6H_2O$ from solid calcium and gaseous chlorine, hydrogen, and oxygen.

In the sixth column of Table XIV are values of the heats of solution of a few compounds. The heat of solution represents the change in enthalpy resulting from the formation of a solution of the specified concentration from 1 gram-mole of the compound and the number of gram-moles of liquid water indicated in the fifth column. By means of these values the total heat of formation of a dissolved material may be calculated. For example, the heat of solution of 1 gram-mole of $CaCl_2 \cdot 6H_2O$ in 400 gram-moles of water is $+9050$ calories. The heat of formation of the undissolved $CaCl_2 \cdot 6H_2O$ is $-623,450$ calories. The combined heat of formation and solution of 1 gram-mole of $CaCl_2 \cdot 6H_2O$ in 400 gram-moles of water is the algebraic sum of these two values, $-623,450 + 9,850 = -614,400$ calories.

If an element normally exists in more than one allotropic form at 18°C and atmospheric pressure, one of these forms is selected to serve as the basis of heats of formation throughout the table. This is equivalent to assigning a heat of formation of zero to the element in the particular form selected. For example, carbon may exist as graphite, diamond, or amorphous carbon. The β-graphite form has been selected as the basis of the heats of formation of all carbon compounds. This is indicated by presenting the heat of formation of β-graphite as zero in the table. On this basis all other forms of elementary carbon have positive heats of formation. The heat of formation, for example, of barium carbonate is the heat of reaction accompanying the formation of the compound from graphite and the other necessary elements. When more than one allotropic form of an element exists, the particular form on which the tables are based can be identified as the form for which the heat of formation is given as zero.

Laws of Thermochemistry. At a given temperature and pressure the quantity of energy required to decompose a chemical compound into its elements is precisely equal to that evolved in the formation of that compound from its elements.   This principle was first formulated by Lavoisier and Laplace in 1780.   For example, the heat of formation of sodium chloride is −98,330 calories.   Theoretically the same amount of energy is required to decompose sodium chloride into sodium and chlorine.

A corollary of this first principle of thermochemistry is known as the law of constant heat summation, which states that the net heat evolved or absorbed in a chemical process is the same whether the reaction takes place in one or several steps.   The total change in enthalpy of a system is dependent upon the temperature, pressure, state of aggregation, and state of combination at the beginning and at the end of the reaction and is independent of the number of intermediate chemical reactions involved.   This principle is known as the law of Hess, formulated in 1840.

By means of this principle it is possible to calculate the heat of formation of a compound from a series of reactions not involving the direct formation of the compound from the elements.   The majority of chemical compounds cannot be prepared in the pure state directly from the elements.   For example, the heat of formation of carbon monoxide cannot be measured directly because it cannot be prepared in a pure state from the elements without the concomitant formation of carbon dioxide.   However, pure carbon dioxide may be formed from its elements and the heat of reaction measured.   Also, pure carbon monoxide may be oxidized to form carbon dioxide and the heat of this reaction measured.   Thus, at 18°C,

$$C(\beta) + O_2(g) = CO_2(g), \qquad \Delta_{H_f} = -94,030 \text{ calories} \qquad (4)$$

$$CO(g) + \tfrac{1}{2}O_2(g) = CO_2(g), \qquad \Delta_{H_f} = -67,410 \text{ calories} \qquad (5)$$

From the law of Hess, the heat of formation of carbon monoxide is the same as the net heat of reaction accompanying (a) the formation of carbon dioxide from the elements and (b) the decomposition of this carbon dioxide into carbon monoxide and oxygen.   The first step is represented by Equation (4) with a heat of reaction, −94,030.   The second step is the reverse of the reaction of Equation (5) with a heat of reaction of +67,410.   The net heat of reaction of the two processes is −94,030 + 67,410 = −26,620 calories, the heat of formation of carbon monoxide.

The result of this application of the law of Hess is exactly the same as that obtained by treating Equations (4) and (5) as algebraic equalities and combining them as such.   If Equation (5) is subtracted from (4),

$$C \text{ (graphite)} + \tfrac{1}{2}O_2 = CO; \qquad \Delta_{H_f} = -26,620 \text{ calories} \qquad (6)$$

All thermochemical equations may be treated in this manner and combined with each other according to the rules of algebra. In this way it is possible to use the principle of Hess effectively in calculating the heat of a reaction for a complicated series of reactions. The heat of formation of any compound may be calculated if the heat of any one reaction into which it enters is known together with the heat of formation of each of the other compounds present in the reaction.

For example, it is practically impossible to measure directly the heats of formation of the hydrocarbons. However, a hydrocarbon may be oxidized completely to carbon dioxide and water and this heat of reaction measured. The heat of formation of the hydrocarbon may then be calculated from the heats of formation of the other compounds present in the reaction, namely, carbon dioxide and water. Thus,

$$CH_4(g) + 2O_2(g) = CO_2(g) + 2H_2O(l); \qquad \Delta H_a = -212{,}805 \text{ cal} \qquad (7)$$

$$C(\beta) + O_2(g) = CO_2(g); \qquad \Delta H_b = -94{,}030 \text{ cal} \qquad (8)$$

$$2H_2(g) + O_2(g) = 2H_2O(l); \qquad \Delta H_c = -136{,}640 \text{ cal} \qquad (9)$$

Equation (8) + Equation (9) − Equation (7) gives

$$C(\beta) + 2H_2(g) = CH_4(g) \qquad (10)$$

$$\Delta H_f = \Delta H_b + \Delta H_c - \Delta H_a = -94{,}030 - 136{,}640 - (-212{,}805)$$
$$= -17{,}865 \text{ cal}$$

By the principle of constant heat summation it is thus possible to calculate heats of reaction which cannot be determined by direct measurement. For example, the oxidation of linseed oil and the souring of milk are reactions which proceed very slowly. By measuring the heats of combustion of the initial reactants and final products it is possible to calculate the desired heat of reaction. Similarly, the heat of transition of a compound to an allotropic form such as graphite to diamond may be impossible to measure directly. However, the difference between the heats of combustion of these two forms of carbon will give the desired heat of transition. Since this method involves taking the difference between two large numbers, small errors in either number may result in a large error in the difference.

Standard Heat of Combustion. The heat of combustion of a substance is the heat of reaction resulting from the oxidation of the substance with elementary oxygen. All experimental thermochemical data on organic compounds are ordinarily expressed in terms of the heats of combustion of the respective compounds. These data are not necessarily the results of direct combustion experiments but may be indirectly obtained from measurements of other heats of reaction which lead to greater accuracy. Calculated heats of combustion data are available for many substances which are ordinarily noncombustible, for example, carbon tetrachloride.

The usually accepted *standard heat of combustion* is that resulting from the combustion of a substance, in the state which is normal at 18°C and atmospheric press're, with the combustion beginning and ending at a temperature of 18°C. The data ordinarily presented for standard heats of combustion correspond to final products of combustion which are in their normal states at a temperature of 18°C and at a pressure of 1 atmosphere. The major final products are generally gaseous carbon dioxide and liquid water.

The standard heat of combustion of a substance is dependent on the extent to which oxidation is carried. Unless otherwise specified, a value of standard heat of combustion corresponds to complete oxidation of all carbon to carbon dioxide and of all hydrogen to liquid water. Where other oxidizable elements are present it is necessary to specify the extent to which the oxidation of each is carried in designating a heat of combustion. For example, in oxidizing an organic chlorine compound, either gaseous chlorine or hydrochloric acid may be formed, depending on the conditions of combustion. If sulfur is present its final form may be SO_2, SO_3, or the corresponding acids.

The situation is further complicated by the fact that such products may form aqueous solutions with the water. Standard heats of combustion of compounds which contain elements such as S, Cl, I, Br, N, and F must always be accompanied by complete specification of the final state of each product.

In Table XV are a few selected values of standard heats of combustion of organic compounds. These values, in all cases, correspond to the formation of gaseous carbon dioxide from all carbon present in the compound. The hydrogen in the original compound forms liquid water or may be utilized, in part, to form mineral acids when such elements as Cl, S, or N are present. The final products of combustion which are formed from other elements are specifically designated. For example, the heat of combustion of gaseous chloroform is given as −96,250 calories per gram-mole. In the heading of this section of the table dealing with halogen derivatives the final state of chlorine is specified as HCl in dilute aqueous solution. Therefore, the heat of combustion of chloroform corresponds to the heat of the following reaction:

$$CHCl_3(g) + \tfrac{1}{2}O_2 + H_2O(aq) = CO_2 + 3HCl(aq)$$

$$\Delta_{Hc} = -96,250 \text{ calories} \tag{11}$$

TABLE XV

STANDARD HEATS OF COMBUSTION*

Reference conditions: 18°C, 1 atmosphere pressure

ΔH_c = heat of combustion in kilocalories per gram-mole

Multiply values by 1000 to obtain kilocalories per kilogram-mole and by 1800 to obtain Btu per pound-mole.

Abbreviations

s = solid l = liquid g = gaseous

Hydrocarbons

Final Products: $CO_2(g)$, $H_2O(l)$.

Compound	Formula	State	Heat of Combustion $q_c = \Delta H_c$
Carbon (graphite)	C	s	−94.030
Carbon (coke)	C	s	−96.630
Carbon monoxide	CO	g	−67.410
Hydrogen	H_2	g	−68.320
Methane	CH_4	g	−212.805
Acetylene	C_2H_2	g	−310.61
Ethylene	C_2H_4	g	−337.26
Ethane	C_2H_6	g	−372.83
Allylene	C_3H_4	g	−464.78
Propylene	C_3H_6	g	−492.01
Trimethylene	C_3H_6	g	−496.80
Propane	C_3H_8	g	−530.62
Isobutylene	C_4H_8	g	−646.20
Isobutane	C_4H_{10}	g	−683.37
n-Butane	C_4H_{10}	g	−688.00
Amylene	C_5H_{10}	l	−811.75
Cyclopentane	C_5H_{10}	l	−783.60
Isopentane	C_5H_{12}	g	−843.40
n-Pentane	C_5H_{12}	g	−845.33
Benzene	C_6H_6	l	−782.00
Hexene-1	C_6H_{12}	l	−963.90
Cyclohexane	C_6H_{12}	l	−940.00
n-Hexane	C_6H_{14}	l	−1002.40
Toluene	C_7H_8	l	−934.20
Cycloheptane	C_7H_{14}	l	−1087.30
n-Heptane	C_7H_{16}	l	−1150.77
o-Xylene	C_8H_{10}	l	−1090.90
m-Xylene	C_8H_{10}	l	−1090.90
p-Xylene	C_8H_{10}	l	−1087.10
n-Octane	C_8H_{18}	l	−1306.8
Mesitylene	C_9H_{12}	l	−1242.80
Naphthalene	$C_{10}H_8$	s	−1231.6
Decane	$C_{10}H_{22}$	l	−1619.4
Diphenyl	$C_{12}H_{10}$	s	−1493.30
Anthracene	$C_{14}H_{10}$	s	−1684.75
Phenanthrene	$C_{14}H_{10}$	s	−1674.95
Hexadecane	$C_{16}H_{34}$	s	−2556.1

* Values up to n-pentane taken from F. D. Rossini, Bureau of Standards *Journal of Research*. Other values on this page taken from M. P. Doss, *Physical Constants of Principal Hydrocarbons* Texas Co. (1942), with permission. Values on following pages taken from International Critical Tables, Vol. V, 162 (1929).

TABLE XV — *Continued*

Alcohols

Final Products: $CO_2(g)$, $H_2O(l)$.

Compound	Formula	State	Heat of Combustion $q_c = \Delta H_c$
Methyl alcohol	CH_4O	l	-170.9
Ethyl alcohol	C_2H_6O	l	-328.0
Glycol	$C_2H_6O_2$	l	-281.9
Allyl alcohol	C_3H_6O	l	-442.4
n-Propyl alcohol	C_3H_8O	l	-482.0
Isopropyl alcohol	C_3H_8O	l	-474.8
Glycerol	$C_3H_8O_3$	l	-397.0
n-Butyl alcohol	$C_4H_{10}O$	l	-639.0
Amyl alcohol	$C_5H_{12}O$	l	-787.0
Methyl-diethyl carbinol	$C_6H_{14}O$	l	-927.0

Acids

Final Products: $CO_2(g)$, $H_2O(l)$.

Compound	Formula	State	Heat of Combustion
Formic	CH_2O_2	l	-62.80
Oxalic	$C_2H_2O_4$	s	-60.15
Acetic	$C_2H_4O_2$	l	-208.00
Acetic anhydride	$C_4H_6O_3$	l	-431.90
Glycolic	$C_2H_4O_3$	s	-166.60
Propionic	$C_3H_6O_2$	l	-365.00
Lactic	$C_3H_6O_3$	s	-326.00
d-Tartaric	$C_4H_6O_6$	s	-275.10
n-Butyric	$C_4H_8O_2$	l	-520.00
Citric (anhydrous)	$C_6H_8O_7$	s	-474.50
Benzoic	$C_7H_6O_2$	s	-771.84
o-Phthalic	$C_8H_6O_4$	s	-771.00
Phthalic anhydride	$C_8H_4O_3$	s	-781.50
o-Toluic	$C_8H_8O_2$	s	-928.90
Palmitic	$C_{16}H_{32}O_2$	s	-2380.00
Stearolic	$C_{18}H_{32}O_2$	s	-2629.00
Elaidic	$C_{18}H_{34}O_2$	s	-2664.00
Oleic	$C_{18}H_{34}O_2$	l	-2669.00
Stearic	$C_{18}H_{36}O_2$	s	-2698.00

Carbohydrates — Cellulose — Starch, etc.

Final Products: $CO_2(g)$, $H_2O(l)$.

Compound	Formula	State	Heat of Combustion
d-Glucose (dextrose)	$C_6H_{12}O_6$	s	-673.0
l-Fructose	$C_6H_{12}O_6$	s	-675.0
Lactose (anhydrous)	$C_{12}H_{22}O_{11}$	s	-1350.8
Sucrose	$C_{12}H_{22}O_{11}$	s	-1349.6
			Calories per gram
Starch			-4179
Dextrin			-4110
Cellulose			-4181
Cellulose acetate			-4496

TABLE XV — *Continued*

Other CHO Compounds

Final Products: $CO_2(g)$, $H_2O(l)$.

Compound	Formula	State	Heat of Combustion $q_c = \Delta H_c$
Formaldehyde	CH_2O	g	-134.0
Acetaldehyde	C_2H_4O	g	-280.5
Acetone	C_3H_6O	l	-427.0
Methyl acetate	$C_3H_6O_2$	g	-397.7
Ethyl acetate	$C_4H_8O_2$	g	-544.4
Ethyl acetate	$C_4H_8O_2$	l	-538.5
Diethyl ether	$C_4H_{10}O$	l	-651.7
Diethyl ketone	$C_5H_{10}O$	l	-736.0
Phenol	C_6H_6O	s	-732.0
Pyrogallol	$C_6H_6O_3$	s	-639.0
Amyl acetate	$C_7H_{14}O_2$	l	-1040.0
Camphor	$C_{10}H_{16}O$	s	-1411.0

Nitrogen Compounds

Final Products: CO_2, $N_2(g)$, $H_2O(l)$.

Compound	Formula	State	$q_c = \Delta H_c$
Urea	CH_4N_2O	s	-152.0
Cyanogen	C_2N_2	g	-260.0
Trimethylamine	C_3H_9N	l	-578.6
Pyridine	C_5H_5N	l	-660.0
(1, 3, 5) Trinitrobenzene	$C_6H_3N_3O_6$	s	-664.0
(2, 4, 6) Trinitrophenol	$C_6H_3N_3O_7$	s	-620.0
o-Dinitrobenzene	$C_6H_4N_2O_4$	s	-703.2
Nitrobenzene	$C_6H_5NO_2$	l	-739.0
o-Nitrophenol	$C_6H_5NO_3$	s	-689.0
o-Nitroaniline	$C_6H_6N_2O_2$	s	-766.0
Aniline	C_6H_7N	l	-812.0
(2, 4, 6) Trinitrotoluene	$C_7H_5N_3O_6$	s	-821.0
Nicotine	$C_{10}H_{14}N_2$	l	-1428.0

Halogen Compounds

Final Products: $CO_2(g)$, $H_2O(l)$, dil. sol. of HCl.

Compound	Formula	State	$q_c = \Delta H_c$
Carbon tetrachloride	CCl_4	g	-90.08
Carbon tetrachloride	CCl_4	l	-82.18
Chloroform	$CHCl_3$	g	-96.25
Chloroform	$CHCl_3$	l	-89.20
Methyl chloride	CH_3Cl	g	-164.00
Chloracetic acid	$C_2H_3ClO_2$	s	-171.00
Ethylene chloride	$C_2H_4Cl_2$	g	-271.00
Ethyl chloride	C_2H_5Cl	g	-321.80

Sulfur Compounds

Final Products: CO_2, $SO_2(g)$, $H_2O(l)$.

Compound	Formula	State	$q_c = \Delta H_c$
Carbonyl sulfide	COS	g	-130.5
Carbon disulfide	CS_2	l	-246.6
Methyl mercaptan	CH_4S	g	-297.6
Dimethyl sulfide	C_2H_6S	g	-455.6
Ethyl mercaptan	C_2H_6S	g	-452.0

It may seem strange to assign negative values to heat of combustion, contrary to common usage in engineering. This practice is, however, consistent with the use of changes of enthalpy as synonymous with heats of formation, heats of vaporization, and so forth. Since combustion proceeds with a reduction in the enthalpy of the system, the value of ΔH must be negative and hence also the heat of combustion when the term is used synonymously with the change of enthalpy.

From the heat of combustion of a substance its heat of formation may be calculated if the heat of formation of each of the other products entering into the combustion reaction is known. Thus, in order to calculate the heat of formation of chloroform from Equation (11) it is necessary to know the heats of formation of CO_2, $H_2O(l)$, and $HCl(aq)$. These values may be obtained from Table XIV.

$$H_2(g) + \tfrac{1}{2}O_2(g) = H_2O(l); \qquad \Delta H_b = -68,320 \text{ cal} \qquad (12)$$

$$C(\beta) + O_2(g) = CO_2(g); \qquad \Delta H_c = -94,030 \text{ cal} \qquad (13)$$

$$\tfrac{1}{2}H_2(g) + \tfrac{1}{2}Cl_2(g) = HCl(aq); \qquad \Delta H_d = -39,687 \text{ cal} \qquad (14)$$

Equation (13) + 3 Equation (14) − Equation (11) − Equation (12) gives

$$C(\beta) + \tfrac{1}{2}H_2(g) + 1\tfrac{1}{2}Cl_2(g) = CHCl_3(g) \qquad (15)$$

or

$$\Delta H_f = \Delta H_c + 3\Delta H_d - \Delta H_a - \Delta H_b$$

$$\Delta H_f = (-94,030) + 3(-39,687) - (-96,250) - (-68,320)$$

$$= -48,521 \text{ cal}$$

Thus the heat of formation of $CHCl_3(g)$ is $-48,521$ calories per gram-mole.

In this manner a general equation may be derived for use in calculating the heat of formation of a compound $C_aH_bBr_cCl_dF_eI_fN_gO_hS_i$ from its heat of combustion. If ΔH_c is the heat of combustion of this compound corresponding to the final products, $CO_2(g)$, $H_2O(l)$, $Br(l)$, $Cl_2(g)$, $HF(aq)$, $I(s)$, $N_2(g)$, $SO_2(g)$, and ΔH_f its heat of formation, then,

$$\Delta H_f = -\Delta H_c - 94,030a - 34,160b - 41,540e - 70,920i \qquad (16)$$

If $\Delta H'_c$ is the heat of combustion of this compound corresponding to the final products $CO_2(g)$, $H_2O(l)$, $Br_2(g)$, $HCl(aq)$, $HF(aq)$, $I(s)$, $HNO_3(aq)$, $H_2SO_4(aq)$, then,

$$\Delta H_f = -\Delta H'_c - 94,030a - 34,160b + 3,825c - 5,527d - 41,540e$$
$$- 15,030g - 147,480i \qquad (17)$$

The coefficient of c in Equation (17) is the heat of formation of bromine vapor from liquid bromine, which is taken as the basis for the heats of formation of bromine compounds.

If neither Equation (16) nor (17) is applicable, the heat of formation of any compound may be calculated from its heat of combustion by the method demonstrated above in the algebraic combination of Equations (11), (12), (13), and (14).

Calculation of the Standard Heat of Reaction from Heats of Formation. The standard heat of reaction accompanying any chemical change may be calculated if the heats of formation of all compounds involved in the reaction are known. If the reference state of enthalpy for a compound at 18°C and 1 atmosphere pressure is taken as its separate component elements at 18°C and 1 atmosphere pressure and in their normal states of aggregation then the relative enthalpy of the compound is equal to its heat of formation. Thus, the standard heat of reaction, or enthalpy change is equal to the algebraic sum of the standard heats of formation of the products less the algebraic sum of the standard heats of formation of the reactants. When an element enters into a reaction its heat of formation is zero if its state of aggregation is that selected as the basis for the heats of formation of its compounds.

Illustration 1. Calculate the standard heat of reaction of the following:

$$HCl(g) + NH_3(g) = NH_4Cl(s)$$

From Table XIV the standard heats of formation are for

$$HCl(g), \quad \Delta_{Hf} = -22,060 = H_a$$
$$NH_3(g), \quad \Delta_{Hf} = -11,000 = H_b$$
$$NH_4Cl(s), \Delta_{Hf} = -74,950 = H_r$$

Substituting these values in the above reaction,

$$\Delta H_{18} = H_r - H_a - H_b$$
$$\Delta H_{18} = + (-74,950) - (-22,060) - (-11,000)$$
$$\text{or } \Delta H_{18} = -41,890 \text{ cal}$$

Illustration 2. Calculate the standard heat of reaction, ΔH_{18} of the following:

$$CaC_2(s) + 2H_2O(l) = Ca(OH)_2(s) + C_2H_2(g)$$

From Table XIV

$$CaC_2(s), \quad \Delta_{Hf} = \quad -14,100 \text{ cal} \quad = H_a$$
$$H_2O(l), \quad \Delta_{Hf} = \quad -68,320 \text{ cal} \quad = H_b$$
$$Ca(OH)_2(s), \Delta_{Hf} = \quad -236,000 \text{ cal} = H_r$$

The heat of formation of acetylene $C_2H_2(g)$, is calculated from the heat of combustion data of Table XV by means of Equation (17)

$$C_2H_2(g), \quad \Delta_{Hf} = +54,230 \text{ cal} = H_s$$

The enthalpy change ΔH_s is then equal to $H_r + H_s - H_a - 2H_b$

$$\Delta H_{18} + (-14,100) + 2(-68,320) = +(-236,000) + (54,230)$$
$$\Delta H_{18} = -236,000 + 54,230 - (-14,100) - 2(-68,320)$$
$$\text{or } \Delta H_{18} = -31,030 \text{ cal}$$

Illustration 3. Calculate the standard heat of reaction ΔH_{18} of the following:

$$2FeS_2(s) + 5\tfrac{1}{2}O_2(g) = Fe_2O_3(s) + 4SO_2$$

The standard heats of formation of the compounds are obtained from Table XIV, for

$$\begin{array}{lll}
FeS_2(s), & \Delta H_f = & -35,500 \text{ cal} = H_a \\
Fe_2O_3(s), & \Delta H_f = & -198,500 \text{ cal} = H_r \\
SO_2(g), & \Delta H_f = & -70,930 \text{ cal} = H_s
\end{array}$$

$$\Delta H_{18} = H_r + 4H_s - 2H_a$$
$$= -198,500 + 4(-70,920) - 2(-35,500) = -411,180 \text{ cal}$$

Calculation of the Standard Heat of Reaction from Heats of Combustion.

For a reaction between organic compounds the basic thermo-chemical data are generally available in the form of standard heats of combustion. The standard heat of reaction where organic compounds are involved can be conveniently calculated by using directly the standard heats of combustion instead of standard heats of formation. An energy balance is again employed, but in this case the standard reference state is not the elements but the products of combustion at 18°C and 1 atmosphere pressure and in the state of aggregation specified by the heat of combustion data. For example, the enthalpy of methane relative to its products of combustion, gaseous CO_2 and liquid H_2O, is equal to the negative value of its standard heat of combustion, or $+212,790$ calories per gram-mole. Therefore, in any equation involving combustible materials the formula of a compound may be replaced by its enthalpy relative to its products of combustion. The enthalpy of the products minus the enthalpy of the reactants is then equal to the standard heat of reaction, or the standard heat of combustion of the reactants minus the standard heat of combustion of the products is equal to the standard heat of reaction.

Illustration 4. Calculate the standard heat of reaction, ΔH_r, of the following:

$$\underset{\text{(ethyl alcohol)}}{C_2H_5OH(l)} + \underset{\text{(acetic acid)}}{CH_3COOH(l)} = \underset{\text{(ethyl acetate)}}{C_2H_5OOCCH_3(l)} + H_2O(l)$$

From Table XV heats of combustion are as follows:

$$\begin{array}{lll}
C_2H_5OH & \Delta H_c = & -328,000 \text{ cal} = -H_a \\
CH_3COOH & \Delta H_c = & -208,000 \text{ cal} = -H_b \\
C_2H_5OOCCH_3 & \Delta H_c = & -538,500 \text{ cal} = -H_r
\end{array}$$

Since the heat of reaction is the difference between the enthalpies of the products and the reactants,

$$\Delta H_{18} = \text{H}_r - \text{H}_a - \text{H}_b = 538,500 - 328,000 - 208,000 = 2500 \text{ cal}$$

In general, the heats of formation of organic compounds are small in comparison to the heats of combustion. Similarly, the heats of reaction are small in systems involving only combinations of organic compounds. Since these relatively small quantities can be determined only by the differences between the large heats of combustion it follows that they are rarely known with a high degree of accuracy. For example, the small heat of reaction which is the final result of Illustration 4 is of uncertain accuracy. An error of only 0.2 per cent in determining the heat of combustion of ethyl acetate results in an error of 40 per cent in the value of this heat of reaction.

When both organic and inorganic compounds appear in a reaction it is best to obtain the heat of reaction by means of heat of formation data. The heats of formation of the organic compounds may be calculated from their heats of combustion by means of Equation (16) or (17). The procedure is then the same as that demonstrated in Illustration 2.

Illustration 5. Calculate the standard heat of reaction, ΔH_{18}, of the following:

$$CH_3Cl(g) + KOH(s) = CH_3OH(l) + KCl(s)$$

The heats of formation of the inorganic compounds are obtained from Table XIV.

$$KOH(s); \quad \Delta \text{H}_f = -102,020 \text{ cal} = \text{H}_b$$
$$KCl(s); \quad \Delta \text{H}_f = -104,361 \text{ cal} = \text{H}_s$$

The heats of combustion of the organic compounds are obtained from Table XV and their heats of formation calculated by means of Equation (17); thus for $CH_3Cl(g)$

$$\Delta \text{H}_c = -164,000 \text{ cal}$$

$$\Delta \text{H}_f = 164,000 - 94,030 - 3(34,160) - 5527 = -38,037 \text{ cal} = \text{H}_a$$

for $CH_3OH(l)$, $\Delta \text{H}_c = -170,900 \text{ cal}$

$$\Delta \text{H}_f = 170,900 - 94,030 - 4(34,160) = -59,770 \text{ cal} = \text{H}_r$$

$$\Delta H_{18} = \text{H}_r + \text{H}_s - \text{H}_a - \text{H}_b = -59,770 + (-104,361) - (-38,037) - (-102,020)$$

$$\Delta H_{18} = -24,074 \text{ cal}$$

Heats of Neutralization of Acids and Bases. The neutralization of a dilute aqueous solution of NaOH with a dilute solution of HCl may be represented by the following thermochemical equation:

$$NaOH(aq) + HCl(aq) = NaCl(aq) + H_2O(l) \tag{18}$$

The heat of reaction, ΔH_r, may be calculated from the respective heats of formation in Table XIV. Thus,

$$NaOH(aq); \Delta H_f = -112,139 \text{ cal} = H_a$$
$$HCl(aq); \Delta H_f = -39,687 \text{ cal} = H_b$$
$$NaCl(aq); \Delta H_f = -97,166 \text{ cal} = H_r$$
$$H_2O(l); \Delta H_f = -68,320 \text{ cal} = H_s$$

From an energy balance

$$\Delta H_{18} = -97,166 + (-68,320) - (-112,139) - (-39,687)$$
$$\Delta H_{18} = -13,660 \text{ cal}$$

Heats of neutralization may be determined by direct calorimetric measurements in series of solutions of finite concentrations and these results extrapolated to correspond to infinite dilution. Following are heats of neutralization based, at least in part, on such measurements:

$$HCl(aq) + LiOH(aq) = LiCl(aq) + H_2O; \qquad \Delta H = -13,660$$
$$HNO_3(aq) + KOH(aq) = KNO_3(aq) + H_2O; \qquad \Delta H = -13,660$$
$$\tfrac{1}{2}H_2SO_4(aq) + KOH(aq) = \tfrac{1}{2}K_2SO_4(aq) + H_2O; \qquad \Delta H = -13,660$$

It will be noted that the heat of neutralization of strong acids with strong bases in dilute solution is practically constant when 1 mole of water is formed in the reaction. The explanation of this fact is that strong acids and bases and their salts are completely dissociated into their respective ions when in dilute aqueous solution. From this viewpoint, a dilute solution of hydrochloric acid consists of only hydrogen and chlorine ions in aqueous solution, and a dilute solution of sodium hydroxide consists of sodium and hydroxyl ions. Upon neutralization the resulting solution contains only sodium and chlorine ions. The reaction may be looked upon as ionic, and Equation (18) rewritten in ionic form.

$$Na^+(aq) + OH^-(aq) + H^+(aq) + Cl^-(aq)$$
$$= Na^+(aq) + Cl^-(aq) + H_2O(l) \qquad (19)$$

Canceling the similar terms from Equation (19),

$$OH^-(aq) + H^+(aq) = H_2O \qquad (20)$$
$$\Delta H_{18} = -13,660 \text{ cal}$$

Thus, the actual net result of the neutralization of dilute solutions of strong acids and bases is the production of water from hydrogen and hydroxyl ions. The accepted average value of the heat of neutralization is $-13,660$ calories per gram-mole of water formed. This corresponds to the heat of reaction of Equation (20), representing the formation of water from its ions.

In the neutralization of dilute solutions of weak acids and weak bases, the heat given off is less than 13,660 cal. For example, in the neutralization of hydrocyanic acid with sodium hydroxide, the heat evolved is only 2661 calories per gram-mole of water formed. The unevolved heat may be considered as the heat required to complete the dissociation of hydrogen cyanide into hydrogen and cyanide ions as neutralization proceeds. As a hydrogen ion from hydrogen cyanide is neutralized by a hydroxyl ion, more hydrogen cyanide ionizes until neutralization is complete. This ionization of hydrogen cyanide requires the absorption of heat at the expense of the heat evolved in the union of hydrogen and hydroxyl ions.

Thermoneutrality of Salt Solutions. When dilute aqueous solutions of two neutral salts are mixed there is no thermal effect provided there is no precipitation, or evolution of gas. However, upon evaporation of a mixture of such solutions four crystalline salts will be found, indicating that double decomposition or metathesis has taken place. For example,

$$NaCl(aq) + KNO_3(aq) = NaNO_3(aq) + KCl(aq); \quad \Delta H = 0 \quad (21)$$

In dilute aqueous solutions it may be considered that each of these four salts is completely ionized and Equation (21) written in ionic form,

$$Na^+(aq) + Cl^-(aq) + K^+(aq) + NO_3^-(aq) = Na^+(aq)$$
$$+ NO_3^-(aq) + K^+(aq) + Cl^-(aq); \quad \Delta H = 0 \quad (22)$$

From this viewpoint it is evident that mixing such systems actually produces no change, the initial and final solutions consisting of the same four ions. It is only upon concentration of the solution and reassociation of the ions that the metathesis leads to a definite change in the nature of the system.

The experimentally observed fact that dilute solutions of neutral salts of strong acids and bases may be mixed without thermal effect is termed the *thermoneutrality of salt solutions.*

Heats of Formation of Ions. Equation (20) represents the formation of one mole of water from the combination of hydrogen and hydroxyl ions. The average value of this heat of reaction has been determined as −13,660 calories per gram-mole. The heat of formation of water from gaseous oxygen and hydrogen is given in Table XIV as 68,320 calories per gram-mole.

$$H_2(g) + \tfrac{1}{2}O_2(g) = H_2O(l); \quad \Delta H_a = -68,320 \text{ cal} \quad (a)$$

$$OH^-(aq) + H^+(aq) = H_2O(l); \quad \Delta H_b = -13,660 \text{ cal} \quad (b)$$

Equation (a) − Equation (b) gives

$$H_2(g) + \tfrac{1}{2}O_2(g) = H^+(aq) + OH^-(aq)$$
$$\Delta_{Hf} = -68,320 - (-13,660) = -54,660 \text{ cal} \qquad (23)$$

Thus the combined heats of formation of the hydrogen and hydroxyl ions from elementary hydrogen and oxygen is −54,660 cal. By arbitrarily assigning a zero value to the heat of formation of the hydrogen ion the heat of formation of the hydroxyl ion is obtained. Thus, by definition,

$$\tfrac{1}{2}H_2(g) = H^+; \quad \Delta_{Hf} = 0 \qquad (24)$$

$$\tfrac{1}{2}H_2(g) + \tfrac{1}{2}O_2(g) = OH^-(aq); \quad \Delta_{Hf} = -54,660 \text{ cal} \qquad (25)$$

On this basis the relative heats of formation of the other ions of highly dissociated acids and bases may be calculated. For example, from Table XIV, the heat of formation of NaOH (aq) is −112,139 calories per gram-mole. Since sodium hydroxide is completely dissociated into sodium and hydroxyl ions when in dilute solution, the formation of 1 mole of NaOH (aq) from the elements is, in actual effect, the formation of 1 mole of sodium and 1 mole of hydroxyl ions. Thus,

$$Na(s) + \tfrac{1}{2}O_2(g) + \tfrac{1}{2}H_2(g) = Na^+(aq) + OH^-(aq)$$
$$\Delta_{Hf} = -112,139 \text{ calories (gram-moles)} \qquad (26)$$

Combining (26) and (25),

$$Na(s) = Na^+; \quad \Delta_{Hf} = -57,479 \text{ cal} \qquad (27)$$

Therefore, the heat of formation of the sodium ion is −57,479 calories per gram-mole. In a similar manner the heats of formation of other ions may be calculated, based on the assignment of a value of zero to hydrogen. Heats of formation of a few common ions, taken from the data of Bichowsky and Rossini,[1] are given in Table XVI.

The heat of formation in dilute aqueous solution of any compound which is completely dissociated under these conditions is equal to the sum of the heats of formation of its ions. From the data of Table XVI the heats of formation of such compounds may be predicted.

Illustration 6. Calculate, from the data of Table XVI, the heat of formation of barium chloride in dilute solution.

Since BaCl₂ may be assumed to be completely disassociated in dilute solution its heat of formation in dilute solution is equal to the sum of the heats of formation of

---

[1] *Thermochemistry of Chemical Substances*, Reinhold Publishing Co. (1936). With permission.

TABLE XVI

HEATS OF FORMATION OF IONS

Δ_{Hf} = heat of formation, kilocalories per gram-mole

Cations			Anions		
Ion	Formula	Δ_{Hf}	Ion	Formula	Δ_{Hf}
Aluminum	Al^{+++}	-126.3	Acetate	CH_3COO^-	-117.506
Ammonium	NH_4^+	-31.455	Bicarbonate	HCO_3^-	-164.6
Barium	Ba^{++}	-128.36	Bisulfate	HSO_4^-	-213.3
Calcium	Ca^{++}	-129.74	Bisulfite	HSO_3^-	-149.0
Hydrogen	H^+	0	Bromide	Br^-	-28.67
Iron	Fe^{++}	-20.6	Carbonate	CO_3^{--}	-160.3
Iron	Fe^{+++}	-9.3	Chloride	Cl^-	-39.687
Lithium	Li^+	-66.628	Fluoride	F^-	-78.20
Magnesium	Mg^{++}	-110.23	Hydroxide	OH^-	-54.66
Manganese	Mn^{++}	-49.2	Iodide	I^-	-13.37
Manganese	Mn^{+++}	-25.0	Nitrite	NO_2^-	-25.3
Potassium	K^+	-60.27	Nitrate	NO_3^-	-49.19
Sodium	Na^+	-57.479	Phosphate	PO_4^{---}	-297.6
Zinc	Zn^{++}	-36.3	Sulfate	SO_4^{--}	-215.8
			Sulfite	SO_3^{--}	-148.5

one barium ion and two chlorine ions. From Table XVI,

$$Ba^{++}; \quad \Delta_{Hf} = -128,360 \text{ cal}$$

$$Cl^-; \quad \Delta_{Hf} = -39,687 \text{ cal}$$

$$Ba^{++} + 2Cl^- = BaCl_2(aq)$$

$$BaCl_2(aq); \quad \Delta_{Hf} = (-128,360) + 2(-39,687)$$

$$\text{or} \quad \Delta_{Hf} = -207,734 \text{ cal}$$

The heat of formation of a substance which is soluble and highly ionized in water may be calculated from its heat of formation in dilute solution from the heat of formation of its ions provided its standard heat of solution is known. Since experimental determinations of the heats of solution are relatively easy, a simple method is provided for estimating the heat of formation of many inorganic compounds. The heat of formation in infinitely dilute solution is calculated from the ionic heats of formation and the heat of solution subtracted from this value. For example, the standard heat of solution of $BaCl_2(s)$ in an infinite amount of water is $-2,450$ calories per gram-mole. From the result of Illustration 6, the heat of formation of $BaCl_2(s)$ is $-207,734 + 2,450$ or $-205,284$ calories per gram-mole.

Chemical reactions which take place between strong acids and bases and their salts in dilute aqueous solution may be treated as ionic, and heats of reaction may be calculated directly from ionic heats of formation. This method is particularly desirable when dealing with analytical data for complex solutions. By treating the reactions as ionic it is unnecessary to formulate hypothetical combinations of the analytically determined ionic quantities.

Heats of Gaseous Dissociation. At high temperatures the elementary gases are known to dissociate into their atomic states with absorption of great amounts of energy. Upon cooling, these monatomic gases rapidly recombine to form the original elementary gas. From an industrial standpoint, the most interesting phenomenon of this type is the dissociation of normal gaseous hydrogen, $H_2(g)$, to form a monatomic gas, $H(g)$. Such molecular dissociations in the gaseous state are accompanied by large absorptions of energy. Conversely, the reassociation of the dissociated products is accompanied by a great evolution of energy. In Table XVII are data for the heats of reaction of several gaseous dissociations.

TABLE XVII

HEATS OF GASEOUS DISSOCIATION AT 25°C

$$\tfrac{1}{2}O_2(g) \;= O(g); \qquad \Delta H_{18} = 59{,}100 \text{ cal}$$
$$\tfrac{1}{2}H_2(g) \;= H(g); \qquad \Delta H_{18} = 51{,}900 \text{ cal}$$
$$\tfrac{1}{2}N_2(g) \;= N(g); \qquad \Delta H_{18} = 85{,}100 \text{ cal}$$
$$\tfrac{1}{2}Cl_2(g) = Cl(g); \qquad \Delta H_{18} = 28{,}900 \text{ cal}$$
$$\tfrac{1}{2}I_2(g) \;= I(g); \qquad \Delta H_{18} = 18{,}140 \text{ cal}$$
$$\tfrac{1}{2}Br_2(g) = Br(g); \qquad \Delta H_{18} = 26{,}880 \text{ cal}$$

THERMOCHEMISTRY OF SOLUTIONS

The enthalpy change accompanying the dissolution of a substance is termed its *heat of solution*, or better, its heat of dissolution. If chemical combination takes place between the solvent and the substance being dissolved, the heat of dissolution will include the heat of solvation or the heat of hydration accompanying this combination. If ionization takes place the heat of solution will also include the energy of ionization. The heat of solution of a neutral, non-dissociating salt is generally positive, that is, heat is absorbed from the surroundings in the isothermal formation of the solution, or the solution cools if dissolution proceeds adiabatically. The dissolution of such a material is analogous to the evaporation of a liquid in that the result of each process is the breaking down of a condensed structure into a state of great dispersion. In each process energy is absorbed in overcoming the attractive forces between the structural units of the condensed state.

Heats of solvation, especially in aqueous systems, are generally negative and relatively large. For this reason the heat of solution of a substance which forms a solvate or hydrate has generally a large negative value, indicating the evolution of heat when the unhydrated substance is dissolved.

The enthalpy change when a substance is dissolved depends upon the amounts and natures of the solute and solvent, upon the temperature, and upon the initial and final concentrations of the solution. The numerical value of the heat effect, therefore, requires an exact and complete statement of all reference conditions.

Standard Integral Heats of Solution. Arbitrarily the *standard integral heat of solution* is defined as the change in enthalpy of the system when one mole of solute is dissolved in n_1 moles of solvent with the temperature maintained at 18°C and the pressure at one atmosphere.

The numerical value of the integral heat of solution depends upon the value of n_1. As successive equal increments of solvent are added to a given mass of solute the heat evolved with each addition progressively diminishes until a high dilution is attained, usually about 100 or 200 moles of solvent per mole of solute, following which no further heat effect is perceptible. The integral heat of solution approaches a maximum numerical value at infinite dilution. This limiting value is termed the integral heat of solution at infinite dilution.

In Figs. 47 through 51 are presented the standard integral heats of solution of common acids, bases, and salts in water. Integral heats of solution are determined calorimetrically by measuring the heat of solution of a solute in enough solvent to form a relatively concentrated solution. The heat of dilution accompanying the addition of solvent to this concentrated solution is then measured. The heat of solution at any desired concentration is obtained by adding, algebraically, the observed heats of solution and dilution.

It is evident that the integral heat of solution as defined above is the enthalpy of the solution containing 1.0 mole of solute, relative to the pure components at the same temperature and pressure. Thus, the enthalpy of a solution relative to any selected reference state is expressed by

$$H_s = n_1 H_1 + n_2 H_2 + n_2 \Delta H_{si} \tag{28}$$

where

H_s = enthalpy of $n_1 + n_2$ moles of solution of components 1 and 2

H_1, H_2 = molal enthalpies of pure components 1 and 2 at the temperature of the solution

ΔH_{si} = integral heat of solution of component 2

Heat of Formation of a Compound in Solution. By combination of the data of Table XIV with those of Figs. 47 to 51 it is possible to calculate the heat of formation of a compound in an aqueous solution of specified concentration. This total standard heat of formation is the sum of the standard heat of formation of the solute, ΔH_f, and its standard integral heat of solution, ΔH_{si}, at the specified concentration.

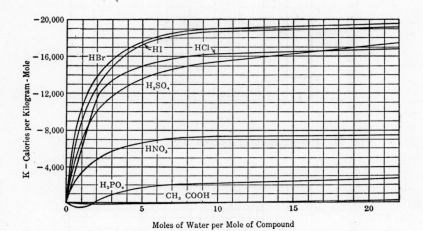

FIG. 47.   Integral heats of solution of acids in water at 18°C.

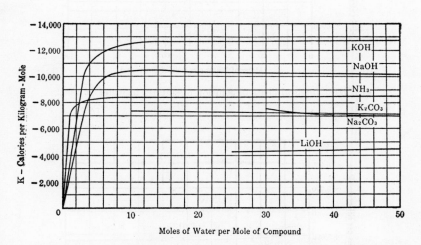

FIG. 48.   Integral heats of solution of alkalies in water at 18°C.

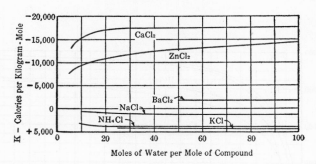

FIG. 49.   Integral heats of solution of chlorides in water at 18°C.

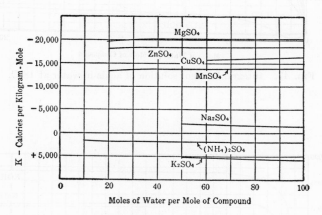

FIG. 50.   Integral heats of solution of sulfates in water at 18°C.

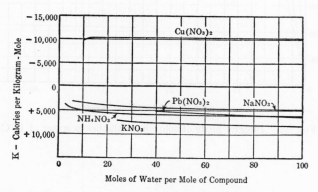

FIG. 51.   Integral heats of solution of nitrates in water at 18°C.

Illustration 7. (a) Calculate the heat of formation of H_2SO_4 to form an aqueous solution containing 5 moles of water per mole of H_2SO_4.

(b) Calculate the heat of dilution of the solution of part (a) to a concentration of 1 mole of H_2SO_4 in 20 moles of water.

(a) From Table XIV,

$$H_2SO_4(l); \quad \Delta H_f = -193,750 \text{ cal}$$

From Fig. 47,

$$H_2SO_4(n_1 = 5.0); \quad \Delta H_s = -13,600 \text{ cal}$$
$$H_2(g) + S(s) + 2O_2(g) + 5H_2O = H_2SO_4(n_1 = 5)$$
$$\Delta H_{18} = -193,750 - 13,600 = -207,350 \text{ cal}$$

Since the same amount of water appears on both sides of the reaction its heat of formation does not appear in the equation.

(b) From Fig. 47,

$$H_2SO_4(n_1 = 20); \quad \Delta H_s = -17,200 \text{ cal} = H_a$$
$$H_2SO_4(n_1 = 5); \quad \Delta H_s = -13,600 \text{ cal} = H_r$$
$$H_2SO_4(n_1 = 5) + 15H_2O = H_2SO_4(n_1 = 20)$$
$$\Delta H_{18} = H_r - H_a = -17,200 - (-13,600)$$
$$\Delta H_{18} = -3600 \text{ cal}$$

Heat of Solution of Hydrates. If a solute forms a hydrate, the standard heat of solution of the hydrate is the difference between the heat of solution of the anhydrous substance and its heat of hydration. The heat of hydration is calculated from the data of Table XIV as the difference between the heat of formation of the hydrated compound and the sum of the heats of formation of the anhydrous substance and of the water of hydration.

Illustration 8. Calculate the standard heat of solution of $CaCl_2 \cdot 6H_2O$ to form a solution containing 10 moles of water per mole of $CaCl_2$.

From Table XIV,

$$CaCl_2; \qquad \Delta H_f = -190,600 = H_a$$
$$H_2O; \qquad \Delta H_f = -68,320 = H_b$$
$$CaCl_2 \cdot 6H_2O; \qquad \Delta H_f = -623,450 = H_r$$

The enthalpy change accompanying hydration is then $H_r - H_a - 6H_b$ or

$$\Delta H_{18} = -623,450 - (-190,600) - 6(-68,320)$$

or

$$\Delta H_{18} = -22,930 \text{ calories per gram-mole} = H_a'$$

This heat of reaction represents the molal enthalpy of $CaCl_2 \cdot 6H_2O$ relative to $CaCl_2$ and H_2O at 18°C. The enthalpy of $CaCl_2(n_1 = 10)$ relative to the same reference substances and state is obtained from Fig. 49,

$$CaCl_2(n_1 = 10); \quad \Delta H_s = -15,500 \text{ calories per gram-mole} = H_r'$$

For the reaction

$$CaCl_2 \cdot 6H_2O + 4H_2O \rightarrow CaCl_2(n_1 = 10)$$

$$\Delta H_{18} = H_r' - H_a' = -15,500 - (-22,930)$$

$$\Delta H_{18} = 7430 \text{ cal per gram-mole}$$

Heats of Mixing. Heats of solution in a system in which both solute and solvent are liquids are termed heats of mixing. Heats of mixing are frequently expressed on a unit weight rather than a molal basis. If 10 grams of glycerin are mixed with 90 grams of water, the heat evolved is 191 calories. Therefore, the heat of mixing water and glycerin to form a solution containing 10 per cent glycerin is -1.91 calories per gram of solution formed. The integral heat of solution of either component may be calculated from heat of mixing data.

Illustration 9. The heat of mixing of water and glycerin to form a solution containing 40% glycerin is -4.50 calories per gram of solution. Calculate the integral heats of solution of glycerin and of water at this concentration.

Basis   100 grams of solution.

Heat of mixing $= -450$ calories

Heat of solution of glycerin in water $= \dfrac{-450}{40} = -11.25$ calories per gram

Heat of solution of water in glycerin $= \dfrac{-450}{60} = -7.5$ calories per gram

Enthalpy-Concentration Charts. W. L. McCabe[2] has called attention to the usefulness of the enthalpy-concentration diagram for binary solutions. In this chart the enthalpy per unit weight of solution is plotted against concentration for a series of constant-temperature lines. Once such a diagram for a given binary solution has been constructed, calculations of the heat effects involved in changing the concentrations and temperatures of the solution become simple and rapid. The expenditure of the time required in constructing such a diagram is well justified in case of specialization on a given system.

In constructing an enthalpy-concentration chart it is convenient to choose as a reference state the pure components at $t_0°C$ and their respective vapor pressures. With heats of solution measured at 18°C it is convenient to establish the first enthalpy curve at this temperature and subsequently from this base line derive all other constant-temperature lines. This is done directly from Equation (28), employing standard integral heats of solution together with heat capacity data on both the pure components and the solutions.

In Figs. 52, 53, and 54 are enthalpy-concentration charts for systems of hydrochloric acid, sulfuric acid, and calcium chloride in water.

[2] *Trans. Am. Inst. Chem. Eng.*, **31**, 129–162 (1935).

Illustration 10.   Calculate the final temperature when 5.0 lb of 10% HCl at 60°F are mixed with 8 lb of 30% HCl at 100°F.   From Fig. 52

$$\text{Enthalpy of } 10\% \text{ HCl at } 60°F = (5) \, (-60) \quad = - \; 300$$
$$\text{Enthalpy of } 30\% \text{ HCl at } 100°F = (8) \, (-175) = -1400$$
$$\text{Total enthalpy of mixture} \ldots \ldots \ldots = -1700$$
$$\text{Enthalpy per pound of final mixture} = \frac{-1700}{13} = -131$$

Final concentration $= (0.5 + 2.4)/13 = 22.3\%$ HCl
From Fig. 52 this enthalpy corresponds to a temperature of 90°F

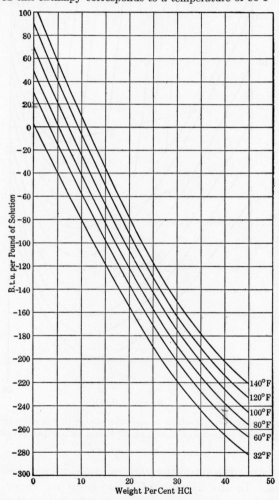

Fig. 52.   Enthalpy-concentration chart of hydrochloric acid solutions relative to pure $HCl(g)$ and pure $H_2O(l)$ at 32°F and 1 atmosphere.

In the enthalpy-concentration diagrams for the systems sulfuric acid-water, and calcium chloride-water, Figs. 53 and 54, steep sloping lines are shown in the two-phase region of vapor and solution. These lines represent the enthalpy of the entire system in Btu per pound, including the liquid and vapor phases for isothermal conditions at one atmosphere

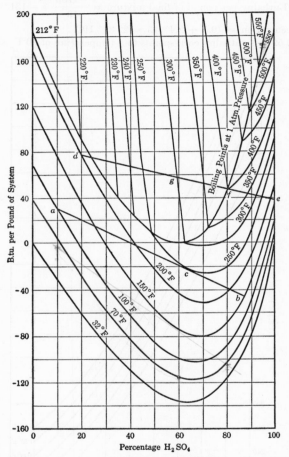

FIG. 53. Enthalpy-concentration of sulfuric acid-water system relative to pure components (water and H_2SO_4 at 32°F and own vapor pressures).

(Reproduced in " C.P.P. Charts ")

pressure and meet the saturation curves at the normal boiling points of the solutions. In the absence of other gases no vapor phase exists when the total vapor pressure of the system is less than one atmosphere.

For the illustrated temperature-concentration range of the sulfuric acid system the vapors are essentially pure water. Thus, a point on the chart in the two-phase region indicates a mixture of water vapor and solution saturated at the indicated temperature and atmospheric pressure. The enthalpy of the mixture is the sum of the enthalpies of the

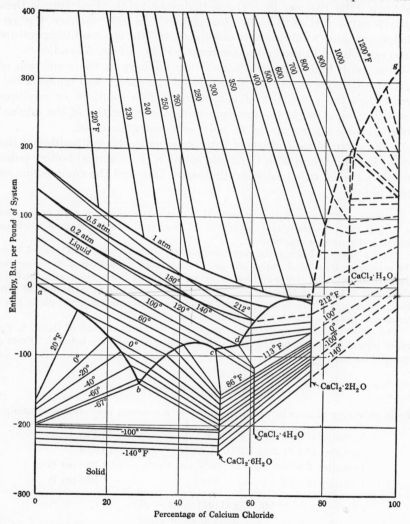

FIG. 54. Enthalpy of calcium chloride-water system. Modified from data of Bošnjakovic, *Technische Thermodynamik*, Theodor Steinkopff, Dresden and Leipzig (1937).

saturated solution and the water vapor. For example, a mixture at 250°F which has an overall composition of 45% H_2SO_4 consists of .06 lb of water vapor plus .94 lb of solution saturated at 250°F and containing 48% H_2SO_4. The enthalpy of the mixture is the sum of the enthalpy of the saturated solution at 250°F which is 8 Btu per lb plus the enthalpy of the superheated water vapor which from the steam tables is 1169 Btu per lb. The enthalpy of 1.0 lb of mixture is then $(.06)(1169) + (.94)(8.0) = 78$ Btu per lb. In this manner the enthalpies of other mixtures were calculated and the constant temperature lines established in the two-phase regions of both Figs. 53 and 54.

Since both enthalpies and masses are additive in the formation of mixtures it follows from the principles of the energy and material balances that the properties of a mixture of two solutions or mixtures must lie on a straight tie-line connecting the properties of the original solutions or mixtures on the enthalpy-concentration chart.

If the vapor phase consists of two components, the composition of the vapor in equilibrium with the liquid solution at its normal boiling point must be known in establishing the vapor lines and the composition of both phases.

Illustration 11. One pound of pure H_2SO_4 at 150°F is mixed with one pound of 20% H_2SO_4 solution initially at 200°F. Calculate the temperature of adiabatic mixing and the water evaporated.

From a material balance the resultant mixture contains 60% H_2SO_4 based upon the combined liquid and vapor phases. By constructing a tie-line de connecting the enthalpies of the two initial solutions and noting its intersection with the 60% abscissa it will be seen that the resultant enthalpy per pound of mixture is 58 Btu. The temperature of the mixture is 300°F and the corresponding liquid phase has a composition of 63% H_2SO_4 with a boiling point of 300°F.

The mass of water evaporated, y, (neglecting the small amount of H_2SO_4 in the vapor phase) can be calculated from either a material or energy balance. From a material balance,

$$0.80 = y + 0.37 \ (2 - y)$$
$$y = 0.095 \text{ lb water vapor}$$

From an energy balance in which enthalpies at a constant pressure are equated,

Enthalpy of 1 lb H_2SO_4 at 150°F	=	40 Btu
Enthalpy of 1 lb 20% H_2SO_4 at 200°F	=	76 Btu
Enthalpy of water vapor at 300°F and 1 atm	=	1194 Btu per lb
Enthalpy of 63% solution at 300°F	=	1 Btu per lb

$$116 = y \ (1194) + (2 - y)$$
$$y = 0.095 \text{ lb water vapor}$$

Maximum Temperature in Mixing Solutions. The maximum temperature attainable in mixing two solutions of the same components but of different temperatures and concentrations may be readily obtained from an enthalpy-concentration diagram by constructing a tie-line connecting the enthalpies corresponding to the two initial solutions. The point of meeting or tangency of this line with the highest isotherm on the diagram is the desired maximum temperature for adiabatic mixing. For example, in Fig. 53, a tie-line ab is constructed joining the enthalpy a of a 10% H_2SO_4 solution at 100°F and the enthalpy b of an 87% solution at 150°F. The maximum temperature attainable occurs at point c, corresponding to the 250°F isotherm, and at a composition of 63%. The relative weights of the two solutions to give this concentration are obtained from a material balance. Thus, for 1 pound of 10% H_2SO_4 solution 2.21 pounds of an 87% solution are required to produce 3.21 pounds of a 63% solution.

If the tie-line crosses the region of partial vaporization the maximum temperature corresponds to the higher temperature of intersection with the saturation line. Thus in Fig. 53 the tie-line de connects enthalpy d of a 20% solution at 200°F with the enthalpy e of pure H_2SO_4 at 150°F. The maximum temperature occurs at point f corresponding to 400°F and a solution of 80% H_2SO_4. The relative amount of the two solutions to give this concentration is obtained from a material balance. Thus, for 1 pound of 20% solution, 3.0 pounds of H_2SO_4 are required to give 4.0 pounds of solution having a concentration of 80%.

Solid-Liquid Systems. The enthalpy composition chart of a complex system is shown in Fig. 54 where the isothermal lines are drawn for calcium chloride, covering the entire solubility diagram and a region of partial vaporization. In this diagram line $abcde$ represents the solubility of the various hydrates of calcium chloride at various temperatures with corresponding enthalpies. Line efg represents the melting points and corresponding enthalpies of the various solid hydrates.

a = freezing point of water, 32°F
b = eutectic of ice and $CaCl_2 \cdot 6H_2O$, −67°F
c = transition of $CaCl_2 \cdot 6H_2O$ to $CaCl_2 \cdot 4H_2O$ at 86°F
d = transition of $CaCl_2 \cdot 4H_2O$ to $CaCl_2 \cdot 2H_2O$ at 113°F
e = transition of $CaCl_2 \cdot 2H_2O$ to $CaCl_2 \cdot H_2O$
f = transition of $CaCl_2 \cdot 1H_2O$ to $CaCl_2$
g = melting point of $CaCl_2$

In the two-phase region the sloping lines starting at the line marked 1 atmosphere represent the enthalpies of the liquid-vapor system at the stated temperatures and compositions of combined phases. For example, one pound of a vapor-liquid system containing 40% $CaCl_2$ at 250°F has an enthalpy of 95 Btu and consists of $(0.40/0.42)$ lb of a 42% solution, boiling at 250°F at 1 atmosphere pressure and $\left(1 - \dfrac{0.40}{0.42}\right)$ lb of water vapor at 250°F in equilibrium with the boiling solution. The value of 42% is obtained as the intersection of the 250°F with the 1 atmosphere line.

The sloping lines running upwards from the dotted melting point line *efg* represent the enthalpies of the liquid-vapor system at the stated temperature and composition of the combined phases. For example, one pound of liquid-vapor system containing 80% $CaCl_2$ at 800°F has an enthalpy of 220 Btu and consists of (0.8/0.84) lb of liquid containing 84% $CaCl_2$ and $\left(1 - \dfrac{0.8}{0.84}\right)$ lb. of water vapor at 800°F in equilibrium with the solid. The vapor pressures corresponding to the latter sloping lines in the two-phase region are not given.

Illustration 12. To 200 pounds of anhydrous $CaCl_2$ at 100°F are added 500 pounds of a solution containing 20% $CaCl_2$ at 80°F.

(*a*) What is the temperature of the final mixture?

(*b*) How much heat must be removed to start crystallization?

From Fig. 54,

Enthalpy of $CaCl_2$ at 100°F = (200) (12) = 2400 Btu

Enthalpy of $CaCl_2$ solution at 80°F = (500) (−15) = −7500 Btu

Total enthalpy = −5100 Btu

Final composition = $\dfrac{(200 + 100)\,(100)}{700}$ = 42.8% $CaCl_2$

Final enthalpy = $\dfrac{-5100}{700}$ −7.3 Btu/lb

(*a*) Final temperature is hence 190°F.

(*b*) This solution must be cooled to 70°F before crystallization will begin.

$$\Delta H = 700\,[-83 - (-7.3)] = -53,000 \text{ Btu}$$

This problem can also be solved by the construction of a tie-line as demonstrated in Illustration 11.

Partial Properties. In dealing with the extensive properties of solutions it is convenient to consider separately the contribution attributable to each component present. For example, the total volume of a solution represents the sum of the contributions of all of the components present. In ideal systems, such as a gas at low pressure, these contributions may be the same as the properties of the pure components existing separately at the temperature and pressure of the solution. However, in many systems this additivity of properties does not exist and the volume of a solution may be quite different from the sum of the pure component volumes. That portion of the total volume of a solution which is attributable to the presence of a particular component is termed the *partial volume* of that component.   Other partial extensive properties may be similarly defined such as *partial enthalpy* and *partial heat capacity*.

The partial volume of a component in a solution of given composition may be experimentally determined by adding a unit mass of the pure component to a quantity of solution so large that only an infinitesimal or negligible change in composition results.   The increase in total volume under these conditions is termed the *partial specific volume* of the component added and in nonideal solutions may differ considerably from the specific volume of the pure component.   The partial volume of that component in the solution is then the product of the partial specific volume and the mass of the component present.   Similarly, the *partial molal volume* of the component may be determined as the increase in volume resulting from the addition of one mole of the component to a quantity of solution so large that no appreciable change in composition results.

If the composition of the solution is varied to approach 100% of the component under consideration it is evident that the partial specific volume must approach the specific volume of the pure component. Thus, in nonideal solutions the partial properties vary as functions of the composition.   The concept and use of partial properties have been developed by Lewis and Randall[3] whose general methods are followed in this discussion.

Mathematically, the partial molal volume of a component of a solution may be defined as

$$\bar{V}_1 = \left(\frac{\partial V}{\partial n_1}\right)_{P,T,n_2,n_3\cdots} \tag{29}$$

where

$\bar{V}$ = partial molal volume of component 1

V = total volume of solution

n_1 = moles of component 1

The partial form of the derivative indicates the requirement of constancy of all intensive properties such as temperature, pressure, and composition.   Thus, Equation (29) is merely the formal expression of the result of the experimental measurement, described above, of the change in volume per mole of component added at conditions of constant composition, temperature, and pressure.

Where u is a single valued, continuous function of several independent variables x, y, z, the total differential of u can be expressed in terms of its

[3] G. N. Lewis and M. Randall, *Thermodynamics*, McGraw-Hill Book Co. (1923).

partial derivative with respect to its independent variables, thus,

$$du = \frac{\partial u}{\partial x}\,dx + \frac{\partial u}{\partial y}\,dy + \frac{\partial u}{\partial z}\,dz \tag{30}$$

This principle is illustrated in Fig. 55 for two independent variables x and y. Applying this mathematical principle to the volume of a solution in terms of the independent variables of composition,

$$dV = \frac{\partial V}{\partial n_1}\,dn_1 + \frac{\partial V}{\partial n_2}\,dn_2 + \frac{\partial V}{\partial n_3}\,dn_3 + \cdots \tag{30a}$$

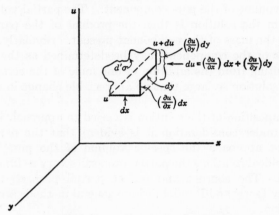

FIG. 55. Partial and total differentials.

or, combining Equations (29) and (30a)

$$dV = \bar{v}_1 dn_1 + \bar{v}_2 dn_2 + \bar{v}_3 dn_3 + \cdots \tag{31}$$

If the composition of the solution and its temperature and pressure are held constant, the partial molal volumes of Equation (28) are constant and the equation may be integrated:

$$V = \bar{v}_1 n_1 + \bar{v}_2 n_2 + \bar{v}_3 n_3 + \cdots \tag{32}$$

Thus, the total volume of a solution is equal to the sum of the partial volumes of the components.

Equation (32) may be differentiated to obtain another useful relationship of partial properties. Thus

$$dV = \bar{v}_1 dn_1 + n_1 d\bar{v}_1 + \bar{v}_2 dn_2 + n_2 d\bar{v}_2 + \bar{v}_3 dn_3 + n_3 d\bar{v}_3 + \cdots \tag{33}$$

Combining Equations (31) and (33):

$$n_1 d\bar{v}_1 + n_2 d\bar{v}_2 + n_3 d\bar{v}_3 + \cdots = 0 \tag{34}$$

This equation relates the changes in the partial molal volume of one component to changes in the partial molal volumes of the others.

Equations (29) to (34) were developed for volumes because of the ease with which this property may be visualized. However, by parallel reasoning similar equations may be developed relating any extensive property to the contributions of the separate components. Thus, the heat capacity of a solution is equal to the sum of the partial heat capacities of its components, derived by an equation similar to (32). Similarly, the total molal enthalpy of a solution is equal to the sum of the products of the partial enthalpies of the components times their mole fractions, or the total enthalpy per gram is equal to the sum of the products of the partial enthalpies per gram times the weight fractions.

It is evident that partial molal volumes or partial specific volumes or any of the other partial extensive properties per mole or unit weight of component are themselves intensive properties, independent of the mass of solution under consideration but varying as functions of composition as well as temperature and pressure.

Partial Enthalpies.   For the establishment of energy balances the enthalpy of a solution may be calculated either from integral heat of solution data and the enthalpies of its pure components according to Equation (28), or directly as the sum of the partial enthalpies of the components.   Thus,

$$H_s = n_1 H_1 + n_2 H_2 + n_2 \Delta H_{si2} = n_1 \bar{H}_1 + n_2 \bar{H}_2 \qquad (35)$$

The latter method is frequently the more convenient and more nearly accurate, particularly where small changes in composition are involved. For this reason it is desirable to derive partial from total enthalpy data.

Lewis and Randall[4] present several methods of calculating partial properties, two of which are illustrated by determining in one case partial enthalpies per pound and in the other partial molal enthalpies of components in solutions.   The same procedures are involved in calculating any other partial extensive properties such as volumes or heat capacities.

Method of Tangent Slope.   If the total enthalpy of a solution is plotted as ordinates against the moles of solute, component 2, per fixed quantity of all other components, it follows from an equation similar to Equation (29) that the partial molal enthalpy of component 2 is represented by the slope of this curve.   If the solution is binary the partial molal enthalpy of the solvent component then may be calculated from an equation similar to (31).

For most accurate results covering the entire range of composition of a binary solution two such plots should be constructed, one covering the concentration range of solute from 0 to 50% on the basis of

[4] *Thermodynamics*, p. 36, McGraw-Hill Book Co. (1923).

one mole or unit weight of solute and the other covering the concentration range of solvent from 0 to 50% constructed on the basis of one mole or unit weight of solvent.

Illustration 13. From the following data for the heats evolved when water and glycerin are mixed to form 1 gram of solution, calculate the partial enthalpies at 18°C of water and glycerin, per gram of each component, at each of the designated concentrations. Plot curves relating the partial enthalpies of glycerin and of water to percentage of glycerin by weight. As the reference state of zero enthalpy use the pure components at 18°C, the temperature of the solutions.

Solution: Method of Tangent Slope.

Percentage Glycerin by Weight	$\Delta H_m = H$	w'	H'	w''	H''	$\bar{H}'$	$\bar{H}''$
10	−1.91	111	−2120	....	......	−13.5	−0.6
20	−3.21	250	−4020	....	......	−10.5	−1.4
30	−3.90	429	−5560	....	......	−8.1	−2.1
40	−4.50	668	−7500	....	......	−6.2	−3.4
50	−4.50	1000	−9000	1000	−9000	−4.2	−4.8
60	−4.21	....	......	668	−7000	−2.65	−6.5
70	−3.70	....	......	429	−5290	−1.7	−8.3
80	−2.61	....	......	250	−3660	−0.8	−11.4
90	−1.79	....	......	111	−1990	−0.4	−14.3

ΔH_m = heat of mixing, = H, the total enthalpy, calories per gram of solution.

w' = grams of glycerin per 1000 grams of water.

w'' = grams of water per 1000 grams of glycerin.

H' = total enthalpy of solution per 1000 grams of water.

H'' = total enthalpy of solution per 1000 grams of glycerin.

$\bar{H}'$ = partial enthalpy of glycerin, calories per gram.

$\bar{H}''$ = partial enthalpy of water, calories per gram.

The values of H', the enthalpy of solution per 1000 grams of water, are obtained by multiplying the heat of solution, per gram of solution, by the number of grams of the solution ($w' + 1000$). For a 40% solution of glycerin

$$H' = (668 + 1000)(-4.50) = -7500 \text{ calories}$$

Values of w' are plotted as abscissas and values of H' as ordinates in Fig. 56. The slope of a tangent to this curve is the partial enthalpy per gram, of glycerin H' in a solution of concentration corresponding to the abscissa of the point of tangency. For a 40% solution:

$$\text{Slope of tangent} = \bar{H}' = \frac{-9600 + 3400}{1000} = -6.2$$

The corresponding values of $\bar{H}''$, the partial enthalpy of water, are calculated from the following equation, which is similar to (35):

$$(w_1 + w_2)H = w_1\bar{H}' + w_2\bar{H}''$$

For a 40% solution:

   Basis: 100 grams of solution.

$$\bar{H}'' = \frac{100H}{w_2} - \frac{w_1\bar{H}'}{w_2} = \frac{-450 + (40)\,(6.2)}{60} = -3.4$$

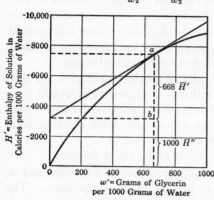

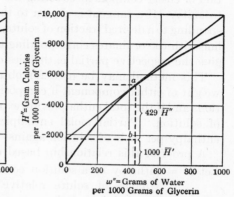

Fig. 56.   Calculation of partial enthalpies   Fig. 57.   Calculation of partial enthalpie
   by the tangent slope method.   by the tangent slope method.

The concept of partial enthalpies may be clearly visualized by inspection of Fig. 56. Point *a* represents the total enthalpy of a solution containing 668 grams of glycerin and 1000 grams of water. This total enthalpy is the sum of the partial enthalpy of the 668 grams of glycerin, equal to $a - b$ on the diagram, plus the partial enthalpy of the 1000 grams of water, equal to *b*.

It will be noted that at the higher concentrations of glycerin the slope of the curve of Fig. 56 becomes small and it is difficult to determine a value of $\bar{H}'$ from it with sufficient accuracy to permit a reliable calculation of $\bar{H}''$. For this reason it is inadvisable to use this curve in the range of concentrations above 50% glycerin. In this range more accurate graphical results may be obtained by plotting total enthalpies for solutions containing 1000 grams of glycerin against the weight of water dissolved. From this curve values of $\bar{H}''$ are determined directly. The corresponding values of $\bar{H}'$ are calculated from Equation (35). In Fig. 57 values of w'', the grams of water per 1000 grams of glycerin, are plotted as abscissas against H'', the total enthalpy per 1000 grams of glycerin. The slope of a tangent to this curve is the partial enthalpy of water per gram in a solution of concentration corresponding to the abscissa at the point of tangency. For a solution containing 70% glycerin:

$$\text{Slope of tangent} = \bar{H}'' = \frac{-10,000 + 1700}{1000} = -8.3 \text{ calories}$$

From Equation (35), on the basis of 100 grams of solution:

$$\bar{H}' = \frac{100H}{w_1} - \frac{w_2\bar{H}''}{w_1} = \frac{-370 - (-8.3 \times 30)}{70} = -1.7$$

In Fig. 57 point *a* represents the total enthalpy of a solution containing 429 grams of water and 1000 grams of glycerin. This total is the sum of the partial enthalpy of the water, equal to $a - b$ plus the partial enthalpy of the glycerin, equal to *b*.

Method of Tangent Intercepts. In the method of tangent intercepts the total enthalpy of the solution based upon unit quantity of solution instead of unit quantity of either component is plotted against the fraction of either component covering the range of concentrations from zero to unity. If a tangent is drawn to this curve at a concentration corresponding to a desired fraction of solute, then the intercepts of this tangent line with the ordinates corresponding to zero and unit fractions of solute give the respective partial enthalpies of the solvent and solute.

This method may be used for obtaining partial enthalpies per unit weight of either component if compositions are expressed in weight fractions and if the total enthalpies are expressed on the basis of a unit weight of solution. Partial molal enthalpies are obtained by plotting total enthalpy per mole of solution against mole fraction.

A proof of this relationship based upon molal units follows:

Let H = enthalpy of a solution containing n_1 moles of solvent and n_2 moles of solute relative to the pure components. The enthalpies of the pure solvent and pure solute are each arbitrarily taken as zero at the temperature of the solution.

H = enthalpy per mole of solution

N_1 = mole fraction of solvent

N_2 = mole fraction of solute

$$\text{H} = \frac{H}{n_1 + n_2} \tag{36}$$

$$N_2 = \frac{n_2}{n_1 + n_2} \tag{37}$$

In Fig. 58, Curve III, values of enthalpy H in Btu per pound mole of solution are plotted against mole fraction N_2 of the solute. A tangent is drawn to Curve III at point E corresponding to a mole fraction of N_2 equal to C. This tangent intercepts the ordinate $N_2 = 0$ at point A and the ordinate $N_2 = 1$ at point F. Ordinate OB represents the value of H at a mole fraction $N_2 = C$. The slope of the tangent at E is equal to $\dfrac{d\text{H}}{dN_2}$; also

$$OA = OB - AB = \text{H} - N_2 \frac{d\text{H}}{dN_2} \tag{38}$$

Variation in the mole fraction of solute, N_2, may be produced by addition or removal of solvent, keeping n_2 constant. Differentiating Equation (36) with respect to n_1 keeping n_2 constant,

$$d\text{H} = \frac{dH}{n_1 + n_2} - H \frac{dn_1}{(n_1 + n_2)^2} \tag{39}$$

Differentiating (37) with respect to n_1 keeping n_2 constant,

$$dN_2 = -\frac{n_2 dn_1}{(n_1 + n_2)^2} \tag{40}$$

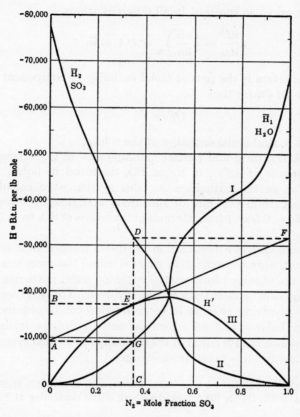

FIG. 58. Calculation of differential heats of solution by the method of tangent intercepts, SO_3–H_2O system.

Combining (39) and (40)

$$\frac{d\text{H}}{dN_2} = \frac{-dH(n_1 + n_2)}{n_2 dn_1} + \frac{H}{n_2} \tag{41}$$

Substituting (41) in (38),

$$OA = \text{H} + \frac{N_2 dH(n_1 + n_2)}{n_2 dn_1} - \frac{N_2 H}{n_2} \tag{42}$$

Substituting (36) and (37) in (42)

$$OA = \text{H} + \frac{dH}{dn_1} - \text{H} = \frac{dH}{dn_1} \tag{43}$$

Since n_2 was taken as constant in all differentiations

$$\frac{dH}{dn_1} = \left(\frac{\partial H}{\partial n_1}\right)_{n_2} = OA = \bar{\text{H}}_1 \tag{44}$$

which by definition is the partial molal enthalpy of component 1. Similarly, it may be shown that

$$OF = \bar{\text{H}}_2 \tag{45}$$

which is the partial molal enthalpy of the solute.

In Fig. 58 the total and partial enthalpies at 18°C, referred to the pure components at 18°C, of liquid SO_3 dissolved in liquid H_2O are plotted. The partial enthalpies for this system were calculated by R. A. Morgen[5] from the data of Bichowski and Rossini[6] for values of SO_3 up to $N_2 = 0.5$ and from Hermann[7] for values of SO_3 from $N_2 = 0.5$ to 1.0.

For large changes in concentration the use of integral heat of solution data or an enthalpy concentration chart is sound but more nearly accurate results are obtained from partial enthalpy data, particularly where slight changes in concentration are involved. The heat evolved in any process involving changes in concentration can be obtained by the usual energy balance method where the enthalpy of the initial system plus the heat absorbed is equal to the enthalpy of the final system. The following illustration is taken from Morgen.[8]

Illustration 14. It is desired to increase the strength of a 23.2% H_2SO_4 solution (19% SO_3) to 80.6% H_2SO_4 (65.6% SO_3) with an oleum containing 41.2% free SO_3 (89.2% SO_3).

Calculate the heat evolved at 18°C on the basis of 100 pounds of weak acid.

From a material balance of SO_3

$$\text{Weak Acid} + \text{Oleum} = \text{Strong Acid}$$
$$(0.19)(100) + x(0.892) = (100 + x)(0.656)$$
$$x = 197.5 \text{ pounds oleum}$$
$$100 + x = 297.5 \text{ pounds strong acid}$$

[5] *Ind. Eng. Chem.*, **34**, 571, (1942), with permission.

[6] *Thermochemistry of Chemical Substances*, Reinhold Publishing Company (1936). With permission.

[7] *Ind. Eng. Chem.*, **33**, 898 (1941).

[8] Morgen, *ibid.*

	Weak Acid	Oleum	Strong Acid
Pounds SO₃	19	176	195
Pounds H₂O	81	21	102
Pounds total	100	197.5	297.5
Pound moles SO₃	0.238	2.202	2.44
Partial molal enthalpy, $SO_3 \times 10^{-3}$	−67.1	−4.68	−35.82
Pound moles H₂O	4.50	1.185	5.686
Partial molal enthalpy of $H_2O \times 10^{-3}$	−0.14	−35.10	−7.16

From an energy balance of the process using partial enthalpies the following results are obtained.

$$\frac{\Delta H}{1000} + \overbrace{0.238(-67.1) + 4.50(-0.14)}^{\substack{\text{Weak Acid} \\ SO_3 \qquad\qquad H_2O}} + \overbrace{2.202(-4.68) + 1.185(-35.10)}^{\substack{\text{Oleum} \\ SO_3 \qquad\qquad H_2O}}$$

$$= \overbrace{2.440(-35.82) + 5.686(-7.16)}^{\substack{\text{Strong Acid} \\ SO_2 \qquad\qquad H_2O}}$$

$$\Delta H = -59,620 \text{ Btu/100 pounds weak acid}$$

or   Heat evolved = 59,620 Btu

The partial enthalpy of a component in a solution referred to the pure components at the temperature of the solution is termed the *differential heat of solution* of the component.

HEAT OF WETTING

When a solid surface is brought into contact with a liquid in which it is insoluble the liquid will spread in a thin film over the surface of the solid provided the solid is wetted by the liquid. This implies that the surface tension of the liquid relative to air is less than the adhesion tension between the liquid and the solid. The liquid film may be highly compressed as a result of attractive forces or a chemical bond may occur. The formation of such films of liquids is accompanied by an evolution of heat.

Because of the small amount of liquid affected by the interfacial forces of wetting, the heat of wetting per square centimeter of interfacial surface is a small quantity and is negligible unless a large area of interface is formed as in wetting a fine powder or a porous material. However, the thermal effects may then be of considerable importance, their magnitude depending upon the nature of both solid and liquid and on the area of interface formed.

Experimental measurements have been made of heats of wetting in many systems. Because of the uncertainty of the surface areas of materials whose heats of wetting are important, such data are ordinarily expressed as the change in enthalpy when 1 gram of the solid is wetted with sufficient liquid so that further addition of liquid produces no thermal effect. This is termed the heat of *complete wetting*. Its magnitude is roughly proportional to the surface area exposed by a given mass of solid and is dependent on the quantity of the liquid originally present in the solid material. In Table XVIII are experimental values of heats of wetting of dry clay, silica gel, starch, and charcoal by various liquids. The chemical and physical structures of the substances are not specified, and it must be recognized that these data may be used only for prediction of the order of magnitude of the effect to be expected in a specific case.

TABLE XVIII

HEATS OF COMPLETE WETTING OF POWDERS DRIED AT 100°C

Experiments at 12° to 13°C.
Data from International Critical Tables, Vol. V, page 142.
Unit: calories per gram of the dry material.

Liquid	ΔH Clay	ΔH Amorphous Silica	ΔH Starch	ΔH Sugar Charcoal
H_2O (water)	−12.6	−15.3	−20.4	−3.9
CH_3OH (methyl alcohol)	−11.0	−15.3	−5.6	−11.5
C_2H_5OH (ethyl alcohol)	−10.8	−14.7	−4.9	−6.8
$C_5H_{11}OH$ (amyl alcohol)	−10.0	−13.5	−7.0	−5.6
CH_3COOH (acetic acid)	−9.3	−13.5	−3.0–4.0	−6.0
CH_3COCH_3 (acetone)	−8.0	−13.5	−2.0	−3.6
$C_2H_5OC_2H_5$ (ether)	−5.8	−8.4	−2.2	−1.2
C_6H_6 (benzene)	−5.8	−8.1	−1.2	−4.2
CCl_4 (carbon tetrachloride)	−1.8	−8.1	−1.7	−1.5
CS_2 (carbon disulfide)	−1.7	−3.6	−0.5	−4.0
C_5H_{12}, C_6H_{14} (pentane, hexane)	−1.2	−3.1	−0.3	−0.4

As previously mentioned, the heat given off in wetting a material is diminished if it originally contains liquid. This effect is shown graphically in Fig. 59 by plotting heats of complete wetting, per gram of dry material, against the milligrams of liquid originally present per gram of dry solid. The points at which the extrapolated curves approach the axis corresponding to zero heat of wetting indicate the minimum quantities of liquid required to produce complete wetting. The heat of wetting of an already partially wetted solid is the difference between the heats of complete wetting at the final and the initial concentrations.

A more convenient method of presenting heat of wetting data is as *integral* and *differential* heats *of wetting*. These terms are analogous to integral and differential heats of solution and are similarly defined.

Values of integral and differential heats of wetting of silica gel with water are shown in Fig. 60 from the data of Ewing and Bauer.[9] The relationships involved are

$$\Delta H = (w_2 + 1)H - w_2 H_2 - H_1 = w_2 \overline{H}_2 + \overline{H}_1 - w_2 H_2 - H_1 \quad (46)$$

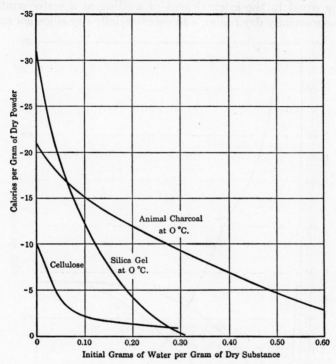

Fig. 59.   Heats of complete wetting with water.

where

H = total enthalpy of wetted gel, Btu per pound of wet gel

ΔH = integral heat of wetting, Btu per pound of commercially dry gel

$\overline{H}_1$ = partial enthalpy of gel, Btu per pound of commercially dry gel

$\overline{H}_2$ = partial enthalpy of water, Btu per pound of water

w_2 = pounds of water per pound of commercially dry gel

w_1 = pounds of dry gel equals unity

H_2, H_1 = enthalpy of pure water and dry gel, respectively, at the temperature of the system, in Btu per pound. Both values are given as zero in Fig. 60.

[9] *J. Am. Chem. Soc.* **59**, 1548 (1937).

It will be observed that the value of $\bar{H}_2$ is -400 Btu per pound of water when adsorbed upon dry gel but rapidly falls off when the water content rises, reaching zero at 40% water (commercially dry basis). The value of $\bar{H}_1$ changes in a reverse manner being zero for dry gel and becoming equal to the integral heat of wetting at a water content of 40% (commercially dry basis). In commercially dry silica gel approxi-

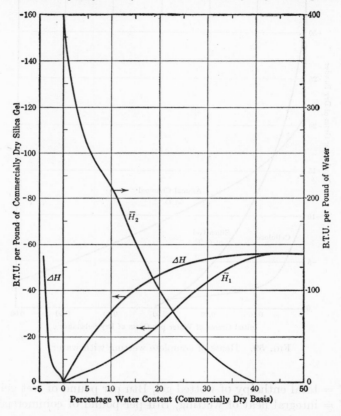

Fig. 60. Integral and differential heats of wetting of silica gel at 25°C. From Ewing and Bauer, *Jour. Am. Chem. Soc.* **59**, 1548 (1937).

mately 5% of water of constitution is present. This water of constitution is always present in active gel and represents the minimum content after regeneration. If this last trace of water were removed the gel would lose its capacity for adsorption. It is also of interest to note that the removal of this form of water as liquid would result in evolution of heat rather than in absorption of heat, as indicated by that portion of the ΔH curve of Fig. 60 below zero useful water content.

Illustration 15. One hundred pounds of commercially dry silica gel are wetted with water from a 5% to a 20% content (commercially dry basis). Estimate the heat evolved from (a) integral and (b) differential heat of wetting data.

(a) From integral heat of wetting data (Fig. 60) the enthalpy change is equal to

$$100[-46 - (-18)] = -2800 \text{ Btu}$$

(b) From the partial enthalpy data (Fig. 60) the enthalpy change is equal to

$$100[-26 + 0.20(-100)] - 100[-5 + 0.05(-260)] = -2800 \text{ Btu}$$

Therefore, the heat evolved is 2800 Btu.

When a liquid is evaporated from a wetted solid adsorbent the heat required is greater than the normal heat of vaporization of the liquid removed by the amount of the heat of wetting of the dried material finally produced. This additional heat which must be supplied corresponds to the work which must be done in removing the adsorbed molecules away from the attractive forces holding them to the surface. The heat of wetting is negligible in drying and calcining most solids.

HEAT OF ADSORPTION

In Chapter V two types of gas adsorption on solid surfaces are discussed, that caused by van der Waals forces and that by activated adsorption. These two types are also characterized by differences in their heats of adsorption. The heat of adsorption of a gas caused by van der Waals forces of attraction and capillarity is the sum of the heat of normal condensation plus the heat of wetting. In activated adsorption the gas is adsorbed by formation of a surface compound at temperatures even above the critical temperature of the gas. The corresponding heat of adsorption greatly exceeds that of normal condensation. The difference between heat of adsorption and heat of normal condensation represents the heat of wetting in van der Waals adsorption and the heat of formation of a surface compound in activated adsorption.

Data on heats of adsorption are presented either as integral or as differential values. The integral heat of adsorption is the change in enthalpy per unit weight of adsorbed gas when adsorbed on gas-free or " out-gassed " adsorbent to form a definite concentration of adsorbate. The integral heat of adsorption varies with the concentration of the adsorbate, in general diminishing with an increase in concentration.

The differential heat of adsorption of a gas is the change in enthalpy when a unit quantity of the gas is adsorbed by a relatively large quantity of adsorbent on which a definite concentration of the adsorbed gas already exists. The differential heat of adsorption is also a function of concentration, diminishing with an increase in concentration. As complete saturation of an adsorbent is approached the differential heat of adsorption approaches that of normal condensation. In activated adsorption the differential heat of adsorption decreases as the more active centers of the surface become covered.   In capillary adsorption the heat of adsorption decreases as the capillaries become progressively filled.   When the gas begins to condense at its normal saturation pressure the heat of adsorption assumes the value of its normal heat of condensation.

Integral and differential heats of adsorption bear the same relationship to each other as do integral and differential heats of solution and may be calculated in a similar manner from total and partial enthalpy relationships.   Heat effects in adsorption are calculated more accurately from differential heat of adsorption data or partial enthalpies because concentration changes are usually small and average values of the differential heat of adsorption may be used.

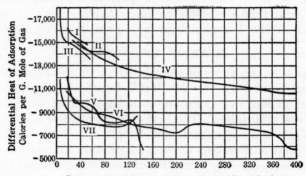

Concentration of Gas, cc. (S.C.) per Gram of Dry Adsorbent

I. C_6H_6, 0°C on inactive cocoanut carbon, out-gassed at 350°C.
II. C_6H_6 0°C on active cocoanut carbon, out-gassed at 350°C.
III. C_2H_5OH, 0°C on active cocoanut carbon, out-gassed at 350°C.
IV. H_2O, 0°C on silica gel dried at 300°C for 2 hr and out-gassed at 250°C, containing 3.5–5.5% H_2O.
V. SO_2, 0°C on same silica gel.
VI. SO_2, −10°C on blood carbon (puriss. Merck), out-gassed at 450°C ($d = 1.63$).
VII. NH_3, 0°C on cocoanut carbon, heated to 550°C, out-gassed at 400°C ($d = 1.86$).

FIG. 61.   Differential heats of adsorption.

Experimental measurements have been made of the heats of adsorption of many of the more common gases on the important adsorbents such as charcoal, silica gel, and various solid catalysts.   In Fig. 61 are values of differential heats of adsorption in calories per gram-mole adsorbed at 0°C, plotted against concentrations in cubic centimeters of adsorbed gas, measured at standard conditions of 0°C and 760 mm of mercury, per gram of adsorbent.   Similar data for other systems may be found in the International Critical Tables (Vol. V, page 139).

Illustration 16.   An adsorber for the removal of water vapor from air contains 250 lb of silica gel on which is initially adsorbed 28.0 lb of water.   Calculate the heat evolved per pound of water adsorbed at this concentration, assuming that the characteristics of the gel are similar to that on which the data of Fig. 61 are based.

Solution: Concentration of H_2O in gel $= \dfrac{28}{250} = 0.112$ gram per gram of gel.

or $\qquad \dfrac{0.112}{18} \times 22{,}400 = 139$ cc per gram of gel

From Fig. 61 $\qquad \Delta H_a = -12{,}500$ calories per gram-mole

or $\qquad\qquad -22{,}500$ Btu per lb-mole

Heat evolved per pound of water adsorbed $= \dfrac{22{,}500}{18} = 1250$ Btu

The effect of temperature on heat of adsorption is given by the Clapeyron equation (page 59) and may be evaluated by an equal temperature reference substance method of plotting equilibrium adsorption pressure similar in principle to that used for normal vapor pressures and heats of vaporization. By referring the heat of adsorption of a vapor to the normal heat of condensation of the same vapor at the same temperature Equation (III–5) (page 65) becomes

$$\lambda_1 = \lambda \frac{d \ln p_1}{d \ln p} \tag{47}$$

where at the equal temperatures for each system

λ = normal heat of condensation
λ_1 = heat of adsorption
p = normal vapor pressure of the condensed vapor
p_1 = actual equilibrium pressure of the adsorbed vapor

Enthalpies of Adsorbed Systems. The enthalpy diagram of a solid-adsorbed gas or solid-liquid system has special advantages in problems dealing with adsorption. In Fig. 62 the enthalpy of the silica gel-water system is shown at various temperatures and compositions relative to dry silica gel at 32°F and liquid water at 32°F. For convenience, the abscissas represent pounds of water per pound of commercially dry gel. In constructing this chart the 70°F isotherm was derived first since heats of wetting are known at this temperature. Thus,

$$(w_2 + 1)H = H_1 + H_2 w_2 + \Delta H_w \tag{48}$$

where

w_2 = pounds of water per pound of dry gel
H_1 = enthalpy of dry gel at 70°F, Btu/lb
H_2 = enthalpy of liquid water at 70°F, Btu/lb
ΔH_w = heat of wetting when w_2 pounds of water are added
  to one pound of dry gel at 70°F
H = enthalpy of wet gel at 70°F, Btu/lb

The enthalpy at any other temperature is then obtained from

$$H = H_{70} + \int_{70}^{t} C_p dt \qquad (49)$$

where C_p is the specific heat of the system.

From a combination of the enthalpy chart for air and that of silica gel it is easy to calculate the thermal effects in drying air by silica gel.

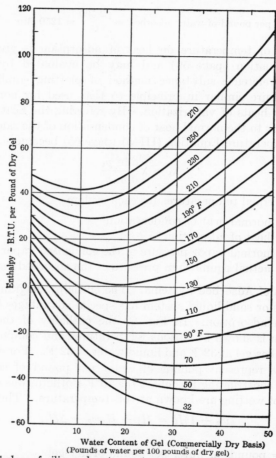

FIG. 62.   Enthalpy of silica gel-water system per pound of commercially dry gel. (Reference state: Commercially dry gel and liquid water at 32°F.)

Illustration 17.   Calculate the heat to be removed when 100 lb of air (dry basis) at 70°F and 0.010 humidity are dried to a final humidity of 0.001 at 70°F and 1 atmosphere pressure starting with 10 pounds of dry gel at 70°F.   The average temperature of the gel is initially at 70°F and finally at 80°F.

Average composition of gel after adsorption =

$$\frac{100(0.010 - 0.001)}{10} = 0.09$$

From Fig. 46 the heat removed from the air = $100(27.7 - 18) =$ 970 Btu
From Fig. 62 the heat removed from the gel = $10[6 - (-14)] =$ 200 Btu

Total heat removed = 1170 Btu

INCOMPLETE AND SUCCESSIVE REACTIONS

In the preceding sections consideration was given to the thermal effects accompanying chemical reactions which went to completion and in which the reactants were present in stoichiometric proportions. In industrial processes excess reactants are nearly always present and the reactions are seldom complete.

Frequently the chemical transformations of the same reactant proceed in successive steps or in divergent stages, and the quantities of different products which are formed bear no stoichiometric relationship to one another. For example, in the combustion of carbon, both carbon monoxide and carbon dioxide are formed in variable proportions which depend upon the conditions of the reaction. In such reactions the standard heat of reaction is calculated by exactly the same general procedure followed when all materials are present in stoichiometric proportions, namely by subtracting the heats of formation of the quantities of reactants which actually react from the heats of formation of the products actually formed. Or the heats of combustion of the products actually formed are subtracted from the heats of combustion of the reactants transformed to give the standard heat of reaction.

Illustration 18. In the production of metallic manganese, 10 kg of manganese oxide, Mn_3O_4, are heated in an electric furnace with 3.0 kg of amorphous carbon (coke). The resulting products are found to contain 4.8 kg of manganese metal and 2.6 kg of manganous oxide, MnO, as slag. The remainder of the products consists of unconverted charge and carbon monoxide gas. Calculate the standard heat of reaction of this process for the entire furnace charge.

Solution:

Initial Mn_3O_4 = 10.0 kg or 10/229 = ...   0.0437 kg-mole
Initial C = 3.0 kg or 3.0/12 = ...   0.250 kg-atom
Mn formed = 4.8 kg or 4.8/55 = ...   0.0874 kg-atom
MnO formed = 2.6 kg or 2.6/71 = ...   0.0366 kg-mole

Unconverted Mn_3O_4 = $0.0437 - \dfrac{0.0874 + 0.0366}{3}$ = ...   0.0024 kg-mole

O in final CO = $4(0.0437 - 0.0024) - 0.0366$ = ...   0.1286 kg-atom
CO formed = ...   0.1286 kg-mole
Unconverted C = 0.250 - 0.1286 = ...   0.1214 kg-mole

Material balance

Materials entering		Materials leaving	
Mn₃O₄........	10.0 kg	Mn....................	4.8 kg
C.............	3.0 kg	MnO....................	2.6 kg
	———	Mn₃O₄ = 0.0024 × 229 =	0.55 kg
Total.......	13.0 kg	CO = 0.1286 × 28 =	3.6 kg
		C = 0.1214 × 12 =	1.45 kg
			13.00 kg

Heats of formation of active reactants (data from Table XIV, page 253)

$$\text{Mn}_3\text{O}_4 = -345{,}000(0.0437 - 0.0024) = \dots\dots\dots\dots\dots\dots \quad -14{,}200 \text{ kcal}$$
$$\text{C(coke)} = +2600(0.250 - 0.1214) = \dots\dots\dots\dots\dots \quad +335 \text{ kcal}$$
$$\text{Total}\dots\dots\dots\dots\dots\dots\dots\dots\dots\dots\dots\dots\dots\dots\dots\dots \quad -13{,}865 \text{ kcal}$$

Heats of formation of products actually formed:

$$\text{MnO} = -96{,}500 \times 0.0366 = \dots\dots\dots\dots\dots\dots\dots\dots \quad -3530 \text{ kcal}$$
$$\text{CO}   = -26{,}620 \times 0.1286 = \dots\dots\dots\dots\dots\dots\dots\dots \quad -3430 \text{ kcal}$$
$$\text{Mn}   = \dots\dots\dots\dots\dots\dots\dots\dots\dots\dots\dots\dots\dots\dots\dots\dots\dots \quad 0 \text{ kcal}$$
$$\text{Total}\dots\dots\dots\dots\dots\dots\dots\dots\dots\dots\dots\dots\dots\dots\dots\dots \quad -6{,}960 \text{ kcal}$$

Standard heat of reaction = $-6960 + 13{,}865 = +6905$ kcal

EFFECT OF PRESSURE ON HEAT OF REACTION

In Chapters II and VII it was pointed out that the internal energy and enthalpy of an ideal gas are dependent only on temperature and independent of pressure. Moreover, changes in pressure have negligible effects on the internal energy contents and enthalpies of solids and liquids.

Therefore, the heat of reaction at constant pressure is independent of the pressure provided that in a reaction where gases are involved these gases behave ideally, and provided that when liquids or solids are present the changes in volume of the condensed phases are negligible.

Where these assumptions are not justified the effect of pressure on heat of reaction must be calculated from an energy balance, taking into account the effects of pressure on enthalpy by methods described later.

Reactions at Constant Volume. From the above section it follows that the changes in enthalpy and internal energy accompanying a chemical reaction are the same whether the reaction proceeds under conditions of constant pressure or under conditions of constant volume provided the gases present behave ideally. If a reaction proceeds in a calorimeter at 18°C at constant volume the heat evolved $-q$ can be measured and is equal to the loss in internal energy.

$$-q_v = -\Delta U_v \tag{50}$$

If there is no change in the product pV during the reaction then the loss in enthalpy will also equal the heat evolved.   If, however, the reaction proceeds with an increase in number of gaseous moles then there will be an increase in pV equal to $V\,\Delta p$ at constant volume.   The change in enthalpy then becomes:

$$\Delta H = \Delta U + V\,\Delta p \tag{51}$$

or
$$\Delta H = q_v + V\,\Delta p \tag{52}$$

If Δn represents the increase in number of gaseous moles, then for ideal gas behavior

$$V\,\Delta p = \Delta n R T \tag{53}$$

or
$$\Delta H = q_v + \Delta n R T \tag{54}$$

Therefore, the heat evolved in a reaction at constant volume is ~~less~~ *greater* than the enthalpy decrease by the heat equivalent of the increase in pV, or where ideal gases are involved, by the heat equivalent of the product $\Delta n R T$.

The difference between heat evolutions at constant pressure and constant volume is of particular value in correcting experimental determinations of heats of reaction to constant pressure conditions.   Calorimetric reactions are for the most part conducted at conditions of constant volume.   Standard heats of reaction and formation are calculated from such data by means of Equation (54).

Illustration 19.   In the combustion of 1 gram-mole of benzoic acid C_6H_5COOH at constant volume and a temperature 18°C, forming liquid water and gaseous CO_2, 771,550 calories are evolved.   Calculate the standard heat of combustion of benzoic acid.

$$C_6H_5COOH(s) + 7\tfrac{1}{2}O_2(g) = 7CO_2(g) + 3H_2O(l)$$
$$\Delta n = 7 - 7\tfrac{1}{2} = -\tfrac{1}{2}$$

From Equation 54

$$\Delta H = q_v + \Delta n R T = -771,550 - (\tfrac{1}{2} \times 1.99 \times 291) =$$
$$-771,840 \text{ calories (gram-moles)}$$

EFFECT OF TEMPERATURE ON HEAT OF REACTION

Standard heats of reaction represent the enthalpy changes during a reaction at constant pressure in which all reactants are initially at a selected standard temperature and all products are finally existent at that same temperature.   Such conditions are rarely encountered in industrial reactions.   Various reactants may enter at different temperatures and the various products may each leave at still different temperatures.   The heat effects of such reactions may be calculated from data on standard heats of reaction and thermophysical properties.

Kirchhoff's Equation. A useful analytical relation for the effect of temperature on the heat of reaction may be derived for the special case of reactions which begin and end with all materials at the same temperature.

Let ΔH_T = heat of reaction at temperature T

$\Delta H_T + d\Delta H$ = heat of reaction at temperature $T + dT$

Since the enthalpy change in going from reactants at T to products at $T + dT$ is independent of path,

$$dH_R + \Delta H_T + d\Delta H = \Delta H_T + dH_P \qquad (55)$$

where

dH_R and dH_P are the changes in enthalpies of the reactants and products, respectively, corresponding to temperature change dT

Where no latent heat changes are involved, that is, where there are no changes in phase of reactants or products,

$$dH_R = C_p dT \qquad (56)$$

and

$$dH_P = C'_p dT \qquad (57)$$

where

C_p and C'_p are the heat capacities of the reactants and products, respectively

Equation (55) becomes

$$d\Delta H = (C'_p - C_p)dT = \Delta C_p dT \qquad (58)$$

where

ΔC_p = change in heat capacity of entire system at constant pressure.

Equation (58), known as Kirchhoff's equation, may be applied to the following general reaction at constant pressure:

$$n_b B + n_c C + \cdots = n_r R + n_s S + \cdots$$

where

n_b, n_c, n_r, n_s = number of moles of components $B, C, R, S \cdots$

For this reaction,

$$\Delta C_p = n_r c_{pr} + n_s c_{ps} + \cdots - n_b c_{pb} - n_c c_{pe} \cdots \qquad (59)$$

Where the molal heat capacity of a substance is represented by an empirical equation of the form

$$c_p = a + bT + cT^2 \qquad (60)$$

each heat capacity term in Equation (59) may be replaced by an empirical equation of type (60). Then,

$$\Delta C_p = \Delta a + \Delta bT + \Delta cT^2 \qquad (61)$$

where

$$\Delta a = n_r a_r + n_s a_s \cdots - n_b a_b - n_c a_c \cdots \qquad (62)$$

$$\Delta b = n_r b_r + n_s b_s \cdots - n_b b_b - n_c b_c \cdots \qquad (63)$$

$$\Delta c = n_r c_r + n_s c_s \cdots - n_b c_b - n_c c_c \cdots \qquad (64)$$

Substituting (61) in (58) and integrating,

$$\Delta H_T = I_H + \Delta aT + \tfrac{1}{2}\Delta bT^2 + \tfrac{1}{3}\Delta cT^3 \qquad (65)$$

The constant of integration I_H of Equation (65) may be evaluated from a single value of ΔH_T. This is usually done at the standard conditions of 18°C. All other constants of Equation (65) are obtained directly from the empirical heat capacity equations for the reactants and products.

Equation (65) is of thermodynamic importance in determining the effect of temperature upon chemical equilibria. The utility of this equation is limited to reactions which begin and end at the same temperature where no changes in phase are involved, and where heat capacities are expressed by the given empirical equations.

Enthalpy Changes in Reactions with Varying Temperatures. The enthalpy change accompanying any reaction may be expressed in terms of an overall energy balance. Thus, in constant pressure or flow processes where changes in kinetic, potential, and surface energies are negligible and no work is performed,

$$\Delta H = \sum H'_P - \sum H_R = q \qquad (66)$$

where

$\sum H_R$ = sum of the enthalpies of all materials entering the reaction relative to the reference state for standard heats of reaction, conveniently 18°C and 1.0 atmosphere

$\sum H'_P$ = sum of the enthalpies of all materials leaving the reaction, referred to the form of chemical combination in which they entered the reaction at the standard reference conditions

A substance which is produced by chemical reaction has an enthalpy at 18°C referred to the reactants at 18°C which is by definition equal to the standard heat of reaction. Thus,

$$H'_p = H_p + \Delta H_{18} \qquad (67)$$

where

H_p = enthalpy of the product, referred to its standard state at 18°C
ΔH_{18} = standard heat of reaction

Combining (66) and (67)

$$\Delta H = \sum H_p + \sum \Delta H_{18} - \sum H_R \qquad (68)$$

In using Equation (68) standard heats of reaction are included in the summations of the equation only for the products actually resulting from reactions taking place in the process and to the extent that they are formed in the process. This term becomes zero for all materials passing through the process without chemical change.

Illustration 20. Carbon monoxide at 200°C is burned under atmospheric pressure with dry air at 500°C in 90% excess of that theoretically required. The products of combustion leave the reaction chamber at 1000°C. Calculate the heat evolved in the reaction chamber in kcals per kilogram-mole of CO burned, assuming complete combustion.

Basis: 1.0 kg-mole of CO.

$$CO(g) + \tfrac{1}{2}O_2(g) = CO_2(g)$$

From the data of Table XV, page 262,

ΔH_{18}	−67,410 kcal (kg-moles)
O_2 required =	0.5 kg-mole
O_2 supplied = 0.5 × 1.90 =	0.95 kg-mole
Air supplied = 0.95/0.21 =	4.52 kg-moles
N_2 present = 4.52 − 0.95 =	3.57 kg-moles
Unused O_2 = 0.95 − 0.50 =	0.45 kg-mole

Enthalpy of reactants (H_R) relative to standard state at 18°C

CO:

  $c_p(18 - 200°C) = 7.00$
  Enthalpy = (7.00)(1.0)(200 − 18) = 1270 kcal

Air:

  $c_p(18 - 500°C) = 7.21$
  Enthalpy = (7.21)(4.52)(500 − 18) = 15,710 kcal

$\sum H_R$ = 15,710 + 1270 = 16,980 kcal

Enthalpy of products (H_p) relative to standard state at 18°C

CO_2:

  $c_p(18 - 1000°C) = 11.94$
  Enthalpy = (11.94)(1.0)(1000 − 18) = 11,720 kcal

O_2:

  $c_p(18 - 1000°C) = 7.94$
  Enthalpy = (7.94)(0.45)(1000 − 18) = 3510 kcal

N_2:

$$c_p(18 - 1000°C) = 7.50$$
$$\text{Enthalpy} = (3.57)(7.50)(1000 - 18) = 26,290 \text{ kcal}$$
$$\sum H_p = 11,720 + 3510 + 26,290 = 41,520 \text{ kcal}$$

From Equation (68)

$$\Delta H = 41,520 - 67,410 - 16,980 = -42,870 \text{ kcal}$$

Since the reaction proceeds at constant pressure the enthalpy change is equal to the heat absorbed, or there is an evolution of heat of 42,870 kcal.

Enthalpy Terms in Energy Balances. As has been pointed out in Chapter VII, page 207, enthalpy is a convenient property combining internal energy and the product pV. Although enthalpy has energy units it cannot in the general case be considered as solely energy. For example, the enthalpy of an incompressible liquid may be increased by merely increasing the pressure without doing any work or changing the energy of the system. In such a case enthalpy is not a measure of energy and there can be no generalization of " conservation of enthalpy " parallel to the principle of conservation of energy. In the general case the enthalpy input and output items of a process do not necessarily balance and true energy balances should include only energy items.

It is proper to include enthalpy terms in energy balances under two conditions:

1. In a flow process the enthalpy is the sum of the internal energy and the flow work and as such is properly included in the general energy balance represented by Equation (VII–1), page 205.

2. In a non-flow process at constant pressure where work is performed only by expansion enthalpy is properly included as a term in the general energy balance of the form of Equation (VII–8), page 206.

Fortunately the great majority of processes of industrial importance fall in one or the other of these classifications.

General energy balances are of great value in interpreting process results and a formal tabulation of such a balance serves as a means of verifying calculations and presenting a compact summary of results. For such purposes it is desirable to tabulate all energy input items in one column and all output items in another. For convenience in addition all entries in each column are arranged to be of positive sign. Thus, in such a tabulation any work done on the system will appear as an input item while work done by the system will appear as an output item. Similarly, heat absorbed is an input, and heat evolved an output, item.

For all flow processes and for non-flow processes at constant pressure the enthalpies of all entering materials are entered as input items and those of the product materials, referred to their own standard states, as output items. In such processes enthalpy changes equal heat absorptions or evolutions and may be entered in the energy balance as such. Standard heats of reaction are input items if they are negative, representing a contribution of heat to the process and output items if positive. The overall enthalpy changes of the process are entered as output heat items if negative and as input items if positive.

Since energy balances of this type are entirely valid for the specified types of processes, all forms of energy may be properly included along with the work, heat, and enthalpy items. Where potential, kinetic, and surface energies are not negligible, input and output items are entered as part of the general balance.

Illustration 21. From the results of Illustration 20 prepare an energy balance for the combustion process under consideration.

Since this is a flow process, enthalpies may be included in the energy balance and enthalpy changes are equal to heat absorptions and liberations.

Input		Output	
Enthalpy of CO	1,270 kcal	Enthalpy of CO_2	11,670 kcal
Enthalpy of Air	15,690	Enthalpy of O_2	3,510
Standard Reaction		Enthalpy of N_2	27,750
Heat Added	67,410	Heat Evolved	41,440
	84,370 kcal		84,370 kcal

TEMPERATURE OF REACTION

Adiabatic Reactions. If a reaction proceeds without loss or gain of heat and all the products of the reaction remain together in a single mass or stream of materials, these products will assume a definite temperature known as the *theoretical reaction temperature*. In this particular case, for a flow process or a non-flow process at constant pressure the enthalpy change is zero, and it follows from Equation (68) that the sum of the enthalpies of the reactants must equal the sum of the standard heat of reaction and the total enthalpy of all the products. The temperature of the products which corresponds to this total enthalpy may be calculated by mathematically expressing the enthalpy of the products as a function of their temperature. This requires data on the heat capacities and latent heats of all products.

The products considered in calculating a theoretical reaction temperature must include all materials actually present in the final system: inerts and excess reactants as well as the new compounds formed. If the reaction is incomplete, only the standard heat of reaction resulting from the degree of completion actually obtained is considered and the products will include some of each of the original reactants.

The enthalpy, H, of n moles of any material at a temperature $T°K$, referred to a temperature of 291°K is expressed by:

$$H = n \int_{291}^{T} c_p \, dT + n\lambda \qquad (69)$$

where

λ = sum of the molal latent enthalpy changes in heating from 291°K to T°K at constant pressure

Expressing the molal heat capacity, c_p, as a quadratic function of temperature, $c_p = a + bT + cT^2$,

$$H = n \int_{291}^{T} (a + bT + cT^2) \, dT + n\lambda \qquad (70)$$

Integrating:

$$H = n \left[aT + \tfrac{1}{2}bT^2 + \tfrac{1}{3}cT^3 \right]_{291}^{T} + n\lambda \qquad (71)$$

The enthalpy of each product of a reaction may be expressed by an equation of the form of Equation (71) and all these added together to represent $\sum H_p$ of Equation (68), page 306. The theoretical reaction temperature is then obtained by solution of this equation for T. The solution is generally best carried out graphically.

Illustration 22. For the production of sulfuric acid by the contact process iron pyrites, FeS_2, is burned with air in 100% excess of that required to oxidize all iron to Fe_2O_3 and all sulfur to SO_2. It may be assumed that the combustion of the pyrites is complete to form these products and that no SO_3 is formed in the burner. The gases from the burner are cleaned and passed into a catalytic converter in which 80% of the SO_2 is oxidized to SO_3 by combination with the oxygen present in the gases. The gases enter the converter at a temperature of 400°C.

Assuming that the converter is thermally insulated so that heat loss is negligible, calculate the temperature of the gases leaving the converter.

$$4FeS_2 + 11O_2 = 2Fe_2O_3 + 8SO_2$$
$$SO_2 + \tfrac{1}{2}O_2 = SO_3$$

Basis: 4.0 gram-moles of FeS_2

Oxygen supplied for 100% excess = 11.0 × 2.0 = ...	22 gram-moles
Air introduced = 22/0.21 = ...	104.8 gram-moles
N_2 introduced = 104.8 − 22 = ...	82.8 gram-moles
Excess O_2 in burner gases = 22 − 11 = ...	11.0 gram-moles
SO_2 in burner gases = ...	8.0 gram-moles

Gases entering converter:

SO_2......................	8.0 gram-moles
O_2......................	11.0 gram-moles
N_2......................	82.8 gram-moles
Total...............	101.8 gram-moles

SO$_3$ formed in converter = 8.0 × 0.8 = ... 6.4 gram-moles
O$_2$ consumed in converter = 6.4/2 = ... 3.2 gram-moles

Gases leaving converter:

SO$_3$......................	6.4 gram-moles
SO$_2$ = 8.0 − 6.4...........	1.6 gram-moles
O$_2$ = 11.0 − 3.2...........	7.8 gram-moles
N$_2$........................	82.8 gram-moles
Total................	98.6 gram-moles

Heat capacity of gases entering converter
(Mean molal heat capacities between 18°C and 400°C are taken from Table VI, page 216.)

$$SO_2 = (8.0)(10.91) = \quad 87$$
$$O_2 = (11.0)(7.40) = \quad 81$$
$$N_2 = (82.8)(7.09) = \underline{587}$$
$$\text{Total} \qquad 755$$

Enthalpy, $\sum H_R$, of gases entering converter = (755)(400 − 18) = 288,000 cal
Standard heat of reaction, ΔH_{291}
Heats of formation from Table XIV, page 253,

$$SO_2; \quad \Delta_{HF} = -70,920 \text{ cal}$$
$$SO_3; \quad \Delta_{HF} = -93,900 \text{ cal}$$

$$\sum \Delta H_{291} = (6.4) \times (-93,900) - (6.4) \times (-70,920) = -147,000 \text{ cal}$$

From Equation (68), since $\Delta H = 0$,

$$\sum H_p = \sum H_R - \sum \Delta H_{291} = 288,000 - (-147,000) = 435,000 \text{ cal}$$

In order to solve for the temperature of the products $\sum H_p$ is expressed as a function of temperature by the equations of Table V, page 214. Integrating these equations between the limits of 291 and $T°K$ gives expressions for the enthalpies of the individual components.

O$_2$:

$$H = 7.8 \left[6.13(T - 291) + \frac{0.00299}{2}(T^2 - \overline{291}^2) - \frac{0.806}{3}(10^{-6})(T^3 - \overline{291}^3) \right]$$
$$= 47.8(T - 291) + 0.0117(T^2 - \overline{291}^2) - 2.095(10^{-6})(T^3 - \overline{291}^3)$$

N$_2$:

$$H = 82.8 \left[6.30(T - 291) + \frac{0.001819}{2}(T^2 - \overline{291}^2) - \frac{0.345}{3}(10^{-6})(T^3 - \overline{291}^3) \right]$$
$$= 522(T - 291) + 0.0753(T^2 - \overline{291}^2) - 9.53(10^{-6})(T^3 - \overline{291}^3)$$

SO$_2$:

$$H = 1.6 \left[8.12(T - 291) + \frac{0.00683}{2}(T^2 - \overline{291}^2) - \frac{2.103}{3}(10^{-6})(T^3 - \overline{291}^3) \right]$$
$$= 13.0(T - 291) + 0.00546(T^2 - \overline{291}^2) - 1.122(10^{-6})(T^3 - \overline{291}^3)$$

SO_3:

$$H = 6.4\left[8.20(T - 291) + \frac{0.01024}{2}(T^2 - \overline{291}^2) - \frac{3.156}{3}(10^{-6})(T^3 - \overline{291}^3)\right]$$

$$= 52.5(T - 291) + 0.0328(T^2 - \overline{291}^2) - 6.73(10^{-6})(T^3 - \overline{291}^3)$$

Adding these equations

$$\Sigma H_p = 635.3(T - 291) + 0.1253(T^2 - \overline{291}^2) - 19.48(10^{-6})(T^3 - \overline{291}^3)$$

$$= 635.3T + 0.1253T^2 - 19.48(10^{-6})T^3 - 195,030$$

Equating the heat input to the enthalpy of the products

$$635.3T + 0.1253T^2 - 19.48(10^{-6})T^3 - 195,030 = 435,000$$

This equation is best solved graphically or by substituting values of T until the equation is satisfied.

Solving,

$$T = 865°K, \text{ or } 592°C$$

This calculation can be verified or can be estimated more readily by assuming average values of the heat capacities of the products. A slight error in the estimated final temperature will not alter the values of mean heat capacities appreciably. For example, let it be assumed that the final temperature is 590°C.

Mean molal heat capacities are obtained from Table VI, page 216,

$$O_2 = 7.60$$
$$N_2 = 7.20$$
$$SO_2 = 11.40$$
$$SO_3 = 12.46 \text{ (Table V, page 214)}$$

$$\Sigma H_p = [(7.8)(7.60) + (82.8)(7.20) + (1.6)(11.40) + (6.4)(12.46)](T - 291)$$
$$= (753.3)(T - 291) = 435,000$$
$$T = 869°K(596°C).$$

This value is better than 592°C because experimental values of heat capacities were used in deriving the mean values of Table VI rather than the empirical equations of Table V. If the calculated final temperature were considerably different from that assumed as a first approximation for obtaining mean heat capacities, the resultant error may be reduced by repeating the calculation with heat capacity data based on the temperature calculated as a first approximation. Since mean heat capacities do not vary greatly with temperature successive approximations of this type rapidly approach the correct solution.

Non-Adiabatic Reactions. If a reaction does not proceed adiabatically the amount of heat gained or lost from the system during the reaction must be known in order to calculate the temperature.

If a flow process or a non-flow process at constant pressure is under consideration the heat absorbed by the system is equal to ΔH of Equation (68). Using this value of ΔH the enthalpy and temperature of the products are calculated as in Illustration 22.

Theoretical Flame Temperatures. The temperature attained in the *complete, adiabatic* combustion of a fuel which is thoroughly admixed with air or oxygen is termed the *theoretical flame temperature.* The methods developed in the preceding sections may be used to calculate the theoretical flame temperature of a gaseous, atomized liquid, or powdered solid fuel when burned with air in any desired proportions. The maximum theoretical flame temperature of a fuel corresponds to combustion with only the theoretically required amount of pure oxygen. The maximum flame temperature *in air* corresponds to combustion with the theoretically required amount of air and is obviously much lower than the maximum flame temperature in pure oxygen. Because of the necessity of using excess air in order to obtain complete combustion, the theoretical flame temperatures of actual combustions are always less than the maximum values.

Illustration 23. Calculate the theoretical flame temperature of a gas containing 20% CO and 80% N_2 when burned with 100% excess air, both air and gas initially being at 18°C.

Basis: 1.0 gram-mole of CO.

N_2 in original gas $= \dfrac{1}{0.20} \times 0.80 =$ .... 4.0 gram-moles

O_2 supplied $= 0.5 \times 2 =$ .... 1.0 gram-mole

N_2 from air $= \dfrac{1.0}{0.21} \times 0.79 =$ .... 3.76 gram-moles

Total $N_2 = 3.76 + 4.0 =$ .... 7.76 gram-moles

Moles of original N_2, O_2, CO $= 7.76 + 1.0$
$+ 1.0 =$ .... 9.76 gram-moles

Combustion products:

CO$_2$ formed $=$ .... 1.0 gram-mole

O_2 remaining $= 1.0 - 0.5 =$ .... 0.5 gram-mole

$N_2 =$ .... 7.76 gram-moles

Enthalpy of products, $\sum H_p$ (referred to 18°C)

CO_2:

$$H = 1.0\left[6.85(T - 291) + \frac{0.008533}{2}\,(T^2 - \overline{291}^2) - \frac{2.475}{3}\,(10^{-6})(T^3 - \overline{291}^3) \right]$$

O_2:

$$H = 0.5\left[6.13(T - 291) + \frac{0.00299}{2}\,(T^2 - \overline{291}^2) - \frac{0.806}{3}\,(10^{-6})(T^3 - \overline{291}^3) \right]$$

N_2:

$$H = 7.76\left[6.30(T - 291) + \frac{0.00182}{2}\,(T^2 - \overline{291}^2) - \frac{0.345}{3}\,(10^{-6})(T^3 - \overline{291}^3) \right]$$

Adding:

$$\sum H_p = 58.80(T - 291) + 0.01207(T^2 - \overline{291}^2) - 1.851(10^{-6})(T^3 - \overline{291}^3)$$
$$H_p = 58.80T + 0.01207T^2 - 1.851 \times 10^{-6}T^3 - 18,095$$

Since $\sum H_R = 0$ and $\Delta H = 0$, from Equation (68), $\sum H_p = -\sum \Delta H_{291}$
or $\qquad 58.80T + 0.01207T^2 - 1.851 \times 10^{-6}T^3 - 18,095 = 67,410$

Solving this equation graphically, $T = 1204°$K or $931°$C.

A better value may be obtained more directly by assuming the final temperature and then using the corresponding values of mean heat capacities from Table VI, page 216. Thus the mean heat capacities between 18°C and 930°C are

$$CO_2 = 11.81$$
$$O_2 = 7.89$$
$$N_2 = 7.45$$

Then,

$$\sum H_p = [(1)(11.81) + (0.5)(7.89) + (7.76)(7.45)](T - 291) = 67,410$$
or $\qquad T = 1206°$K $(933°$C$)$

The theoretical flame temperature of a fuel is dependent on the initial temperature of both the fuel and the air with which it is burned. By *preheating* either the fuel or the air the total heat input is increased and the theoretical flame temperature is correspondingly raised.

Illustration 24. Calculate the effect on the theoretical flame temperature of Illustration 23 of preheating both the gas and air to 1000°C before combustion.

Basis: Same as Illustration 23.

Enthalpy of reactants at 1000°C relative to 18°C

$$CO = (1.0)(7.58)(1000 - 18) = 7,450$$
$$O_2 = (1.0)(7.94)(1000 - 18) = 7,800$$
$$N_2 = (7.76)(7.50)(1000 - 18) = \underline{57,100}$$
$$H_R = 72,350$$

Using the values of ΔH_{291} and the equation for $\sum H_p$ from Illustration 23,

$$67,410 + 72,350 = 58.80T + 0.01207T^2 - 1.851 \times 10^{-6}T^3 - 18,095$$
$$58.80T + 0.01207T^2 - 1.851 \times 10^{-6}T^3 = 157,855$$

Solving this equation,

$$T = 2078°\text{K } (1805°\text{C})$$

In Table XXII, page 340, are values of the *maximum* theoretical flame temperatures of various hydrocarbon gases when burned with air at 18°C.

Actual Flame Temperatures. A theoretical flame temperature is always higher than can be obtained by actual combustion under the specified conditions. There is always loss of heat from the flame, and it is impossible to obtain complete combustion reactions at high temperatures. The partial completion of these reactions results from the establishment of definite equilibrium conditions between the products and reactants. For example, at high temperatures an equilibrium is established between carbon monoxide, carbon dioxide, and oxygen, corresponding to definite proportions of these three gases. Combustion of carbon monoxide will proceed only to the degree of completion which will give a mixture of gases in proportions corresponding to these equilibrium conditions. The general subject of reaction equilibria is discussed in Chapter XV.

In Table XXII are experimentally observed values of the maximum flame temperatures of various hydrocarbon gases when burned with air at 18°C. These data are from the work of Jones, Lewis, Friauf and Perrott.[10] These investigators also demonstrated a method of calculating flame temperatures, taking into account the degree of completion actually obtained if the combustion proceeds to equilibrium conditions but neglecting heat loss. The calculated values were ordinarily higher than those experimentally observed. Values of such maximum *calculated* flame temperatures are included in Table XXII.

The maximum values of actual and calculated flame temperatures do not correspond to the air-fuel proportions theoretically required for complete combustion. Because of the incomplete combustion actually produced at high temperatures the maximum flame temperature is obtained with a ratio of air to fuel which is somewhat less than that required for complete combustion.

It will be noted from Table XXII that the maximum flame temperatures of the various gases vary but little. For example, although pentane has twelve times the heating value of hydrogen, its flame temperature is lower by 110°C. This results from the fact that, in the combustion of the gases of high heating values, correspondingly large quantities of combustion products are formed with high total heat capacities.

PROBLEMS

1. Calculate the heat of formation, in calories per gram-mole, of $SO_3(g)$ from the following experimental data on standard heats of reaction:

$PbO(s) + S(s) + \frac{3}{2}O_2(g) = PbSO_4(s)$; $\Delta H = -165,500$ calories (gram-moles)
$PbO(s) + H_2SO_4 \cdot 5H_2O(l) = PbSO_4(s) + 6H_2O(l)$; $\Delta H = -23,300$ calories
$SO_3(g) + 6H_2O(l) = H_2SO_4 \cdot 5H_2O(l)$; $\Delta H = -41,100$ calories

2. From heat of formation data, calculate the standard heats of reaction of the following, in kcal per kilogram-mole:

(a) $SO_2(g) + \frac{1}{2}O_2(g) + H_2O(l) = H_2SO_4(l)$.
(b) $CaCO_3(s) = CaO(s) + CO_2(g)$.

[10] *J. Am. Chem. Soc.*, **53**, 869 (1931).

(c) $CaO(s) + 3C(graphite) = CaC_2(s) + CO(g)$.

(d) $2AgCl(s) + Zn(s) + aq = 2Ag(s) + ZnCl_2(aq)$.

(e) $CuSO_4(aq) + Zn(s) = ZnSO_4(aq) + Cu(s)$.

(f) $N_2(g) + 3H_2(g) = 2NH_3(g)$.

(g) $N_2(g) + O_2(g) = 2NO(g)$.

(h) $2NaClO_3(s) \rightarrow 2NaCl(s) + 3O_2(g)$.

(i) $CuO(s) + H_2(g) \rightarrow H_2O(l) + Cu(s)$.

(j) $Ca_3(PO_4)_2(s) + 3SiO_2(s) + 5C(s) \rightarrow 3CaO \cdot SiO_2(s) + 5CO(g) + 2P(s)$.

(k) $NH_3(g) + 5O_2(g) \rightarrow 4NO(g) + 6H_2O(g)$.

(l) $3NO_2(g) + H_2O(l) \rightarrow 2HNO_3(aq) + NO(g)$.

(m) $CaF_2(s) + H_2SO_4(l) \rightarrow CaSO_4(s) + HF(l)$.

(n) $P_2O_5(s) + 3H_2O(l) \rightarrow H_3PO_4(aq)$.

(o) $CaC_2(s) + 5H_2O(g) \rightarrow CaO(s) + CO_2(g) + 5H_2(g)$.

(p) $AsH_3(g) + 3O_2(g) \rightarrow As_2O_3(s) + 3H_2O(l)$.

(Heat of formation of arsine = $+43,500$ cal per g-mole)

3. Calculate the heats of formation of the following compounds from the standard heat of combustion data:

(a) Benzene (C_6H_6)(l).

(b) Ethylene glycol ($C_2H_6O_2$)(l).

(c) Oxalic acid $(COOH)_2(s)$.

(d) Aniline ($C_6H_5NH_2$)(l).

(e) Carbon tetrachloride (CCl_4)(l).

(f) Propane(g).

(g) Phenol (s).

(h) Carbon disulfide (l).

(i) Urea (s).

(j) Chloroform (l).

Note: The heat of combustion data given in the table for halogen compounds are based on having all chlorine converted into $HCl(aq)$, by hydrolysis. Accordingly, the actual reaction to be considered is one of the combined oxidation and hydrolysis. For CCl_4 it is all hydrolysis and no oxidation.

4. Calculate the standard heats of reaction of the following reactions, expressed in calories per gram-mole:

(a) $(COOH)_2(s) = HCOOH(l) + CO_2(g)$.
    (oxalic acid)   (formic acid)

(b) $C_2H_5OH(l) + O_2(g) = CH_3COOH(l) + H_2O(l)$.
    (ethyl alcohol)   (acetic acid)

(c) $2CH_3Cl(g) + Zn(s) = C_2H_6(g) + ZnCl_2(s)$.
    (methyl chloride)   (ethane)

(d) $3C_2H_2(g) = C_6H_6(l)$.
    (acetylene) (benzene)

(e) $(CH_3COO)_2Ca(s) = CH_3COCH_3(l) + CaCO_3(s)$.
    (calcium acetate)   (acetone)

(f) $CH_3OH(l) + \frac{1}{2}O_2(g) = HCHO(g) + H_2O(g)$.
    (methyl alcohol)   (formaldehyde)

(g) $2C_2H_5Cl(l) + 2Na(s) = C_4H_{10}(g) + 2NaCl(s)$.
    (ethyl chloride)   (n-butane)

(h) $C_2H_2(g) + H_2O(l) = CH_3CHO(g)$.
    (acetylene)   (acetaldehyde)

(*i*) $C_6H_5NO_2(l) + 3Fe(s) + 6HCl(aq) = C_6H_5NH_2(l) + 3FeCl_2(aq) + H_2O(l)$.

(nitrobenzene)   (aniline)

(Heat of formation of $FeCl_2(aq)$ = $-99,950$ g cal/g-mole)

5. The integral heat of solution of LiCl in water to form a solution of infinite dilution is -8665 calories per gram-mole. Calculate the heat of formation of $LiCl(s)$ from the data of Table XVI (page 272).

6. (*a*) Calculate the number of Btu evolved at 18°C when 90 lb of $ZnCl_2$ are added to 150 lb of water.

(*b*) Calculate the number of Btu evolved when 50 lb of $CaCl_2$ are added to 200 lb of an aqueous solution containing 10% $CaCl_2$ by weight at 18°C.

7. Calculate the heat evolved, expressed as Btu, when the following materials are mixed at 18°C:

(*a*) 50 lb H_2SO_4 and 50 lb H_2O.

(*b*) 50 lb H_2SO_4 and 200 lb of a solution of sulfuric acid and water, which contains 50% by weight of H_2SO_4.

(*c*) 50 lb H_2O and 200 lb of a solution of sulfuric acid and water, which contains 50% by weight of H_2SO_4.

(*d*) 60 lb of $Na_2SO_4 \cdot 10H_2O$ and 100 lb of water.

8. An aqueous sulfuric acid solution contains 50% H_2SO_4 by weight. To 100 grams of this solution are added 75 grams of a solution containing 95% H_2SO_4 by weight. Calculate the quantity of heat evolved.

9. 100 pounds of an oleum solution containing 15.4% free SO_3 is to be diluted with water to make a 30.8% solution of H_2SO_4. Calculate the heat evolved.

10. 100 grams of oleum containing 71.0% free SO_3 are diluted with a large volume of water to infinite dilution. Calculate the heat evolved.

11. One ton of 19.1% H_2SO_4 is to be concentrated to 91.6% H_2SO_4. Steam is available to heat the acid to 302°F and a vacuum can be maintained equivalent to the vapor pressure of 91.6% H_2SO_4 at 150°C, i.e., 14 mm. Calculate the heat to be supplied.

12. A batch of dilute sulfuric acid at 90°F, which weighs 250 lb and contains 20% H_2SO_4 is to be brought to 50% strength by the addition of 98% acid, which is at 70°F. How much heat is abstracted by the cooling system if the temperature of the final acid is 100°F?

By means of a sketch, show how Fig. 53 was used in the solution of this problem.

13. In a continuous concentrating system, dilute acid (60% H_2SO_4) is concentrated to 95% strength. The dilute acid enters the system at 70°F, while the water vapor and the concentrated acid leave the system at the boiling temperature of the latter. How many Btu are needed to concentrate 1000 lb of dilute acid?

By means of a sketch, show how Fig. 53 was used in the solution of this problem.

14. Hydrochloric acid, $G(60°/60°F) = 1.20$, is prepared by absorbing HCl gas at 80°F in water that is admitted to the absorption system at 50°F. If the final acid leaves the system at 70°F, how much heat is withdrawn from the equipment per 1000 lb of acid produced?

By means of a sketch, show how Fig. 52 was used in the solution of this problem.

15. Assume that pure sulfuric acid and water, both at 70°F, are mixed under adiabatic conditions. If the acid is gradually added to the water, what is the maximum temperature that can be attained?

By means of a sketch, show how Fig. 53 was used in the solution of this problem.

16. Two sulfuric acid solutions of 10% and 80% concentrations are to be mixed under adiabatic conditions. If the stronger solution is gradually added to the weaker, what is the maximum temperature attained when initial solutions are at 18°C? By means of a sketch, show how Fig. 53 was used in the solution of this problem.

17. Calculate the heat evolved when 5 lb HCl gas at 80°F are dissolved in 20 lb of 10% HCl at 60°F, to form a solution at 60°F (Fig. 52).

18. Calculate the resultant temperature when 10 lb of water at 120°F are added to 10 lb of 40% HCl at 60°F (Fig. 52).

19. Calculate the heat required to concentrate 40 lb of 5% HCl at 120°F to 8 lb of 20% HCl at 120°F with the vapors leaving at 120°F (Fig. 52).

20. Calculate the final temperature when 100% H_2SO_4 at 60°F is diluted with a solution containing 10% H_2SO_4 at 100°F to form a solution containing 50% H_2SO_4 (Fig. 53).

21. In the following table are values of the standard integral heats of solution, ΔH_s, in calories per gram-mole, of liquid acetic acid in water to form solutions containing m moles of water per mole of acid. (From International Critical Tables, Vol. V, page 159.)

m	ΔH_s	m	ΔH_s
0.25	+ 70	5.00	+ 24
0.58	+126	6.19	− 13
1.11	+149	30.00	− 92
1.42	+149	63.3	−107
1.95	+130		

(a) Using the method of tangent slope, calculate the differential molal heat of solution of acetic acid in a solution containing 15% acetic acid by weight.

(b) Using the method of tangent intercepts, calculate the differential molal heats of solution of acetic acid and of water in a solution containing 50% acetic acid by weight.

22. From the data of Problem 21 calculate the heat, in Btu, evolved when 10 lb of acetic acid are added to 1000 lb of a solution containing 50% acetic acid by weight. Perform the calculation both from integral heat of solution data and from differential heat of solution data derived in Problem 21.

23. From the integral heat of solution data of Problem 21 calculate the heat, in calories, which is evolved at 18°C when 1 liter of aqueous acetic acid containing 75% acid by weight is diluted to 2 liters by the addition of water.

24. The heats of mixing, in calories per gram-mole of solution, of carbon tetrachloride (CCl_4) and aniline ($C_6H_5NH_2$) at 25°C are given in the following table. (From the International Critical Tables, Vol. V, page 155.)

Mole fraction CCl_4	ΔH	Mole fraction CCl_4	ΔH
0.0942	98	0.6215	288
0.1848	169	0.7175	270
0.3005	237	0.7888	246
0.4152	282	0.8627	188
0.4827	291	0.9092	149
0.5504	298		

(a) Calculate the integral and differential molal heats of solution of each component in a solution containing 50% CCl_4 by weight.

(b) Calculate the heat evolved, in Btu, when 1 lb of aniline is added to a large quantity of solution containing 40% CCl_4 by weight.

25. (a) For one of the binary systems from the following list, determine the differential heat of solution of each component at each of the concentrations listed.

From the *International Critical Tables*, Vol. V

Carbon Tetrachloride Benzene		Carbon Disulfide Acetone		Chloroform Acetone	
Weight % CCl₄	ΔH	Weight % CS₂	ΔH	Weight % CHCl₃	ΔH
10	0.301	10	5.78	10	− 4.77
20	0.598	20	11.80	20	− 9.83
30	0.816	30	16.45	30	−14.31
40	0.963	40	19.92	40	−19.38
50	1.030	50	20.93	50	−23.27
60	1.030	60	20.80	60	−25.53
70	0.912	70	17.62	70	−25.07
80	0.699	80	16.15	80	−21.55
90	0.452	90	10.80	90	−13.56
$t = 18°C$		$t = 16°C$		$t = 14°C$	

ΔH = heat of mixing, as joules per gram of solution. (1 joule = 0.2389 calorie.)

(b) Prepare a plot showing the differential heats of solution (calories per gram of component) and the heats of mixing (calories per gram of solution) as a function of the composition, expressed as weight per cent.

(c) Calculate the integral molal heat of solution of each component for a solution containing 50 weight per cent of each component.

(d) Calculate the Btu evolved when 1 lb of the first-named component is dissolved in a large volume of solution containing 60% by weight of the first-named component.

26. Activated charcoal is used for the recovery of benzene (C_6H_6) vapors from a mixture of inert gases. Calculate the heat evolved, in Btu, per pound of benzene absorbed on a large quantity of charcoal at 0°C, when the charcoal contains 0.25 lb of benzene per pound of charcoal.

27. Calculate the heat evolved, in Btu, when 1 lb of SO_2 gas is adsorbed on 6.0 lb of outgassed silica gel at 0°C.

28. Estimate the heat evolved, in Btu, in completely wetting 10 lb of dried clay with water.

29. Silica gel contains 12% H_2O by weight. Calculate the heat, in calories, evolved when 2.0 kg of this material at 0°C are completely wetted with water.

30. Calculate the heat evolved, expressed in Btu, when 100 lb of silica gel (commercially dry basis) adsorbs 25 lb of water at 70°F.

31. Using the data in Table XVIII and Fig. 59, calculate the heat of complete wetting, expressed in Btu, for the following solids when completely wetted with the liquid specified.

(a) 100 lb of dry starch, wetted with water.

(b) 100 lb of dry clay, wetted with ethyl alcohol.

(c) 100 lb of animal charcoal containing 8% water when wetted completely with water.

32. From *International Critical Tables* obtain the following data:

   (*a*) Heat of formation of $SnBr_4(c)$; calories per gram-mole.

   (*b*) Heat of fusion of $SnBr_4(c)$; calories per gram-mole.

   (*c*) Integral heat of solution of $SnBr_4(l)$ in a large amount of water; calories per gram-mole.

   (*d*) Integral heat of solution of $O_2(g)$ in water to form a dilute solution; Btu per pound-mole.

   (*e*) Integral heat of solution of 1.0 mole of $HCl(g)$ in 400 moles of water; Btu per pound-mole.

   (*f*) Heat of transition of rhombic to monoclinic sulfur; calories per gram-mole.

   (*g*) Standard heat of combustion of *o*-toluic acid $(C_8H_8O_2)(s)$; calories per gram.

   (*h*) Heat of mixing of $CS_2(l)$ and ethyl acetate $(C_4H_8O_2)(l)$; calories per gram-mole of mixture.

   (*i*) Total heat evolved in the adsorption of 65 cc of CO_2 on 1 gram of outgassed cocoanut charcoal at 0°C; calories.

   (*j*) Differential heat of adsorption of CH_3OH vapors on activated charcoal containing 100 cc of vapor per gram of charcoal at 0°C; calories per gram-mole.

   (*k*) Heat of wetting of dried bone charcoal with gasoline; calories per gram of charcoal.

33. 200 grams of 20% $CaCl_2$ at 40°C and 300 grams of $CaCl_2 \cdot 6H_2O$ at 20°C are mixed.

   (*a*) Estimate the temperature of the final mixture.

   (*b*) How much heat must be removed to solidify completely the mixture?

34. Estimate the final temperature when 100 lb of air (dry basis) initially at 70°F and a humidity of 0.010 lb per pound of dry air are mixed with 10 lb of silica gel, initially commercially dry and at 70°F. The final humidity of the air = 0.001. Assume no heat losses from the system.

35. Calculate the heat removed when 100 lb of air (dry basis) initially at 70°F and a humidity of 0.010 lb per pound of dry air are mixed with 10 lb of silica gel, initially commercially dry and at 70°F. The final humidity of the air is 0.001. The gel and air are kept at 70°F.

36. Calculate the heat removed when 100 lb of air (dry basis) initially at 70°F and 70% relative humidity are mixed with 10 lb of silica gel, initially commercially dry at 70°F. The final relative humidity of the air is 10% and the final temperature of the air and gel is 80°F.

37. Solve Illustration 17, assuming the average temperature of air is 85°F and average temperature of gel is 95°F. (*Note:* Actually beds of silica gel are operated under intermittent-flow adiabatic conditions. The treatment of the adiabatic case presents formidable difficulties because of the fluctuating temperature waves in the bed.)

38. Calculate the standard heat of reaction, in kcal, accompanying the reduction of 20 kg of Fe_2O_3 by carbon (coke) to form 12 kg of $Fe(s)$. The only other products leaving the process are $FeO(s)$ and $CO(g)$.

39. The heat requirements are to be estimated for a low-temperature reduction process applied to a magnetite ore. The ore averages 90% Fe_3O_4; the balance is inert gangue, largely SiO_2. The process is to be conducted in an externally heated retort which is closed except for the two openings which are to serve as entry for the charge and as exit for the product. The opening for admitting the charge also serves as a vent for the escape of the gas formed by the reaction between the magnetite and the reducing agent. On the basis of laboratory tests, it is anticipated that 95% of the iron will be reduced to the metallic state, with the remainder being reduced to FeO. The reducing agent to be used is a metallurgical coke containing 85% carbon and 15% ash, the latter being largely SiO_2. The coke is to be charged 300% in excess of the theoretical demand for complete reduction. The solid discharged is thus a mixture of Fe, FeO, unused coke and SiO_2 from the ore and from the coke used up in the reduction process. The gas escaping from the retort will be practically pure CO, formed by the reaction between the magnetite and the coke.

Analyses

Coke:	Carbon = 85%		Ore:	Fe_3O_4 = 90%
	Ash = 15%			SiO_2 = 10%

Temperatures		*Specific heats*	
Entering ore = 200°C		Fe_3O_4:	0°C ... 0.151
Entering coke = 200°C			100°C ... 0.179
Leaving solids = 950°C			200°C ... 0.203
Leaving gas = 950°C			300°C ... 0.222

Calculate the material balance for the process, on the basis of 2000 lb of metallic iron produced.

Estimate the heat requirements for the process, as Btu per 2000 lb of metallic iron produced.

40. By the combustion at constant volume of 2.0 grams of $H_2(g)$ to form liquid water at 17°C, 67.45 kcal are evolved. Calculate the quantity of heat which would be evolved were the reaction conducted under a constant pressure at 17°C.

41. When 1.0 gram of naphthalene ($C_{10}H_8$) is burned in a bomb calorimeter, the water formed being condensed, 9621 calories are evolved at 18°C. Calculate the heat of combustion at constant pressure and 18°C, the water vapor remaining uncondensed.

42. A fuel oil analyzes 80% C and 20% H_2 by weight. The standard heat of combustion of this oil is determined in an oxygen bomb calorimeter. Calculate the correction that must be applied to get the heat of reaction at constant pressure. Which is the greater, the heat of reaction at constant pressure, or the heat of reaction at constant volume?

43. Calculate the actual heat of reaction in cal per g-mole for each of the following reactions. Reactants and products are at a constant pressure of 1 atm. The temperature of each reactant and product is as indicated. Assume the mean specific heat of $NaHSO_4$ (0 to 250°C) to be 0.23 and Na_2SO_4 (0 to 600°C) to be 0.26.

a) $SO_2(g) + \frac{1}{2}O_2(g) \rightarrow SO_3(g)$
   450°C   450°C   450°C

b) $NaHSO_4(s) + NaCl(s) \rightarrow Na_2SO_4(s) + HCl(g)$
   250°C   30°C   600°C   300°C

c) $SiO_2(s) + 3C$ (coke) $\rightarrow SiC(s) + 2CO(g)$
   18°C   18°C   1800°C   1200°C

d) $2NaCl(s) + SO_2(g) + \frac{1}{2}O_2(g) + H_2O(g) \rightarrow$
   400°C   400°C   400°C   400°C

         $Na_2SO_4(s) + 2HCl(g)$
         400°C   400°C

44. Derive a general equation for each of the following reactions which will express the heat of reaction as a function of temperature, with the reactants and products at the same temperature. Base all equations on the assumption that H_2O is in the vapor state.

$$CO(g) \ + \tfrac{1}{2}O_2(g) \rightarrow CO_2(g)$$
$$H_2O(g) + CO(g) \rightarrow H_2(g) + CO_2(g)$$
$$N_2(g) \ + 3H_2(g) \rightarrow 2NH_3(g)$$
$$SO_2(g) \ + \tfrac{1}{2}O_2(g) \rightarrow SO_3(g)$$
$$H_2(g) \ + Cl_2(g) \rightarrow 2HCl(g)$$

45. Sulfur dioxide gas is oxidized in 100% excess air with 80% conversion to SO_3. The gases enter the converter at 400°C and leave at 450°C. How many kcal are absorbed in the heat interchanger of the converter per kilogram-mole of SO_2 introduced?

46. Calculate the heat of neutralization in calories per gram-mole of $NaOH(n_1 = 5)$ with $HCl(n_1 = 7)$ at 25°C where n_1 = moles H_2O per mole of solute.

47. A bed of petroleum coke (pure carbon) weighing 3000 kg, at an initial temperature of 1300°C, has saturated steam at 100°C blown through it until the temperature of the bed of coke has fallen to 1000°C. The average temperature of the gases leaving the generator is 1000°C. The analysis of the gas produced is CO_2, 3.10%; CO, 45.35%; H_2, 51.55% by volume, dry basis. How many kilograms of steam are blown through the bed of coke to reduce the temperature to 1000°C? Neglect loss of heat by radiation and assume that no steam passes through the process undecomposed.

48. In Problem 47, 20% of the steam passes through the coke undecomposed. How much steam is blown through the bed of coke to reduce its temperature to 1000°C? Neglect loss of heat by radiation.

49. Steam at 200°C, 50° superheat, is blown through a bed of coke initially at 1200°C. The gases leave at an average temperature of 800°C with the following composition by volume on the dry basis:

H_2	53.5%
CO	39.7%
CO_2	6.8%
	100.0%

Of the steam introduced 30% passes through undecomposed. Calculate the heat of reaction in kcal per kilogram-mole of steam introduced. Mean specific heat of coke (0 to 1200°C) = 0.35.

50. Calculate the number of Btu required to calcine completely 100 lb of limestone containing 80% $CaCO_3$, 11% $MgCO_3$, and 9% H_2O. The lime is withdrawn at 1650°F and the gases leave at 400°F. The limestone is charged at 70°F.

51. Limestone, pure $CaCO_3$, is calcined in a continuous vertical kiln by the combustion of producer gas in direct contact with the charge. The gaseous products of combustion and calcination rise vertically through the descending charge. The limestone is charged at 18°C and the calcined lime is withdrawn at 900°C. The producer gas enters at 600°C and is burned with the theoretically required amount of air at 18°C. The gaseous products leave at 200°C. The analysis of the producer gas by volume is as follows:

CO_2	9.21%
O_2	1.62%
CO	13.60%
N_2	75.57%
	100.00%

Calculate the number of cubic meters (0°C, 760 mm Hg) of producer gas, required to burn 100 kg of limestone, neglecting heat losses and the moisture contents of the air and producer gas.

52. A fuel gas of the following composition at 1600°F is burned in a copper melting furnace with 25% excess air at 65°F.

CH₄	40%
H₂	40%
CO	4%
CO₂	3%
N₂	11%
O₂	2%
	100%

The copper is charged at 65°F and poured at 2000°F. The gaseous products leave the furnace at an average temperature of 1000°F. How much copper is melted by burning 4000 cu ft (32°F; 29.92 in. Hg; dry) of the above gas, assuming that the heat lost by radiation is 50,000 Btu and neglecting the moisture contents of the fuel gas and air?

53. One thousand cubic meters of gas, measured at standard conditions (0°C, 760 mm and dry), containing 20 grams of ammonia per cubic meter (as measured above) are passed into an ammonia absorption tower at a temperature of 40°C, saturated with water vapor, and at a total pressure of 740 mm. The gas is passed upward countercurrent to a descending stream of water which absorbs 95% of the incoming ammonia. The gas leaves the tower at 38°C, saturated with water vapor. Six hundred kilograms of water enter the top of the tower at 20°C. What is the temperature of the solution leaving the tower, neglecting heat losses? Assume the mean molal heat capacity of moisture-free, ammonia-free gas to be 7.2.

54. Pure HCl gas comes from a Mannheim furnace at 300°C. This gas is cooled to 60°C in a silica coil and is then completely absorbed by passing it countercurrent to a stream of aqueous hydrochloric acid in a series of Cellarius vessels and absorption towers. The unabsorbed gas from the last Cellarius vessel enters the first absorption tower at 40°C. Fresh water is introduced in the last absorption tower at 15°C and leaves the first absorption tower at 30°C containing 31.45% HCl (20° Bé acid). This acid is introduced into the last Cellarius vessel and leaves the first vessel at 30°C containing 35.21% HCl (22° Bé acid). There are produced in this system 9000 lb of 22° Bé acid in 10 hours. Calculate separately the heat removed in the cooling coil, Cellarius vessels, and absorption towers, neglecting the presence of water vapor in the gas stream and assuming complete absorption of HCl.

CHAPTER IX

FUELS AND COMBUSTION

Because of the universal use of the combustion of fuels for the generation of heat and power, special techniques and methods have been developed for establishing the material and energy balances of such processes. Each problem should be pursued independently and as rigorously as the available experimental data permit, using the chemical principles involved and not empirical equations.

Heating Values of Fuels. The most important property of a fuel is its heating value, which is numerically equal to its standard heat of combustion but of opposite sign. This property is usually determined by direct experimental measurements, although methods are also given for its estimation.

The major products of complete combustion from practically all fuels are carbon dioxide and water. Two methods of expressing heating values are in common use, differing in the state selected for the water present in the system after combustion. The *total heating value* of a fuel is the heat *evolved* in its complete combustion under constant pressure at a temperature of 18°C when all the water formed and originally present as liquid in the fuel is condensed to the liquid state. The *net heating value* is similarly defined except that the final state of the water in the system after combustion is taken as vapor at 18°C. The total heating value is also termed the " higher " or " gross " heating value; the net is often termed the " lower " heating value. The net heating value is obtained from the total heating value by subtracting the latent heat of vaporization at 18°C of the water *formed and vaporized* in the combustion.

Coke and Carbon. The combustible constituents of cokes and charcoals are practically pure carbon. The heating value of such a fuel may be predicted with accuracy sufficient for most purposes by simply multiplying its carbon content by the heating value per unit weight of carbon.

In Table XIV of Chapter VIII, page **253**, it will be noted that heats of formation of carbon compounds are based on a value of zero assigned to the heat of formation of β graphite. On this basis various other forms of elementary carbon have positive heats of formation. Several types of " amorphous " carbon are included in the table, each having a different heat of formation. These differences arise in part from differences in allotropic forms and in the surface energy of carbon in different states of subdivision and porosity, and in part from the presence of hydrocarbon compounds of high molecular weights and low hydrogen contents.

The heats of combustion of the various forms of amorphous carbon differ by the same amounts as do their heats of formation. For combustion calculations the value of the heat of combustion of carbon is taken as $-96,630$ calories per gram-atom or $-14,495$ Btu per pound. This value is the difference between the heat of formation of carbon dioxide and that of carbon in coke, given in Table XIV, page 253, or $-96,630 = -94,030 - 2600$.

Coal Analyses. Coal consists chiefly of organic matter of vegetable origin which has been altered by decomposition, compression, and heating during long ages of inclusion in the earth's crust. In addition to organic matter it contains mineral constituents of the plants from which it was formed and also inclusions of other inorganic materials deposited in it during its geological formation.

Two types of analysis are in common use for expressing the composition of coal. In an *ultimate analysis*, determination is made of each of the major chemical elements. In a *proximate analysis* four arbitrarily defined groups of constituents are determined and termed *moisture, volatile matter, fixed carbon,* and *ash.* Following are the ultimate and proximate analyses of a typical Illinois coal:

Ultimate		*Proximate*	
Moisture...........	9.61	Moisture...........	9.61
Ash (corrected)......	9.19	Ash...............	9.37
Carbon.............	66.60	Volatile matter......	30.68
Net hydrogen.......	3.25	Fixed carbon........	50.34
Sulfur.............	0.49		
Nitrogen...........	1.42		100.00
Combined H_2O......	9.44		
	100.00		

The proximate analysis of coal should be carried out according to an arbitrarily standardized procedure which has been recommended by the United States Bureau of Mines. The details of this method are described in most books on methods of technical analysis. The determinations may be rapidly and easily carried out, and the majority of the contracts and specifications for the purchase of coal are based on this analysis. The tedious methods of ultimate analysis are completely carried out only when necessary to serve as a basis for energy and material balance calculations. However, the sulfur content is of particular interest, and determination of this element frequently accompanies the proximate analysis.

In both schemes of analysis " moisture " represents the loss in weight on heating the finely divided coal at 105°C for one hour. The material termed " ash " in the proximate analysis is the residue from complete oxidation of the coal at a high temperature in air. This quantity is needed for calculating the quantity of refuse formed in the ordinary combustion of the coal. However, the ash determined in this manner does not accurately represent the mineral content of the original coal because of the changes which take place during combustion. An important mineral component of many coals is iron pyrites, FeS_2. In combustion this is oxidized to form Fe_2O_3, which is weighed in the residual ash, and SO_2 gas. In the oxidation of pyrites 4 gram-atoms (128 grams) of sulfur are replaced by 3 gram-atoms (48 grams) of oxygen, a loss in weight equal to $\frac{5}{8}$ times the weight of pyritic sulfur present. Thus, in order to determine the actual mineral content of the coal, including the pyritic sulfur, it is necessary to add to the ash-as-weighed, a correction equal to $\frac{5}{8}$ of the pyritic sulfur content. To determine the actual mineral content, *not* including the pyritic sulfur, a correction equal to $\frac{3}{8}$ of the pyritic sulfur must be subtracted from the ash-as-weighed. Other less important corrections may also be applied to the ash. Unless otherwise designated, " ash " refers to ash-as-weighed.

To obtain the ultimate analysis, direct determinations are made of carbon, sulfur, nitrogen, and hydrogen by the usual analytical methods. The moisture and ash are determined by the standardized procedures of the proximate analysis. The percentage oxygen content is then taken as the difference between 100 and the sum of the percentages of carbon, hydrogen, sulfur, nitrogen, and corrected sulfur-free ash. It is recommended that, for this calculation, the corrected ash be estimated by assuming that all sulfur in the coal is present in the pyritic form. On this basis,

$$\% \text{ corrected ash} = \% \text{ ash-as-weighed} - \tfrac{3}{8}(\% \text{ S}) =$$
$$\% \text{ mineral content} - \% \text{ S}$$

where $\%$ S = percentage sulfur content of coal. This correction represents only an approximation, since not all sulfur is pyritic and other changes in the mineral constituents may take place in combustion. More refined methods for estimating oxygen content are not ordinarily justified.

In reporting the ultimate analysis it is convenient to consider that all oxygen is in combination with hydrogen to form moisture and " combined water." The surplus hydrogen, above that required to combine with the oxygen, is termed " net " or " available " hydrogen. This represents the hydrogen present in the form of hydrocarbons and available for further oxidation.

Rank of Coal. The sum of the fixed carbon and volatile matter of a coal is termed the *combustible*. The Bureau of Mines has published extensive tables[1] of the ultimate analyses of coals representing hundreds of coal deposits throughout the United States. If the source of a coal is known the ultimate analysis of its combustible matter can be obtained with fair reliability from these tables since the composition of combustible material in any one coal bed is nearly constant. In every coal sample it is necessary, however, to make separate determinations of ash and moisture contents.

The *fuel ratio* of a coal is defined as the ratio of its percentage of fixed carbon to that of volatile matter. The *rank* of the coal, whether bituminous, or anthracite, may be estimated from the fuel ratio. The generally accepted classification of coals and the corresponding ranges of fuel ratios are as follows:

TABLE XIX

RANK OF COALS

Rank	Fuel Ratio
Anthracite	between 10 and 60
Semi-anthracite	between 6 and 10
Semi-bituminous	between 3 and 7
Bituminous	between $\frac{1}{2}$ and 3

Fuels of lower rank than bituminous, namely, sub-bituminous and lignite, may have fuel ratios within the bituminous range but are characterized by higher water or oxygen contents.

The classification of coals on the basis of fuel ratio is not entirely satisfactory for many purposes. Several other methods[2] have been developed which give more nearly exact differentiation.

Heating Value of Coal. The total heating value of a coal may be determined by direct calorimetric measurement and is usually expressed in Btu per pound. The net heating value is obtained by subtracting from the total heating value the heat of vaporization at 18°C of the

---

[1] U. S. Bureau of Mines, *Bulletin* 123. With permission.
[2] Haslam and Russell, *Fuels and Their Combustion*, McGraw-Hill Book Co. (1926).

water present in the coal and that formed by the oxidation of the available hydrogen. Thus,

$$\text{Net } H.V. = \text{Total } H.V. - 9 \times H \times 1056 \tag{1}$$

where $H.V.$ = heating value, Btu per pound

H = weight fraction of total hydrogen, including available hydrogen, hydrogen in moisture, and hydrogen in combined water

When the heating value of coal is determined by burning in a calorimeter the sulfur is oxided to form sulfuric acid. Normally the sulfur in coal burns to form sulfur dioxide only, so that a correction should be made to the calorimetric value for the heat evolved in forming sulfuric acid from sulfur dioxide and water.

Many attempts have been made to develop a method of calculating the heating value of coal from its proximate analysis. None of these methods are sufficiently reliable to justify their use except as approximations.

A fair approximation to the heating value of a coal may be obtained by considering that each of the combustible constituents, carbon, available hydrogen, and sulfur, is present in its elementary state. On the basis of this assumption the heating value is the sum of the quantities of heat evolved in the combustion of each of these elements, using for carbon the heating value of amorphous carbon and for sulfur the heating value of FeS_2. The respective heats of combustion, in Btu per pound, may be obtained from the data of Table XIV, page 253. It is assumed that sulfur is burned to SO_2.

Element	Heat of Combustion
Carbon...............................	14,495 Btu per pound
Hydrogen (total).....................	61,000 Btu per pound
Hydrogen (net).......................	51,550 Btu per pound
Sulfur (as FeS_2).....................	5,770 Btu per pound

Then,

$$\text{Total } H.V. = 14{,}495C + 61{,}000H_a + 5770S \tag{2}$$

where

$H.V.$ = heating value, Btu per pound

C, H_a, S = weight fractions of carbon, available hydrogen, and sulfur, respectively

Equation (2) is known as Dulong's formula. It is not theoretically sound because it neglects the heats of formation of the compounds of carbon, sulfur, and hydrogen which exist in the coal. However, as previously pointed out, the heats of formation of hydrocarbon compounds are small in comparison to their heats of combustion, and the results of the above equation are rarely in error by more than 3 per cent. The experimentally observed total heating value of the coal whose analysis is given above, page 324, was 11,725 Btu per pound. Applying Equation (2), the heating value would be predicted as (14,495 × 0.666) + (61,000 × 0.0325) + (5770 × 0.0049) = 11,920 Btu per pound, an error of +1.7 per cent.

Because of the fact that the heating value of coal is more easily determined than its ultimate analysis, the use of Equation (2) for calculation of heating value is rarely advantageous. It is more useful as a means of predicting the available hydrogen content from experimentally determined values of carbon and sulfur contents and heating value.

A useful relationship has been pointed out by Uehling[3] between the heating value of coal *per pound of total carbon* and its rank. It was found that, for each rank of coal, the heating value per pound of total carbon is nearly constant, rarely varying by more than 2 per cent. It was also found that the weight of available hydrogen, per pound of total carbon, varies but little among coals of the same rank. On this basis, standard average heating values and available hydrogen contents, per pound of total carbon, were established for the different ranks of fuel. These values, contained in the following table, were based on the published results of a large number of analyses carried out by the United States Bureau of Mines.

TABLE XX

STANDARD HEATING VALUES AND NET HYDROGEN CONTENTS OF COAL

$H.V.'$ = heating value, Btu per pound of total carbon
H_a' = available hydrogen content, pounds per pound of total carbon

Rank	$H.V.'$	H_a'
Coke	14,495	0.0
Anthracite	16,100	0.029
Semi-bituminous	17,400	0.049
Bituminous	17,900	0.054
Sub-bituminous	17,600	0.045
Lignite	17,100	0.037

From the data of Table XX, the heating value of the bituminous coal whose analysis was given on page 324 would be predicted as 17,900 × 0.666 = 11,920 Btu per pound, in error by only 1.7 per cent as com-

[3] E. A. Uehling, *Heat Loss Analysis*, McGraw-Hill Book Co. (1929). With permission.

pared to the experimentally determined value of 11,725.   The available hydrogen content would be predicted as 0.666 × 0.054 × 100 = 3.6 per cent as compared to the experimentally observed value of 3.25 per cent.

Because of the relative difficulty of the total carbon determination as compared to the calorimetric determination of heating value, the relationships pointed out by Uehling have their greatest value in predicting the ultimate analysis from the experimentally determined heating value. From only the heating value and the data of Table XX, good approximations to the total carbon and available hydrogen content of a coal may be predicted.   In view of the fact that the composition of a coal sample will frequently vary by as much as 5 per cent from the true average composition of the coal from which it was taken, the accuracy of these predictions is often as great as is justifiable for calculations of energy and material balances.

PETROLEUM

Petroleum oils are complex mixtures of hydrocarbons including four important series of compounds: paraffins, naphthenes, olefins, and aromatics.   These compounds differ in hydrogen content in the order listed, paraffins having the highest hydrogen content and aromatics the lowest. In naturally occurring petroleums the first two series predominate; in *cracked* products formed by decomposition of natural oils large quantities of olefins and aromatics may also be present.   In addition to hydrocarbons varying quantities of sulfur, oxygen, and nitrogen compounds are generally present.

Because of the complexity of petroleum fractions determination of the actual compounds present is generally impossible.   Elementary analyses may be made, determining carbon, hydrogen, sulfur, and nitrogen as for coal.   Data of this type are available in the publications of the U. S. Bureau of Mines for many naturally occurring petroleums.   However, such analyses give little indication of the actual character of an oil and its thermal properties.   Approximate methods have been developed whereby much of this information may be estimated from easily determined physical properties such as the distillation or boiling range, the specific gravity, and the viscosity.

Characterization of Petroleum.   For general correlation of the average physical properties of petroleum stocks of different types, it is necessary to develop a means of quantitatively expressing the general character of the oil.   Paraffin hydrocarbons of maximum hydrogen content may be considered as one extreme and aromatic materials of minimum hydrogen content as the other.

To serve as a quantitative index to this property, which may be termed paraffinicity, the U.O.P. characterization factor[4] has been developed and empirically related to six commonly available laboratory inspections. Although this factor is not an exact measure of chemical type and does not show perfect constancy in a homologous series, these disadvantages are, to a considerable extent, offset by its simplicity and convenience of definition and use.

The definition of the U.O.P. characterization factor arose from the observation that, when a crude oil of supposedly uniform character is fractionated into narrow cuts, the specific gravity of these cuts is approximately proportional to the cube roots of their absolute boiling-points. The proportionality factor may then be taken as an index of the paraffinicity of the stock. Thus

$$K = \frac{\sqrt[3]{T_B}}{G} \tag{3}$$

where

K = U.O.P. characterization factor
T_B = average boiling point, degrees Rankine
G = specific gravity at 60°F

When dealing with mixtures of wide boiling range a special method of obtaining the average boiling point as described by Watson and Nelson must be used. For narrower cuts the 50 per cent point of the Engler distillation may be taken as the average boiling point.

The characterization factor shows fair constancy throughout the boiling range of a number of crude oils and for others may either increase or decrease in the higher boiling range. In the paraffin series fair constancy for the average of the reported isomers exists up to a boiling temperature of 700°F. Values of the characterization factor range as follows:

Pennsylvania stocks.............	12.2–12.5
Midcontinent stocks	11.8–12.0
Gulf Coast stocks...............	11.0–11.5
Cracked gasolines...............	11.5–11.8
Cracking plant combined feeds....	10.5–11.5
Recycle stocks.................	10.0–11.0
Cracked residuums.............	9.8–11.0

The characterization factor is readily calculated from Equation (3) from only the specific gravity and average boiling point, or it may be read directly from API gravity and average boiling point by interpola-

[4] Watson, Nelson, and Murphy, *Ind. Eng. Chem.*, **25**, 880 (1933); **27**, 1460 (1935). With permission.

tion between the curves of Fig. 63. In this figure, API gravities are plotted as ordinates and average boiling points as abscissas with lines of constant K from Equation (3). The relationship between specific gravity and degrees API is shown by Fig. B in the Appendix.

It has also been found that a fair empirical correlation exists between

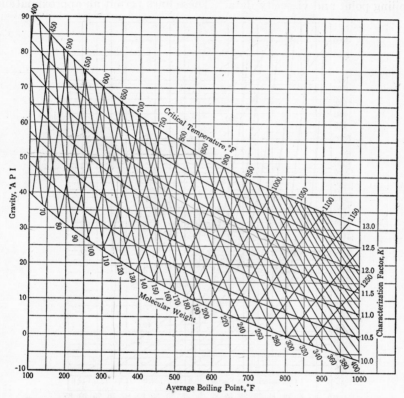

FIG. 63. Molecular weights, critical temperatures and characterization factors of petroleum fractions.

(Reproduced in " C.P.P. Charts ")

the characterization factor and the viscosity-gravity relationship at a given temperature. Paraffinic stocks have high viscosities as compared with aromatic materials of the same gravities.

Because of uncertainties of molecular aggregation at low temperatures, the viscosity measurements used for physical correlations should be made at as high temperature as possible. In Fig. 64[5] viscosity in centistokes at 122°F is plotted against API gravity for stocks of constant characterization factors. By use of the centistoke scale of viscosity

[5] Watson et al., ibid.

the entire range of fractions from light gasolines to heavy residues is covered in a single relationship. A chart for conversion of common viscometer readings into centistokes is included in Fig. C, page 442.

Lines of constant boiling point are plotted on Fig. 64 resulting from combination of the relationships between characterization factor from boiling point and viscosity data. These lines permit an approximation

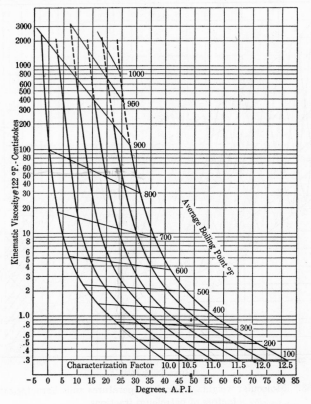

Fig. 64. Characterization factor from viscosity at 122°F.
(Reproduced in " C.P.P. Charts ")

to the boiling point from only viscosity and gravity data. This relationship is particularly useful for heavy stocks on which boiling point data can be obtained only under high vacuum. However, because of the large change in viscosity with a slight change in the gravity of heavy stocks, boiling points estimated in this way may be considerably in error, sometimes as much as 50°F for the heavier residues. Similar charts were developed on the basis of viscosities at other temperatures.[6]

[6] Watson, Nelson, and Murphy, *Ind. Eng. Chem.*, **27**, 1460 (1935).

Molecular Weights of Petroleum Fractions. The average molecular weights of petroleum fractions may be satisfactorily estimated from average boiling point and gravity. Aromatic stocks of low characterization factors have lower molecular weights than paraffinic materials of the same average boiling points.

The relationship between molecular weight, characterization factor, boiling point, and API gravity is included in the curves of Fig. 63. By interpolation between these curves, molecular weights may be estimated, with errors rarely exceeding 5 per cent. If boiling-point data are not available, the boiling point may be estimated from viscosity using Fig. 64.

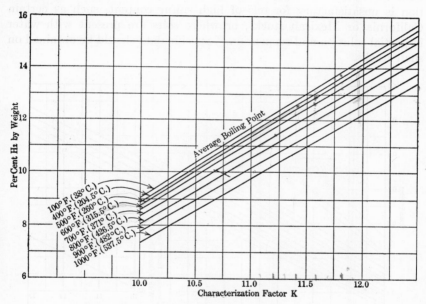

FIG. 65.   Characterization factor *vs.* weight % H_2.

(Reproduced in " C.P.P. Charts ")

Critical Properties. The critical temperature curves of Fig. 63 were calculated directly from Equation (III–8), page 69, but are in satisfactory agreement with the existing data on petroleum.

Critical temperatures estimated from Fig. 63 are applicable with little error to pure hydrocarbons, narrow petroleum cuts, or wide-boiling mixtures if a proper method of obtaining average boiling point is used. Correct methods of averaging have been developed by Smith and Watson.[7] Critical pressures may be estimated by the methods described in Chapter III.

[7] R. L. Smith and K. M. Watson, *Ind. and Eng. Chem.*, **29**, 1408 (1937).

Hydrogen Content. The curves of Fig. 65 represent a relationship between hydrogen content and characterization factor for materials of constant boiling points.

Figure 65 combined with the preceding charts permits estimation of hydrogen content from a knowledge of only the specific gravity and one other property. Ordinarily the error will be less than 0.5 per cent, based on the total weight of the oil, except for highly aromatic, low-boiling materials.

Petroleum oils ordinarily contain little ash and in the absence of specific data may be assumed to be 97 per cent carbon and hydrogen with the remainder oxygen, nitrogen, sulfur, and ash. This assumption is unsatisfactory for oils of high sulfur content, such as certain California or Mexican stocks, or where salts are present with water in partial solution and suspension. Specific data should be obtained on such stocks.

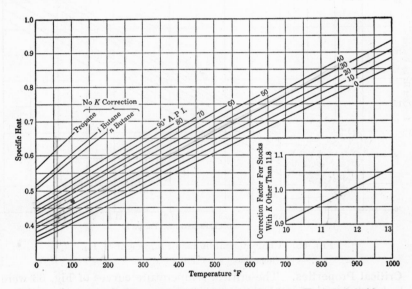

Fig. 66. Specific heat of liquid petroleum oils where $K = 11.8$ (mid-continent stocks). For other stocks multiply by correction factor.

(Reproduced in " C.P.P. Charts ")

Specific Heats of the Liquid State. The following equation was recommended by Fallon and Watson[8] for the specific heats of liquid hydrocarbons and petroleum fractions at temperatures between 0°F

[8] J. Fallon and K. M. Watson, presented before Petroleum Div., Am. Chem. Soc., Pittsburgh meeting, Sept., 1943.

and reduced temperatures of 0.85.

$$C_p = [(0.355 + 0.128 \times 10^{-2}\,°\mathrm{API}) + (0.503 + 0.117$$
$$\times\ 10^{-2}\,°\mathrm{API}) \times 10^{-3}t]\,[0.05K + 0.41]$$

where t is in °F and K is the characterization factor.

Figure 66 is a plot of this relationship together with curves for the individual light paraffin hydrocarbons as recommended by Holcomb and Brown.[9] The curves on the main plot apply directly to Midcontinent stocks whose characterization factors are approximately 11.8. For other stocks the value read from the main plot is multiplied by a correction factor derived as a function of K from the small plot in the lower right-hand corner.

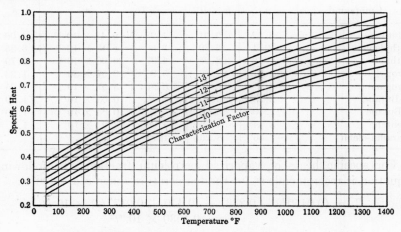

FIG. 67. Specific heats of paraffin gases at atmospheric pressure.
(Reproduced in " C.P.P. Charts ")

Specific Heats of the Vapor State. At temperatures below atmospheric the molal heat capacities of complex gas molecules approach a constant value of 7.95 whereas at high temperatures the curves expressing heat capacity as a function of temperature become concave downward. Since this complicated form of relationship is difficult to express in a single equation, Fallon and Watson[8] proposed the use of the following two forms of equations:

For temperatures from 50 to 1400°F:

$$c_p = a + bT + cT^2$$

For temperatures from -300 to $+200$°F (never above 200°F):

$$c_p = 7.95 + uT^v$$

where T is in °R and c_p is the molal heat capacity.

[9] Holcomb and Brown, *Ind. Eng. Chem.*, **34**, 590 (1942). With permission.

In Table XXI are values of the constants of these equations for the light paraffins and olefins at low pressures. Also included are the constants of an equation expressing the approximate differences between the heat capacities of paraffins of more than three carbon atoms and those of the corresponding olefins.

TABLE XXI

HEAT CAPACITIES OF HYDROCARBON GASES

For 50 to 1400°F, $c_p = a + bT + cT^2$

For −300 to 200°F, $c_p = 7.95 + uT^v$

T = °Rankine

Compound	a	$b \times 10^3$	$-c \times 10^6$	u	v
Methane	3.42	9.91	1.28	6.4×10^{-12}	4.00
Ethylene	2.71	16.20	2.80	8.13×10^{-11}	3.85
Ethane	1.38	23.25	4.27	6.20×10^{-5}	1.79
Propylene	1.97	27.69	5.25	2.57×10^{-3}	1.26
Propane	0.41	35.95	6.97	3.97×10^{-3}	1.25
n-Butane	2.25	45.40	8.83	0.93×10^{-2}	1.19
i-Butane	2.30	45.78	8.89	0.93×10^{-2}	1.19
Pentane	3.14	55.85	10.98	3.9×10^{-2}	1.0
Paraffin-olefin	−1.56	8.26	1.72	....	....

For specific heats of vaporized petroleum fractions the following equation is recommended[10] for the temperature range of 0–1400°F:

$$C_p = (0.0450K - 0.233) + (0.440 + 0.0177K) \times 10^{-3}t \\ - 0.1530 \times 10^{-6}t^2$$

where t is in °F, C_p is specific heat, and K is the characterization factor. This relationship was found to be independent of gravity or boiling point and is plotted in Fig. 67 which is applicable to all petroleum fractions and hydrocarbons containing more than four carbon atoms.

Heat of Vaporization. Heats of vaporization of hydrocarbons and petroleum fractions under atmospheric pressure were calculated by Fallon and Watson[10] by differentiating Equation (III–16), page 73, and substituting in the Clapeyron Equation (III–1). The resultant correlation is plotted in Fig. 68 relating heat of vaporization to boiling point and either API gravity or molecular weight. When working with pure compounds use of the molecular weight is preferable. Heat of vaporization at other pressures may be obtained by means of Equation (VII–32), page 233.

[10] Fallon and Watson, presented before Petroleum Div., Am. Chem. Soc., Pittsburgh meeting, Sept., 1943. To be published *Nat. Petr. News*, 1944.

The values of Fig. 68 are in close agreement with the Kistyakowsky equation for low-boiling compounds but are considerably higher for high-boiling materials. It is believed that these higher values represent a more reliable extrapolation than that of the Kistyakowsky equation.

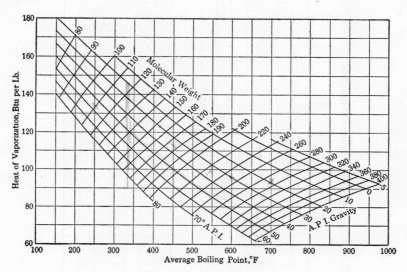

Pressure = 1.0 atmosphere

FIG. 68. Heats of vaporization of hydrocarbons and petroleum fractions. (*Note:* Base on molecular weights rather than API if available.)

(Reproduced in " C.P.P. Charts ")

Heat of Combustion. Average values of heats of combustion of petroleum fractions and hydrocarbons are plotted in Fig. 69 as a function of API gravity and characterization factor. These are total heating values, corresponding to formation of liquid water at 60°F.

Illustration 1. A fuel oil has an API gravity of 14.1 and a viscosity of 150 Saybolt Furol seconds at 122°F. Estimate the characterization factor, average boiling point, hydrogen content, specific heat at 200°F, heating value, and average molecular weight of this oil.

From the conversion chart, Fig. C in the Appendix, it is found that 150 Saybolt Furol seconds is equivalent to 320 centistokes.

(a) Characterization factor, Fig. 64   = 11.35
(b) Average boiling point, Fig. 63 or 64   = 880°F
(c) Hydrogen content, Fig. 65   = 11.5
(d) Specific heat 200°, Fig. 66 = 0.485 × 0.975 = 0.473
(e) Average molecular weight, Fig. 63   = 410
(f) Heating value, Fig. 69   = 18,825 Btu per lb

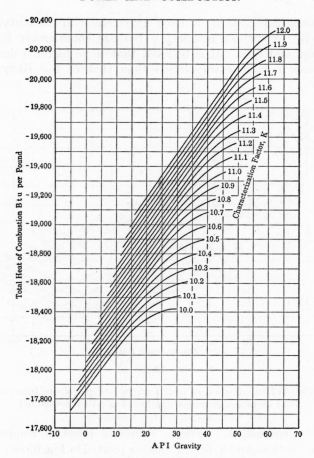

FIG. 69.   Total heats of combustion of liquid petroleum hydrocarbons.

(Reproduced in " C.P.P. Charts ")

FUEL GAS

The standard basis which has been adopted for the expression of the total heating value of a fuel gas is the number of Btu evolved when one cubic foot of the gas, at a temperature of 60°F, a pressure of 30 inches of mercury, and saturated with water vapor, is burned with air at the same temperature, and the products cooled to 60°F, the water formed in the combustion being condensed to the liquid state. Since gas is rarely burned under these standard conditions of temperature and pressure, the heating value per standard cubic foot is not a convenient unit for calculations. However, the unit is widely used as a basis for specifications and legal standards.

Since the vapor pressure of water at a temperature of 60°F is 0.52 inch of mercury, the heating value per standard cubic foot represents the heating value of 1 cubic foot of moisture-free gas under a pressure of $30 - 0.52$ or 29.48 inches of mercury and a temperature of 60°F. The number of moles of moisture-free gas in the standard cubic foot is equal to $1.0 \times \dfrac{492}{520} \times \dfrac{29.48}{29.92} \times \dfrac{1}{359} = 0.002597$ pound-mole. Conversely, $\dfrac{1}{0.002597}$ or 385.5 standard cubic feet of fuel gas contain 1 pound-mole of moisture-free gas if the gas behaves ideally.

The heating value of a fuel gas of known composition may be calculated as the sum of the heats of combustion of its components. The necessary data may be obtained from Table XV, page 262. The total heating values of the common combustible gases are also contained in Table XXII, expressed in Btu per standard cubic foot.

Fuel gases generally contain complex mixtures of both saturated and unsaturated hydrocarbons. The individual analytical determination of each component of these mixtures is not feasible for ordinary industrial purposes. However, Watson and Ceaglske[11] have described a simple scheme of industrial gas analysis which yields data suitable for ordinary combustion calculations. In this scheme carbon monoxide and hydrogen are separately determined and reported as such. The saturated paraffin hydrocarbon gases are reported in terms of a hypothetical compound C_nH_{2n+2}, representing the average composition of the mixture of paraffins in the gas. Similarly, the unsaturated hydrocarbons or illuminants are reported in terms of a hypothetical compound of average composition, C_aH_b. For example, the analysis of a gas might be: CO, 40 per cent; H_2, 42 per cent; $C_{2.5}H_{4.2}$ (illuminants), 7 per cent; $C_{1.2}H_{4.4}$ (paraffins), 11 per cent.

An analysis of this type may be used as effectively for stoichiometric calculations as though all components were individually determined. The heating value of the gas may also be calculated by means of the following approximate formulas for the total heating values of mixtures of paraffins and of unsaturated hydrocarbons.

Paraffin hydrocarbons, C_nH_{2n+2}:

Calories per gram-mole $= 158,100n + 54,700$

Btu per cu ft at 60°F, 30 in., sat. $= 745n + 258$ (4)

Unsaturated hydrocarbons, C_aH_b:

Calories per gram-mole $= 98,200a + 28,200b + 28,800$

Btu per cu ft at 60°F, 30 in., sat. $= 459a + 132b + 135$ (5)

[11] *Ind. Eng. Chem., Anal. Ed.* (January, 1932).

TABLE XXII

Heating Values and Flame Temperatures of Gases

$H.V.$ = total heating value, Btu per standard cubic foot, measured at 60°F, 30 inches of Hg and saturated with water vapor (assuming ideal gas behavior)

Gas	Formula	$H.V.$	Theoretical (assuming complete combustion)	Calculated* (allowing for equilibrium conditions)	Actual*
			Maximum Flame Temperatures with Air at 18°C		
Carbon monoxide	CO	315	2440°C		
Hydrogen	H_2	319	2200		
Paraffins:					
Methane	CH_4	994	1980	1918°C	1880°C
Ethane	C_2H_6	1741	2150	1949	1895
Propane	C_3H_8	2478	2300	1967	1925
n-Butane	C_4H_{10}	3212	2080	1973	1900
n-Pentane	C_5H_{12}	3947	2090		
Olefins:					
Ethylene	C_2H_4	1575	2240	2072	1975
Propylene	C_3H_6	2297	2200	2050	1935
Butylene	C_4H_8	3017	2200	2033	1930
Amylene	C_5H_{10}	3814	2180		
Acetylene	C_2H_2	1544			
Aromatics:					
Benzene	C_6H_6	3686	2240		
Toluene	$C_6H_5CH_3$	4399	2240		
Mesitylene	$C_6H_3(CH_3)_3$	5845	2240		
Naphthalene	$C_{11}H_8$	5795			

* Jones, Lewis, Friauf and Perrott, *J. Am. Chem. Soc.*, **53**, 869 (1931).

If the analysis of a gas is carried out carefully, its heating value may ordinarily be predicted by means of these equations with an error of less than 2 per cent. Larger errors arise if large quantities of acetylene are present in the gas.

Illustration 2. A city gas has the following composition by volume:

CO_2........................... 2.6%
$C_{2.73}H_{4.72}$ (unsaturateds).......... 8.4%
O_2............................. 0.7%
H_2............................. 39.9%
CO............................. 32.9%
$C_{1.14}H_{4.28}$ (paraffins).............. 10.1%
N_2............................. 5.4%
100.0%

(a) Calculate the theoretical number of moles of oxygen which must be supplied for the combustion of 1 mole of the gas.

(b) Calculate the heating value of the gas in calories per gram-mole and Btu per standard cubic foot.

Solution:

(a) *Basis:* 100 gram-moles of gas.

Oxygen required for:

Unsaturateds = $8.4(2.73 + 4.72/4)$ = ... 32.8
Hydrogen = $39.9/2$ = ... 19.95
Carbon monoxide = $32.9/2$ = ... 16.45
Paraffins = $10.1 (1.14 + 4.28/4)$ = ... 22.3
Total................................... 91.5

Oxygen to be supplied per mole of gas = $0.915 - 0.007 = 0.908$ mole.

(b) *Basis:* 1.0 gram-mole of gas.

Heating value of:

Hydrogen = $0.399 \times 68,320$ = ... 27,250 cal
Carbon monoxide = $0.329 \times 67,410$ = ... 22,180 cal
Unsaturateds =
$0.084 [(2.73 \times 98,200) + (4.72 \times 28,200)$
$+ 28,800]$ = ... 36,100 cal
Paraffins =
$0.101 [(1.14 \times 158,100) + 54,700]$ = ... 23,730 cal
Total................................ 109,260 cal

$$\text{Btu per standard cubic foot} = \frac{109,260 \times 1.8}{385.5} = 510.$$

Incomplete Combustion of Fuels. The standard heating values of fuels correspond to conditions of complete combustion of all carbon to carbon dioxide gas, hydrogen to liquid water, and sulfur to sulfur dioxide gas. If a fuel is burned in such a manner that complete combustion does not result, the standard heat of reaction may be calculated by subtracting from its standard heat of combustion the standard heats of combustion of the combustible products formed.

Illustration 3. A coal having a heating value of 12,180 Btu per lb and containing 68.1% total carbon is burned to produce gases having the following composition by volume on the moisture-free basis.

CO_2.......................... 12.4%
CO............................ 1.2%
O_2.............................. 5.4%
N_2............................. 81.0%
100.0%

Calculate the standard heat of reaction in Btu per pound of coal burned.

Basis: 1.0 lb-mole of flue gas.

C in CO_2 = 0.124 lb-atom or......................	1.49 lb
C in CO = 0.012 lb-atom or......................	0.14 lb
Total carbon = 1.49 + 0.14 = ...	1.63 lb
Coal burned = 1.63 ÷ 0.681 = ...	2.39 lb
Heating value of coal = 2.39 × 12,180 = ...	29,100 Btu
Heat of combustion of CO = 0.012(−67,410 × 1.8) =	−1460 Btu
Standard heat of reaction = −29,100 − (−1460) =	
−27,640 Btu or − 27,640 ÷ 2.39 = ...	−11,560 Btu
	per lb of coal

MATERIAL AND ENERGY BALANCES

As previously discussed, in establishing an energy balance all sources of thermal energy are entered on the input side of the balance and all items of heat utilization and dissipation on the output side. It is ordinarily desirable to base all thermal quantities on a reference temperature of 18°C, thus permitting direct use of standard thermochemical data. Other reference temperatures may be used if desired, but in any event it is necessary that each complete balance be based on a single constant-reference temperature.

Where a fuel is used in an industrial reaction two different points of view are emphasized in establishing an energy balance, depending upon whether or not the fuel is intended primarily as a source of heat or principally as a reducing agent. In the first instance the entire heating value of the fuel is listed on the input side of the balance and the entire heating value of the products resulting from the partial combustion of the fuel and its reaction with the charge on the output side. In this instance the utilization of the heating value of the fuel is of principal interest for heating purposes or in producing a fuel gas which is subsequently to be used for heating. In the second instance, where fuel is used primarily as a reducing agent, as in the reduction of ores, the principal interest is in the products of reduction and not in the heating value of the fuel or of the products of reaction. In this latter instance it is customary to include on the input side of the energy balance the heat evolved in the partial combustion of the fuel, which represents the difference between the heating value of the fuel and the heating value of the combustible products resulting from the incomplete combustion of that fuel.

With these two points of view in mind, the input and output items of an energy balance of a chemical process, based upon a reference temperature of 18°C, are distributed in the following classification:

Input Items

Group 1.   The enthalpy, both sensible and latent, of each material entering the process.

Group 2.   Where fuel is used primarily as a source of heat or in the production of fuel gases, the total heating value of the fuel.

Group 3.   The heat evolved at 18°C in the formation of each final product from the initial reactants when such heat effects are exothermic.   Where a fuel is used and its total heating value is included as an input item the products formed from the fuel by combustion or reaction of the fuel with the charge are not considered in this group.

Group 4.   All energy supplied directly to the process from external sources in the forms of heat, electrical or radiant energy or work.

Output Items

Group 1.   The enthalpy of each material leaving the process.

Group 2.   Where fuel is used primarily as a source of heat or in the production of fuel gases the total heating value of each product resulting from the partial combustion of the fuel and its reaction with the charge.

Group 3.   The heat absorbed at 18°C in the formation of each final product from the initial reactants when such heat effects are endothermic.   Where the total heating value of a fuel is included as an input item the products formed from the fuel are not considered in this group.

Group 4.   All heat transferred from the process for useful purposes as for the generation of steam in a boiler furnace.

Group 5.   All energy lost from the process in the form of heat, electrical, or radiant energy or work.

The difference between input group 2 and output group 2 represents the difference between the heating value of the fuel and the heating value of the products of incomplete combustion of that fuel and is considered as one of the items in input group 3 when the fuel is used primarily as a reducing agent.   Separate consideration of the heating values of the fuel and of its products is desirable to produce an energy balance of greater economic significance where recovery of heat is a primary objective.

An energy balance shows how much energy is consumed by necessary endothermic reactions, how much is transferred to a heat interchanger or stored in a fluid used for supplying useful heat or power, and how much heat is wasted owing to incomplete combustion of fuels, to overheating of products, and to inadequate thermal insulation.

For example, in a coal-fired boiler furnace an energy balance indicates the distribution of the chemical energy of the coal into the enthalpy of steam, the heat lost in the gaseous products due to the presence of combustible gases and to sensible heat, the heat loss due to incomplete combustion of coal as represented by the unburned coke and coal in the refuse, and the loss of heat by radiation and conduction through the boiler

setting. The justification of further insulation, of increasing the size of
the combustion space, of increasing the draft, and of using automatic
stoking can be answered, at least in part, from the study of such an
energy balance.

Enthalpy of Water Vapor. The enthalpy of superheated water vapor
referred to the liquid at 18°C is the sum of three separate items:

1. The sensible enthalpy of the water in the liquid state at the saturation temperature of the vapor. This item may be either positive or
negative, depending on whether the saturation temperature is above or
below 18°C.

2. The heat of vaporization of the water at the saturation temperature.

3. The sensible enthalpy of the water vapor referred to the saturation
temperature.

For the sake of consistency the enthalpy of water vapor is determined
in this manner in all energy balances developed in the following pages.
However, where water vapor is highly superheated, as in flue gases, it is
general practice to simplify this calculation by assuming that the
*enthalpy of water vapor is equal to the sum of the heat of vaporization at
18°C plus the sensible enthalpy of the vapor referred to 18°C.* In effect,
this is assuming a saturation temperature of 18°C for the water vapor.
This assumption introduces negligible errors where the partial pressure
of the water vapor is small, of the order of one atmosphere or less.

THERMAL EFFICIENCY

The thermal efficiency of any process is defined as the percentage of
the total heat input which is effectively utilized in the desired manner.
It is evident that the thermal efficiency of a process may be expressed
in a number of different ways, depending on the method of designating
the total heat input and the effectively utilized heat.   In stating a value
of thermal efficiency it is always necessary to specify fully the basis upon
which it was calculated.

The total heat input on which the efficiency is based may be taken as
the total of the input items of the energy balance.   This would seem to be
the most logical basis for general usage, and unless otherwise specified
thermal efficiencies will be considered as on this basis and termed thermal
efficiencies based on *actual heat* input.   However, many special bases
are in common use to fit the needs of particular processes.   In expressing the thermal efficiency of a combustion process it is customary to
obtain the total heat input by deducting from the input items of the
energy balance the heat of vaporization, at 18°C, of the water vapor
present in the air used for combustion.   On this basis the thermal efficiency is termed the thermal efficiency *based on total heating value.*

Because of the fact that in many combustion processes water will not be condensed from the products of combustion, even if they are cooled to 18°C, it is sometimes considered that the thermal efficiency of the apparatus should be based on the net heating value of the fuel. The total heat input is then obtained by considering only the net heating value of the fuel, plus all the sensible heat supplied. A thermal efficiency on this basis is termed the thermal efficiency *based on the net heating value.* This method of expression has as its principal advantage the fact that the percentage efficiencies are higher and appear more encouraging. However, it is undesirable because combustion apparatus is available which is capable of recovering some of the latent heat of the water vapor from the gaseous products. On this basis of expression such apparatus might have an efficiency above 100 per cent. The thermal efficiency based on total heating value is a better general criterion for judging the operation of a combustion process.

Percentage efficiency is also dependent on the quantity of heat which is designated as effectively utilized. Various interpretations frequently may be made of this quantity. For example, a furnace and steam boiler unit used in domestic heating might be considered as effectively utilizing only the heat represented by the enthalpy of the steam produced. On the other hand, it might be logical to include as effective heat the radiation from the furnace itself which is used in heating the room in which it is situated. The efficiency of the unit might be expressed on either basis.

A gas producer or water gas generator produces a combustible gas at a relatively high temperature. If the gas can be utilized while hot its sensible enthalpy as well as its heating value should be included in the effectively utilized heat of the producer unit. The efficiency of the unit on this basis is termed the *hot thermal efficiency.* If the gas must be cooled before use its sensible heat is not useful and only the heating value can be classed as heat effectively utilized in the producer. The efficiency on this basis is termed the *cold thermal efficiency.*

COMBUSTION OF FUELS

In calculating the material and energy balances of processes involving the partial or complete combustion or decomposition of fuels, the same principles are employed whether such fuels are gaseous, liquids, or solid. The material balance of a simple combustion process includes the weights of fuel and air supplied and the weights of refuse and gases produced. This material balance can be calculated fully from a knowledge of the chemical composition of the four items mentioned without any direct measurements of the weights except the weight of fuel consumed. The weights (or volumes) of air and gaseous products are usually not measured because of the great difficulties involved and because these can generally be calculated indirectly with greater accuracy than by direct measurement.

In the burning of coal on a grate as in a boiler furnace the weight and composition of fuel used and composition of gaseous products are measured directly. The chemical analysis of the fuel should include the percentages of carbon, hydrogen, oxygen, moisture, nitrogen, and ash. It may not be necessary to have a complete ultimate analysis, but in any event the carbon, moisture, and ash content should be known. A complete analysis of the dry gaseous products is always necessary. The moisture content in the gaseous products can be calculated, provided the hydrogen content of the fuel is known, or can be determined by measuring the dew point of the gas.

The refuse from the furnace may be considered as consisting of ash, coked carbon, and unchanged combustible matter from the coal. The composition in terms of these constituents may be estimated from a determination of ash, fixed carbon, and volatile matter in the refuse, using the standard scheme of proximate analysis. The weight of refuse actually formed per unit weight of coal should be calculated on the basis of the *ash* contents of refuse and coal, as reported in the proximate analysis, and not on the basis of the *corrected ash* reported in the ultimate analysis.

The air entering the furnace may be assumed to be of average atmospheric composition and its humidity determined by a psychrometric method. The analysis of the flue gases is ordinarily determined by the Orsat type of apparatus, yielding the percentages of carbon dioxide, carbon monoxide, oxygen, and nitrogen in the *moisture-free* gases. For more nearly accurate work, determinations of methane and hydrogen should also be made.

Material and energy balances of combustion processes are based either upon a unit weight of fuel or upon the weight of fuel used in a given cycle or unit time of operation. For example, in boiler furnaces operating continuously the analysis can be based on a period of twenty-four hours or reduced to a basis of one pound of coal consumed. In operating a ceramic kiln of the batch type the analysis should be conducted over a complete cycle of operation including time of preheating and firing and the final results based on the entire cycle of operation.

When sufficient experimental data are collected the same scheme of calculations may be employed for all problems in combustion. However, complete information is seldom available and it becomes necessary to devise methods of circumventing these limitations.   Extensive information can often be built up from but few data, and complete material and energy balances established from a few temperature measurements, the proximate analysis of coal, and an Orsat analysis of gas.

The various calculations which follow illustrate the modifications in procedure necessary to make the best use of data available and also to deal with the special variations in combustion processes represented in four special cases:

Case 1.   Combustion of coal in a boiler furnace where:

    a. Complete ultimate analysis of fuel is known.

    b. No uncoked coal appears in refuse.

    c. Tar and soot are negligible.

    d. Sulfur is negligible.

Case 2.   Combustion of coal where:

    a. Hydrogen and nitrogen contents are unknown.

    b. Uncoked coal drops into refuse.

Case 3.   Combustion of coal where sulfur is not negligible.

Case 4.   Partial combustion of fuel, as in a gas producer, where:

    a. Steam is admitted.

    b. Tar and soot are not negligible.

Case 1.   Combustion of Coal in Boiler Furnace.   The simplest problem in combustion calculations exists where complete information is available or can be directly estimated on the ultimate analysis of coal, where tar and soot in the gases are negligible, where the sulfur content of the fuel is negligible, and where no uncoked coal drops into the refuse. The methods employed in this illustration are general and may be similarly applied to all problems in the combustion or partial combustion of a carbonaceous fuel whether solid, liquid, or gaseous.

The material balance of a furnace is represented by the following items:

Input

    1. Weight of fuel charged.

    2. Weight of dry air supplied.

    3. Weight of moisture in air supplied.

Output

    1. Weight of dry gaseous products.

    2. Weight of water vapor in gaseous products.

    3. Weight of refuse.

The method of calculating each of these items will be discussed in detail.   General methods for such calculations have already been discussed in Chapters I and VI.

Illustration 4. From a 12-hour test conducted on a coal-fired steam generating plant the following data were obtained.

Data on Coal Fired

Ultimate analysis

Carbon...	65.93%
Available hydrogen..................................	3.50%
Nitrogen...	1.30%
Combined water.....................................	6.31%
Free moisture......................................	4.38%
Ash..	18.58%
Total...	100.00%

Total heating value = ...	11,670 Btu per lb
Total weight of coal fired = ...	119,000 lb
Average temperature of coal.........................	65°F

Data on Refuse Drawn from Ash Pit

Ash content = ...	87.4%
Carbon content....................................	12.6%
Average temperature...............................	255°F
Mean specific heat from 65 to 255°F.................	0.23

(Estimated from Fig. 39, page 219)

Data on Flue Gas

1. Orsat analysis

Carbon dioxide....................................	11.66%
Oxygen...	6.52%
Carbon monoxide..................................	0.04%
Nitrogen..	81.78%
Total...	100.00%
Average temperature..............................	488°F

Data on Air

Average dry-bulb temperature.........................	73.0°F
Average wet-bulb temperature........................	59.4°F
Average barometric pressure = 29.08 in. Hg	

Data on Steam Generated

Average feed water temperature....................	193°F
Weight of water evaporated.......................	1,038,400 lb
Average steam pressure = 137.4 lb per sq in. gauge	
Quality of steam.................................	98.3%

Calculate the material and energy balances for the entire plant.

MATERIAL BALANCE

All calculations are based upon 100 lb of coal as fired.

1. Weight of Refuse Formed.

Where the refuse is not weighed directly its weight can be readily calculated from its ash content and that of the coal. The following method is correct where mineral sulfides are not present in the coal.

Ash content of coal............................... 18.58 lb

Ash content per pound of refuse.................... 0.8740 lb

Weight of refuse formed $= \dfrac{18.58}{0.8740} =$ .... 21.2 lb

2. Weight of Dry Gaseous Products.

A direct measurement of the weight of gaseous products from a combustion process is seldom made because of the many difficulties involved. Pitot tubes measure inaccurately because of the low velocities encountered in chimneys and flues.   Orifice and Venturi meters are similarly unreliable because of low-pressure drops encountered and because soot accumulates in the openings.   Electric flow meters read inaccurately if the composition of the gas varies with respect to carbon dioxide or water vapor.   In any case the direct measurement of gas streams is made extremely difficult because of variation in temperature and velocity across each section of the stream.   Any accurate measurement must give a correct integrated value of velocity and temperature over the entire cross section.   Because of these uncertainties and troubles it becomes easier and more nearly accurate to calculate the weight of gaseous products from the stoichiometric relationships of combustion.

The complete analysis of the gaseous products includes the percentages of carbon dioxide, carbon monoxide, oxygen, methane, ethane, hydrogen, and nitrogen present.   Moisture content is not revealed in the usual gas analysis because the entire analysis is conducted with the gas sample saturated with water vapor at a constant temperature and pressure.

The general rule is recommended that the *weight of dry gaseous products should be calculated from a carbon balance.*   A carbon balance is selected as the basis of this calculation for two reasons.   In the first place, carbon is determined with a higher degree of precision in both fuel and gaseous products than any other element present.   Secondly, carbon is the chief constituent in both fuel and gaseous products so that a slight error in its determination will not be magnified in subsequent calculations.   To calculate the weight of gaseous products from a material balance of any other element would invite many additional sources of error.   For example, the hydrogen balance would be out of the question because of the many sources of hydrogen, its relatively low percentage content, its several outlets, and its various methods of combination.

Carbon Balance

Carbon gasified.   *Basis:* 100 lb coal fired.

Carbon in coal $= 100 \times 0.6593 =$ ...... 65.93 lb or 5.49 lb-atoms

Carbon in refuse $= 21.2 \times 0.1260 =$ ..... 2.67 lb or 0.22 lb-atoms

Carbon entering stack gases $=$ .... 63.26 lb or 5.27 lb-atoms

Carbon in stack gases. *Basis:* 1.0 lb-mole of gas.

Carbon in CO_2 = ... 0.1166 lb-atom

Carbon in CO = ... $\underline{0.0004}$ lb-atom

     Total carbon = ... 0.1170 lb-atom

Moles of dry stack gas per 100 lb coal fired
$$= 5.27/0.1170 = 45.1 \text{ lb-moles}$$

Total dry gaseous products. *Basis:* 100 lb coal fired.

CO_2	= 45.1 × 0.1166 =	5.26	lb-moles or × 44 =	....	231.5	lb
CO	= 45.1 × 0.0004 =	0.018	lb-moles or × 28 =	....	0.560	lb
O_2	= 45.1 × 0.0652 =	2.94	lb-moles or × 32 =	...	94.2	lb
N_2	= 45.1 × 0.8178 =	36.90	lb-moles or × 28.2 =	...	1041	lb
	Total..........	45.1	lb-moles	....	1367	lb

Average molecular weight = 1367/45.1 = ... 30.3

3. Weight of Dry Air Supplied.

Direct measurement of the weight or volume of air used in combustion is accompanied by the same difficulties as the direct measurement of gaseous products. Furthermore, air is usually drawn through the grate by chimney draft so that there is no need for confining the supply of air in ducts and there is no opportunity for direct measurement of its flow.

The dry air used in combustion consists of oxygen and inert gases, chiefly nitrogen. These inert gases also include argon and traces of rare gases, but because of the small amount present it is customary to include all the inert gases as nitrogen and assign a *molecular weight of 28.2 to atmospheric " nitrogen."* [12] This nitrogen passes through the furnace unchanged and appears entirely in the gaseous products. Any nitrogen present in the fuel burned will also appear in the flue gases. The nitrogen in ordinary solid and liquid fuels burned will usually be negligible or very small. However, in the combustion of gases a considerable portion of the nitrogen appearing in the flue gases may come from the gaseous fuel.

The composition of dry air may ordinarily be taken as constant, containing 21.0 per cent oxygen and 79.0 per cent nitrogen by volume, the nitrogen content including the argon present. Under certain conditions the air used in combustion may contain appreciable amounts of carbon dioxide, making a separate gas analysis of the air valuable. The moisture content of air is subject to extreme variations depending upon weather conditions so that a separate determination of the moisture content of air is invariably necessary.

Because of the constancy of composition of dry air it is possible to cal-

[12] The molecular weight of atmospheric " nitrogen " is taken as 28.2 because of the argon associated with it.

culate readily the weight of air used in a combustion process from a knowledge of the nitrogen content of the gaseous products and of the fuel used. Accordingly, the general rule is expressed that the *weight of dry air actually used in combustion process is calculated from a nitrogen balance.* The chief objection to the use of the nitrogen balance basis is that in gas analysis, errors resulting from unabsorbed components accumulate on the nitrogen determination which is always found by difference.

Nitrogen Balance.   Basis: 100 lb coal fired.

Nitrogen in gaseous products = ... 36.90 lb-moles
Nitrogen from coal = 1.30/28.0 = ... 0.0464 lb-mole
Nitrogen from air = ... 36.85 lb-moles
Dry air supplied = 36.85/0.79 = ... 46.6 lb-moles
       or 46.6 × 29 = ... 1354 lb

It will be noted that the nitrogen content of the coal might be neglected without introducing a serious error.

4. Weight of Moisture in Air.

The weight of moisture per mole of dry air depends upon the temperature, pressure, and relative humidity of the air.   From the dew point the partial pressure of the water vapor is determined, and the moisture content of the air may be calculated by the methods explained in Chapter IV.

Dry-bulb temperature = ... 73.0°F
Wet-bulb temperature = ... 59.4°F
From Fig. 9, the molal humidity of the air is 0.012
Water supplied with air = 46.6 × 0.012 = ... 0.559 lb-mole

5. Total Volume of Wet Air Introduced.

Basis: 100 lb coal fired.

Total moles of moist air = 46.6 + 0.559 = ... 47.2 lb-moles
Volume at 73°F, 29.08 in. Hg =

$$47.2 \times 359 \times \frac{29.92}{29.08} \times \frac{533}{492} = \ \ldots\ldots\ldots \quad 18{,}870 \text{ cu ft}$$

6. Weight of Moisture in Gaseous Products.

To complete the material balance it is necessary to know the weight of moisture in the gaseous products since this is not obtained by the ordinary gas analysis.   Direct measurement of the moisture content is difficult.   It can also be calculated if the composition of the dry flue gases, the moisture content of the air used, and the hydrogen and moisture in the fuel burned are known.   As a general rule it may be stated that the *moisture content of the gaseous products is calculated from a hydrogen balance.*

Hydrogen Balance. Basis: 100 lb coal fired.

Input

From moisture introduced with dry air = ...	0.559 lb-mole
From combined water in coal = 6.31/18 = ...	0.351 lb-mole
From free moisture in coal = 4.38/18 = ...	0.244 lb-mole
From available hydrogen in coal = 3.50/2.016 = ...	1.738 lb-moles
Total..................................	2.892 lb-moles

Output

The Orsat apparatus used for analyzing the stack gases does not determine hydrogen or hydrocarbon gases. However, in an efficiently operated boiler furnace, hydrogen and hydrocarbons are present only in small quantities. It is therefore assumed that all the hydrogen introduced into the system leaves as water in the stack gases.

Hydrogen in H_2O of stack gases = ...   2.892 lb-moles

The dew point of the stack gases may now be determined.

Partial pressure of $H_2O = \dfrac{2.892}{45.1 + 2.892} \times 29.08 = 1.75$ in. Hg

From Table I, this partial pressure is seen to correspond to a dew point of 36°C or 97°F.

7. Total Volume of Gaseous Products.

Basis: 100 lb coal fired.

Moles of wet gas = 45.1 + 2.892 = ...   48.0 lb-moles

Volume at 488°F and 29.08 in. Hg =

$$48.0 \times 359 \times \frac{29.92}{29.08} \times \frac{948}{492} = \text{.......... } 34{,}150 \text{ cu ft}$$

Summary of Material Balances. To verify the accuracy of experimental data or methods of calculation a summary of all material balances is prepared. The overall material balance is also indicated on the flow chart of Fig. 70.

OVERALL MATERIAL BALANCE

Input		Output	
Coal...................	100 lb	Refuse................	21.2 lb
Dry air (46.6 lb-moles)...	1354 lb	Dry gases (45.1 lb-moles)	1367 lb
H_2O in air (0.559 lb-mole)	10.05 lb	H_2O in stack gases	
		(2.892 lb-moles)......	52.1 lb
Total.............	1464 lb	Total............	1440 lb

CARBON BALANCE

In coal................	65.93 lb	In gases..............	63.26 lb
		In refuse.............	2.67 lb
Total.............	65.93 lb		65.93 lb

NITROGEN BALANCE

In air (36.85 lb-moles)...	1040 lb	In stack gases	
In coal..................	1.30 lb	(36.90 lb-moles)........	1041 lb
Total..............	1041 lb	Total..............	1041 lb

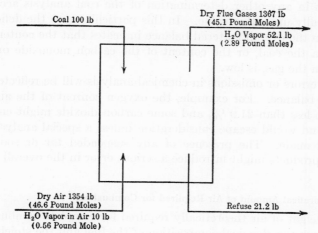

Coal 100 lb

Dry Flue Gases 1367 lb
(45.1 Pound Moles)

H₂O Vapor 52.1 lb
(2.89 Pound Moles)

Dry Air 1354 lb
(46.6 Pound Moles)
H₂O Vapor in Air 10 lb
(0.56 Pound Mole)

Refuse 21.2 lb

FIG. 70. Material balance of steam generating plant (Illustration 4).

HYDROGEN BALANCE

In water vapor of air		In H_2O of stack gases	
(0.559 lb-mole)........	1.127 lb	(2.892 lb-moles)......	5.83 lb
In combined water of coal			
(0.351 lb-mole)........	0.708 lb		
In free moisture of coal			
(0.244 lb-mole)........	0.491 lb		
Available hydrogen of coal			
(1.738 lb-moles).......	3.50 lb		
Total..............	5.83 lb	Total..............	5.83 lb

ASH BALANCE

In coal................	18.58 lb	In refuse................	18.58 lb

OXYGEN BALANCE

In combined water of coal		In CO_2 of stack gases	
$6.31 \times 16/18$.........	5.61 lb	5.26×32..........	168.5 lb
In free moisture of coal		In O_2 of stack gases	
$4.38 \times 16/18$.........	3.89 lb	2.94×32..........	94.2 lb
In dry air		In CO of stack gases	
$36.85 \times 21/79 \times 32$...	313.5 lb	$0.01804/2 \times 32$......	0.289 lb
In water vapor in air.....		In H_2O of stack gases	
$0.559/2 \times 32$.........	8.95 lb	$2.892/2 \times 32$........	46.3 lb
Total..............	332 lb	Total.............	309 lb

It will be seen that there is a deficit of 24 lb or 1.6 per cent on the output side of the overall material balance. This discrepancy falls entirely on the oxygen balance since no direct calculations were made from this basis. The oxygen content of the fuel is obtained by difference, so that all errors in any other determination of the coal analysis accumulate algebraically upon this value. In this particular case the deficit of 24 pounds in the overall material balance indicates that the content of the carbon in the coal, or the content of the carbon monoxide or carbon dioxide in the gas, is low.

Other errors or omissions in chemical analysis will be reflected in the material balance. For example, the oxygen content of the air supply might be less than 21.0 %, and some carbon dioxide might enter with the air and would escape consideration unless a special analysis of the air were made. The presence of any suspended tar or soot in the gaseous products might introduce a serious error in the overall material balance.

8. Theoretical Amount of Air Required for Combustion.

The weight of air theoretically required for complete combustion depends upon the chemical composition of the fuel and the stoichiometric relations involved in combustion. Since the one element in common for all combustion reactions is oxygen, the *weight of air required for combustion must be calculated from an oxygen balance.* The oxygen already in the fuel is assumed to be in combination with hydrogen. Hence only the available hydrogen of the fuel is considered in calculating its oxygen requirement.

Per Cent Excess Air
  Basis: 100 lb coal fired.

OXYGEN BALANCE

Oxygen requirements for combustible constituents of coal charged.

	Oxygen Required
Carbon 65.93 lb = 5.49 lb-atoms..............	5.49 lb-moles
Net hydrogen 3.50 lb = 1.736 lb-moles..........	0.868 lb-mole
Total...................................	6.358 lb-moles
Air required = 6.358/0.21 = ...	30.3 lb-moles
Air supplied = ...	46.6 lb-moles
Excess air = 46.6 − 30.3 = ...	16.3 lb-moles
Per cent excess air = 16.3/30.3 × 100 = ...	53.9%

ENERGY BALANCE

Since it is conventional to utilize total rather than net heating values in energy balances the reference state for all water involved in the process is the liquid state; the enthalpies of reactants and products are evaluated on this basis.

Reference temperature: 65°F.

Basis: 100 lb coal fired.

Input

1. Heating value of the coal = 100 × 11,670 = ...   1,167,000 Btu
2. Enthalpy of the coal = ...   0 Btu
3. Enthalpy of the dry air =
   46.6 × 6.96 × (73 − 65) = ...   2,590 Btu
4. Enthalpy of the water vapor accompanying the dry air.

   Dew point of the air from Fig. 9 = 49°F
   Heat of vaporization = 19,140 Btu per lb-mole at 49°F
   Enthalpy = 0.559[8.00(73 − 49) + 19,140 − 18(65 − 49)] =   10,570 Btu

   Total heat input . . . .   1,180,160 Btu

Output

1. Heating value of refuse.

   Weight of carbon in refuse = 2.67 lb
   Heating value = 2.67 × 14,550 = ...   38,850 Btu

2. Heating value of stack gases.

   Lb-moles of CO in stack gases = 0.01804 lb-mole
   Heating value = 0.01804 × 67,410 × 1.8 = ...   2,190 Btu

3. Enthalpy of refuse.

   Weight of refuse = 21.2 lb
   Enthalpy = 21.2 × 0.23 × (255–65) = ...   1,090 Btu

4. Enthalpy of dry stack gases.

   Mean heat capacities between 65 and 488°F from Fig. 37.
   Enthalpies

$$CO_2 = 5.26 \quad (9.94) \quad (488 - 65) = \quad 22,120 \text{ Btu}$$
$$CO = 0.018 \quad (7.03) \quad (488 - 65) = \quad 50 \text{ Btu}$$
$$O_2 = 2.94 \quad (7.22) \quad (488 - 65) = \quad 8,990 \text{ Btu}$$
$$N_2 = 36.90 \quad (7.02) \quad (488 - 65) = 109,500 \text{ Btu}$$

   Total = 140,660 Btu

5. Enthalpy of the water vapor in stack gases.

   Dew point of stack gases = 97°F
   Heat of vaporization (Fig. 8) =
   18,660 Btu per lb-mole at 97°F
   Enthalpy of the liquid = 2.892 × 18 × (97 − 65) = ...   1,665 Btu
   Heat of vaporization = 2.892 × 18,660 = ...   53,960 Btu
   Superheat = 2.892 × 8.21 × (488 − 97) = ...   9,290 Btu

   Total Enthalpy . . . .   64,920 Btu

6. Heat utilized in generating steam.

$$\text{Pounds of steam generated} = \frac{1,038,400}{119,000} \times 100 = 873 \text{ lb}$$

Enthalpy of 1 lb steam as produced (151.7 lb per sq in. absolute, 1.7% moisture), relative to
32°F =

$331.4 + 862.3 - (0.017 \times 862.3) =$	1179.0 Btu per lb
Enthalpy of feed water at 193°F relative to 32°F =	160.9 Btu per lb
Net heat input into steam produced = ...	1018.1 Btu per lb
Total heat absorbed by steam produced = $873 \times 1018.1 =$	890,000 Btu

7. Undetermined losses (by difference) = ...   42,450 Btu

Summarized Energy Balance of Steam Generating Plant:

Reference temperature: 65°F

Basis: 100 lb coal fired.

Energy Input

	Btu	Per cent
1. Heating value of the coal......................	1,167,000	98.9
2. Enthalpy of the coal...........................	0	0
3. Enthalpy of the dry air.......................	2,590	0.2
4. Enthalpy of the water vapor accompanying the dry air.......................................	10,570	0.9
Total.................................	1,180,160	100.0

Energy Output

	Btu	Per cent
1. Heating value of refuse........................	38,850	3.3
2. Heating value of stack gases....................	2,190	0.2
3. Enthalpy of refuse............................	1,090	0.1
4. Enthalpy of the dry stack gases.................	140,660	11.9
5. Enthalpy of the water vapor in the stack gases....	69,920	5.5
6. Heat utilized in generating steam...............	890,000	75.4
7. Undetermined losses (by difference).............	42,450	3.6
Total.................................	1,180,160	100.0

This energy balance is summarized in Fig. 71.

Thermal Efficiency and Economy. The thermal efficiency of a boiler furnace may be calculated on the total or the net heating value of the coal. The effectively utilized heat is that which is absorbed in steam generation.

Based on total heating value of coal,

$$\text{the thermal efficiency is } \frac{890,000}{1,167,000 + 2590} \text{ or} \qquad 76.2\%$$

Based on net heating value of coal,

$$\text{the thermal efficiency is } \frac{890,000}{1,167,000 + 2590 - (42)(1060)} \text{ or} \qquad 79.1\%$$

Case 2. Combustion Calculations Where Ultimate Analysis of Coal Is Not Completely Known. In the preceding illustration, the calculations were completed without making any assumptions as to the composition of the fuel. However, the determination of the hydrogen content of coal is a difficult procedure, to be avoided if possible. For

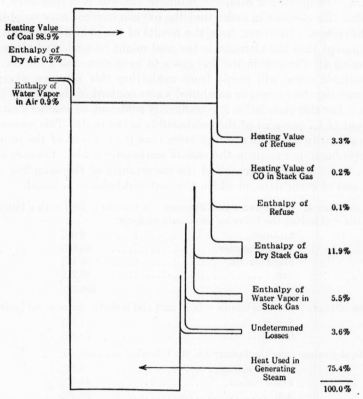

FIG. 71.   Energy balance of steam generating plant (Illustration 4).

this reason it is sometimes desirable to evaluate the material balance of a furnace or gas producer without data on hydrogen content and to calculate this quantity from an oxygen balance. In this illustration an additional complication is introduced in that some of the coal drops through the grate without coking.

Where the hydrogen content of the fuel is calculated from an oxygen balance great care must be taken in analyzing the flue gases. The sampling and determination of oxygen in hot flue gases are particularly uncertain. For this reason it is frequently preferable to estimate the hydrogen content of the fuel by empirical methods such as those illustrated on page 328 for coal, page 334 for gases, and Fig. 65, page 333 for oils. The calculations are then carried out as in the preceding illustration.

In the preceding illustration the nitrogen content of the coal was known. In making a complete ultimate analysis it is necessary to determine this element in order that the oxygen content may be obtained by difference. However, from the results of the previous illustration it is apparent that the nitrogen in the coal might be neglected altogether, assuming all nitrogen in the flue gases to have come from the air. No appreciable error will result from neglecting this nitrogen except in determining the oxygen or combined water content of the coal by difference. For this calculation it is ordinarily sufficient to assume a nitrogen content of 1.7 per cent of the combustible in the coal. This assumption will ordinarily not be in error by more than 0.3 per cent of the weight of the combustible except in the case of anthracite coals. Greater refinement is not justified because of the uncertainty of the sampling of the coal and of other data on which the material balance is based.

Illustration 5. Coal-Fired Boiler Furnace. A furnace is fired with a bituminous material coal having the following proximate analysis:

Moisture.........................	2.9%
Volatile matter....................	33.8%
Fixed carbon......................	53.1%
Ash.............................	10.2%
	100.0%

The ultimate analysis is known only in part and includes (as-received basis):

Sulfur...........................	1.1%
Carbon...........................	73.8%

The dry refuse from the furnace has the following composition:

Volatile matter....................	3.1%
Fixed carbon......................	18.0%
Ash.............................	78.9%
	100.0%

The Orsat analysis of the flue gases is as follows:

Carbon dioxide....................	12.1%
Carbon monoxide..................	0.2%
Oxygen..........................	7.2%
Nitrogen.........................	80.5%
	100.0%

Air enters the furnace at a temperature of 65°F with a percentage humidity of 55%. The barometric pressure is 29.30 in. of Hg. The flue gases enter the stack at a pressure equivalent to 1.5 in. of water less than the barometric pressure and at a temperature of 560°F.

Water is fed to the boiler at a temperature of 60°F and vaporized to form wet steam at a gauge pressure of 100 lb per sq in., quality 98%, at a rate of 790 lb of steam or water per 100 lb of coal charged.

Compute complete material and energy balances, the volumes of air and flue gases per 100 lb of coal charged, and the percentage excess air used.

MATERIAL BALANCE

The calculations are similar to Illustration 4 with special methods in parts 1, 4, 5.

1. Total carbon content of refuse.

Basis: 100 lb of coal charged.

The weight of refuse is calculated from the ash contents of the coal and of the refuse. The weight of ash in 100 lb of coal is 10.2 lb. This weight of ash constitutes but 78.9% of the weight of refuse.

$$\text{Total weight of refuse} = \frac{10.2}{0.789} = \dots\dots\dots\dots \quad 12.9 \text{ lb}$$

Carbon exists in the refuse as fixed carbon and as volatile matter. The volatile matter is due to the dropping of uncoked coal through the grate. The combustible of the uncoked coal in the refuse may be assumed to have the same composition as the combustible of the coal fired. Therefore, the ratio of combustible to volatile matter in the uncoked coal in the refuse will be the same as that in the coal.

Ratio of combustible matter to volatile matter in the coal

$$= \frac{33.8 + 53.1}{33.8} = \dots\dots\dots\dots\dots\dots\dots\dots\dots\dots\dots\dots\dots\dots \quad 2.56$$

Volatile matter in refuse $= 12.9 \times 0.031 = \dots\dots\dots\dots\dots\dots$ 0.40 lb
Unchanged combustible in refuse $= 0.40 \times 2.56 = \dots\dots\dots\dots\dots$ 1.02 lb

Carbon is also present in the refuse as coked coal accompanied by no volatile matter. The amount of carbon as coke is the difference between quantities of total combustible and of unchanged coal combustible in the refuse.

Fixed carbon in refuse $= 12.9 \times 0.18 = \dots\dots\dots\dots\dots\dots\dots$ 2.32 lb
Total combustible in refuse $= 2.32 + 0.40 = \dots\dots\dots\dots\dots\dots$ 2.72 lb
Coked combustible (carbon) $= 2.72 - 1.02 = \dots\dots\dots\dots\dots$ 1.70 lb

The total carbon content of the unchanged combustible in the refuse may be determined from the percentage of carbon in the combustible of the original coal.

$$\text{Total carbon content of combustible of coal} = \frac{73.8}{33.8 + 53.1} = \dots \quad 85\%$$

Carbon in uncoked coal in refuse $= (0.85)(1.02) = \dots\dots\dots\dots$ 0.87 lb
Total carbon in refuse $= 1.70 + 0.87 = \dots\dots\dots\dots\dots\dots$ 2.57 lb

2. Weight of dry flue gases.

Carbon Balance. *Basis:* 100 lb of coal charged.

Carbon gasified $= 73.8 - 2.57 = 71.2$ lb or $\dots\dots\dots\dots\dots$ 5.94 lb-atom

1.0 lb-mole of dry flue gases contains:

Carbon dioxide $\dots\dots\dots\dots\dots\dots\dots\dots\dots\dots\dots\dots$ 0.121 lb-mole
Carbon monoxide $\dots\dots\dots\dots\dots\dots\dots\dots\dots\dots\dots$ 0.002 lb-mole
Total carbon per pound-mole of gas $= \dots\dots\dots\dots$ 0.123 lb-atom

$$\text{Total dry flue gas} = \frac{5.94}{0.123} = \dots\dots\dots\dots\dots\dots\dots\dots \quad 48.3 \text{ lb-moles}$$

Total dry gases:

CO_2 = 48.3 × 0.121 =   5.85 lb-moles or ×44 = ...   258 lb
CO = 48.3 × 0.002 =   0.096 lb-moles or ×28 = ...   3 lb
O_2 = 48.3 × 0.072 =   3.48 lb-moles or ×32 = ...   111 lb
N_2 = 48.3 × 0.805 = 38.87 lb-moles or ×28.2 = ..   1097 lb

Total = ...   48.30 lb-moles or...........   1469 lb

3. Weight of air supplied.

Nitrogen Balance.   Basis: 100 lb of coal charged.

N_2 in flue gases = ...   38.87 lb-moles

Assuming all N_2 to come from the air:

Dry air supplied = $\dfrac{38.87}{0.79}$ = ...   49.2 lb-moles

or 49.2 × 29 = ...   1430 lb

From Fig. 9:

Molal humidity of air = 0.012
Water vapor in air = 0.012 × 49.2 = ...   0.59 lb-mole
  or 0.59 × 18 = ...   10.6 lb
Total wet air = 49.2 + 0.59 = ...   49.8 lb-moles
  or 1430 + 10.6 = ...   1440 lb
Volume of air entering at 65°F, 29.3 in. of Hg =

49.8 × 359 × $\dfrac{525}{492}$ × $\dfrac{29.92}{29.30}$ = ...   19,450 cu ft

4. Hydrogen content of coal.

Oxygen Balance.   Basis: 100 lb of coal charged.

Oxygen in dry flue gas = 5.85 + $\dfrac{0.096}{2}$ + 3.48 =   9.38 lb-moles

Oxygen entering in dry air = 49.2 × 0.21 = ...   10.32 lb-moles

Assuming that the oxygen not accounted for in the dry flue gases was consumed in oxidizing the available hydrogen of the coal:

O_2 oxidizing H_2 = 10.32 − 9.38 = ...   0.94 lb-mole
H_2 burned = 2 × 0.94 = 1.88 lb-moles or......   3.79 lb

The hydrogen burned may be taken as the available hydrogen of the coal, neglecting the small hydrogen content of the uncoked combustible in the refuse.

5. Complete ultimate analysis of coal.
The unknown items of the ultimate analysis are combined water and nitrogen. As pointed out on page 358, the nitrogen content may be assumed to be 1.7 × 0.87 = 1.4%. The combined water may then be determined as the difference between 100 and the sum of the percentages of moisture, carbon, hydrogen, sulphur, nitrogen, and *corrected* ash.

Corrected ash = 10.2 − 3/8(1.1) = ...   9.8%
Combined H_2O = 100 − (2.9 + 73.8 + 3.8 + 1.1 + 1.4
  + 9.8) = ...   7.2%

Ultimate analysis:

Moisture.......................	2.9%
Carbon...........................	73.8%
Available H	3.8%
Sulfur.........................	1.1%
Nitrogen.......................	1.4%
Corrected ash..................	9.8%
Combined H_2O..................	7.2%
	100.0%

6. Water vapor in flue gases.

Hydrogen Balance.   Basis: 100 lb of coal charged.

H_2O from air =............................   0.59 lb-mole

H_2O from coal = $\dfrac{2.9 + 7.2}{18}$ =................   0.56 lb-mole

H_2O formed from H = ...   1.88 lb-moles

Total =.............................   3.03 lb-moles
                                        or 55 lb

7. Volume of wet flue gases.

Moles of wet flue gas = 48.3 + 3.03 =.......   51.33 lb-moles

Pressure in flue = $29.30 - \dfrac{1.5}{13.6}$ =...........   29.19 in. of Hg

Volume at 560°F, 29.19 in. of Hg =

$51.33 \times 359 \times \dfrac{1020}{492} \times \dfrac{29.92}{29.19}$ =...........   39,200 cu ft

8. Complete material balance.

Input

Coal..	100 lb
Dry air (49.2 lb-moles)...........................	1430 lb
Water vapor in air (0.59 lb-mole)..................	10 lb
Total.......................................	1540 lb

Output

Dry flue gases (48.3 lb-moles).....................	1469 lb
Refuse.......................................	13 lb
Water vapor in flue gas (3.03 lb-moles).............	55 lb
Total.......................................	1537 lb

This material balance is summarized in Fig. 72.

9. Percentage excess air.

Basis: 100 lb of coal charged.

Total carbon in coal charged = 73.8 lb =.......   6.15 lb-moles

Available hydrogen in coal charged =..........   1.88 lb-moles

Total O_2 required = $\dfrac{1.88}{2}$ + 6.15 =...........   7.09 lb-moles

Air theoretically required = $\dfrac{7.09}{0.21}$ =...........   33.8 lb-moles

Air actually supplied =......................   49.2 lb-moles

Percentage excess air $= \dfrac{49.2 - 33.8}{33.8} = 45.5$, based on that required for complete combustion of all carbon and available hydrogen in the coal charged and neglecting that required for sulfur.

The percentage excess air may also be calculated directly from the flue-gas analysis:

Basis: 100 lb-moles of flue gas.

Neglecting N_2 from the coal:

O_2 introduced in air $= \dfrac{80.5}{0.79} \times 0.21 = \ldots\ldots\ldots$ 21.4 lb-moles

O_2 in excess $=$ (surplus present in gases minus that required to complete oxidation of the CO) $=$

$7.2 - \dfrac{0.2}{2} = \ldots\ldots\ldots\ldots\ldots\ldots\ldots\ldots\ldots\ldots\ldots$ 7.1 lb-moles

O_2 actually consumed $= \ldots\ldots\ldots\ldots\ldots\ldots\ldots$ $\overline{14.3 \text{ lb-moles}}$

Percentage excess oxygen $=$ percentage excess air $= \dfrac{7.1}{14.3} = 49.6$.

This percentage excess is based on the oxygen required for all combustible substances which were *actually burned*.

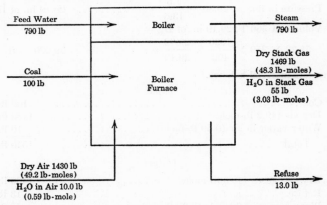

FIG. 72. Material balance of a boiler furnace (Illustration 5).

ENERGY BALANCE

The energy balance of this problem is calculated exactly as in Illustration 4, estimating the heating value of the coal by the empirical method discussed on page 328.

Effect of Sulfur in Coal. In the preceding illustrations the combustion of the sulfur in the coal has been neglected. This does not introduce appreciable error if the sulfur content is low, 1 per cent or less. It is difficult to take into account the combustion of the sulfur by any rigorous method because of the uncertainty of the forms in which it is present in the coal and the difficulty of determining its distribution in the combustion products. A considerable part of the sulfur which is present in the coal in a combustible or available form will appear as sulfur dioxide in the flue gases. The remainder will be present in the refuse.

The ordinary scheme of flue gas analysis by the Orsat apparatus, in which the gas sample is confined over water, does not permit determination of sulfur dioxide. Because of the high solubility of sulfur dioxide in water (about thirty times that of carbon dioxide) the bulk of the SO_2 will be absorbed in the water of the sampling apparatus and burette. Any SO_2 which is not removed in this manner will be absorbed and reported as CO_2. The reported analysis will ordinarily represent approximately the composition of the SO_2-free gases. Methods are available by which separate determination may be made of the SO_2, but this is not ordinarily feasible for performance tests of combustion equipment.

Neglect of the combustion of sulfur in calculations of the type carried out in Illustration 4, in which the oxygen balance is not used, does not introduce any serious error, even if the sulfur content is relatively high. The calculated total weight of dry flue gases will be low by approximately the weight of SO_2 formed. This error is usually negligible. However, in the method of calculation of Illustration 5, in which the net hydrogen in the coal is calculated from an oxygen balance, the errors resulting from neglect of combustion of sulfur may be more serious. In this type of calculation it is assumed that all free oxygen not accounted for as CO_2, CO, or free oxygen in the flue gas was utilized in the oxidation of the net hydrogen of the coal. This assumption neglects the oxygen consumed in oxidation of sulfur. As a result the net hydrogen content calculated by this method will be too high, and the combined water content, calculated by difference, will be too low. These errors are particularly high when the sulfur is originally present as iron pyrites, FeS_2, in which case oxygen is consumed in the oxidation of both the sulfur and the iron. However, because of the relatively high atomic weight of sulfur it is permissible to neglect these errors in dealing with coals containing less than 1 per cent of sulfur.

The probable errors involved in neglecting the combustion of the sulfur may be estimated for a typical case by consideration of the data of Illustration 5. The coal contained 1.1 per cent sulfur. The maximum error would result if all the sulfur were in the form of iron pyrites. In this case, for each pound-atom of sulfur present 11/8 pound-moles of oxygen would be consumed in producing SO_2 and Fe_2O_3. On the basis of 100 pounds of coal charged 0.048 pound-mole of oxygen would be required for combustion of the sulfur of the coal. Introducing this value into the oxygen balance on page 360 will reduce the unaccounted-for oxygen, used in oxidizing hydrogen, from 0.94 to 0.89 pound-mole. The calculated net hydrogen is correspondingly reduced from 3.8 to 3.6 pounds and the combined water content is increased from 7.2 to

7.4 pounds. These errors are not serious and in addition represent maximum errors because actually not all the sulfur will be in the pyritic form and furthermore it will not be completely burned.

Case 3. Method of Calculation Where Sulfur Is High and the Carbon and Hydrogen Contents of the Coal Are Unknown. Where sulfur contents of more than about 1 per cent are encountered it is not ordinarily desirable to use the oxygen balance for computing the net hydrogen content of the coal. In order to develop an accurate oxygen balance it would be necessary to have data on the forms in which the sulfur occurred in the coal and on the sulfur content of the refuse. Without such data the net hydrogen content is better determined analytically or estimated by the empirical method of Uehling, discussed on page 328. This latter method may also be used to estimate the total carbon in the coal from a determination of the heating value.

If the sulfur content of a coal is so high that sulfur dioxide constitutes a considerable part of the flue gas it is necessary to obtain data from which the amount of sulfur actually burned may be calculated. A determination of either the sulfur in the refuse or of the SO_2 in the flue gases will supply this information. The former determination is more easily carried out. It may then be assumed that the ordinary flue gas analysis yields the composition of the SO_2-free gases and the total quantity of gases computed on this basis. Direct determination of the SO_2 content of the flue gases is more reliable but frequently unwarranted.

In calculating an energy balance involving a coal of high sulfur content a sulfur correction should be applied to the heating value directly determined in the oxygen bomb calorimeter. This correction results from the fact that in the calorimeter the available sulfur is burned almost entirely to SO_3 whereas in ordinary combustion the major part of it will form SO_2. The correction may be taken as 1000 calories per gram of sulfur, to be subtracted from the observed heating value.

Case 4. Gas Producers. In the operation of a gas producer, a fuel gas of low calorific value is produced by blowing air, usually accompanied by steam, through a deep incandescent bed of fuel. Carbon monoxide and carbon dioxide are formed by partial combustion of the fuel. Hydrogen and the oxides of carbon are formed from the reduction of water, and volatile combustible matter is distilled from the coal without combustion.

Effect of Soot and Tar. In many combustion processes the gases contain carbon suspended in the form of soot and tar. These forms of carbon do not appear in an ordinary volumetric gas analysis but must be determined separately by absorption or retention on a weighed filter. The tar can then be separated and analyzed for its hydrogen content although this precision is not usually warranted. It is ordinarily sufficient to assume that the combustible of the suspended tar analyzes 90 per cent carbon and 10 per cent hydrogen and that the combustible of the soot consists of 100 per cent carbon. Frequently refuse also appears suspended in the gaseous products so that the ash content of the suspended material should also be measured.

In addition to the tar and soot suspended in the gases a gradual accumulation of these products settles in the flues. The slight correction necessary for this is usually made by measuring the deposition over a long period of time, for example, over the usual time interval elapsed between successive cleanings of the flues. The tar deposit is removed and weighed and, since the tons of fuel distilled during the time interval is known, a rough approximation can be obtained for the carbon deposited per unit weight of coal charged.

The following illustration of gas producer operation is of value in that consideration is given to the tar and soot suspended in the producer gas and also to the live steam which is passed into the producer. The problem is of added interest in that all experimental data were collected with unusual care. Analyses of gas and coal and measurements of temperatures were made at regular intervals over several days of operation and data carefully weighted to give average results. These data were taken in part from a report by C. B. Harrop.[13]

Illustration 6. Air is supplied to a gas producer at 75°F with a percentage humidity of 75% and a barometric pressure of 29.75 in. of Hg. Coal weighing 70,900 lb and having a gross heating value of 11,910 Btu per lb is charged into the producer. Tar weighing 591 lb is deposited in the flues prior to the point of gas sampling and contains 93% carbon and 7% hydrogen.

The water vapor, suspended tar, and soot in the gases were determined experimentally by withdrawing samples of the gas through an absorption train. The results are expressed in grains per cubic foot of wet, hot gas measured at 1075°F and 29.75 in. of Hg.

Water vapor.................	3.43 grains per cu ft
Suspended tar................	3.31 grains per cu ft
Suspended soot...............	1.52 grains per cu ft

It is assumed that the suspended tar is 90% carbon and 10% hydrogen and that the soot is 100% carbon.

Saturated steam at a gauge pressure of 25.3 lb per sq in. is introduced at the bottom of the fuel bed.

Analysis of coal as charged:

Carbon..........	66.31%	Nitrogen...	1.52%	Sulfur.......	1.44%
Available hydrogen	3.53%	Total water	23.16%	Corrected ash	4.04%
				Total........	100.00%
Ash (as weighed)...					4.58%

[13] *J. Am. Ceramic Soc.*, **1**, 35 (1918).

Analysis of refuse:

Moisture	1.10%	Fixed carbon	3.08%	Ash	95.82%
				Total	100.00%

Analysis of dry, tar- and soot-free gas, by volume:

CO_2	7.12%	CO	21.85%	H_2	13.65%
O_2	0.90%	CH_4	3.25%	N_2	53.24%
				Total	100.00%

The gases leave the flue at 1075°F and 29.75 in. of Hg. The refuse leaves the producer at 350°F and may be assumed to have a specific heat of 0.22. The mean specific heat of the tar and soot between 75°F and 1075°F may be assumed to be 0.32. The heating value of the suspended tar was found to be 16,000 Btu per lb, and that of the deposited tar 15,500 Btu per lb.

Calculate complete material and energy balances for the operation of this gas producer.

Solution: Since a majority of the details of this problem are similar to those of Illustration 4, full explanations of all steps will not be repeated. A flow chart of this process is shown in Fig. 73.

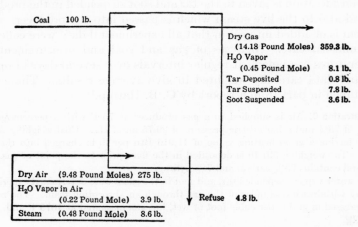

Fig. 73. Material balance of a gas producer (Illustration 6).

MATERIAL BALANCE

Special attention should be given to calculations shown in Parts 1 and 3.

1. Weight of gaseous products.

Carbon Balance:

Carbon gasified: *Basis:* 100 lb. of coal charged.

Carbon in coal = 100×0.6631 = ... 66.31 lb or 5.526 lb-atoms

Carbon in deposited tar = $\dfrac{591}{709} \times 0.93$ = ... 0.78 lb or 0.065 lb-atom

Weight of refuse = 4.58/0.9582 = ... 4.79 lb

(calculated on the basis of the ash as weighed in the proximate analysis).

Carbon in refuse = 4.79 × 0.0308 = ...   0.148 lb or 0.012 lb-atom
Carbon gasified = 5.526 − (0.065 + 0.012) =   5.449 lb-atoms

Carbon in clean gases. *Basis:* 1.0 lb-mole of dry, tar- and soot-free gas.

C in CO_2 = ...   0.0712 lb-atom
C in CO = ...   0.2185 lb-atom
C in CH_4 = ...   0.0325 lb-atom
Total = ...   0.3222 lb-atom

Water, tar and soot in gas. *Basis:* 1000 cu ft of wet gases at 1075°F and 29.75 in. of Hg.

$$\text{Moles of total gas} = \frac{1000}{359 \times \dfrac{1535}{492} \times \dfrac{29.92}{29.75}} = \text{. . . . .}\quad 0.888 \text{ lb-mole}$$

$$H_2O \text{ present} = \frac{3.43 \times 1000}{7000 \times 18} = \text{. . . . .}\quad 0.0272 \text{ lb-mole}$$

Moles of dry gas = ...   0.8608 lb-mole
Water per mole of dry gas 0.0272/0.8608 = ...   0.0316 lb-mole

$$\text{Tar per mole of dry gas } \frac{3.31 \times 1000}{7000 \times 0.8608} = \text{. . . . .}\quad 0.550 \text{ lb}$$

$$\text{Soot per mole of dry gas } \frac{1.52 \times 1000}{7000 \times 0.8608} = \text{. . . . .}\quad 0.252 \text{ lb}$$

Total carbon in gases. *Basis:* 1.0 lb-mole of dry, clean gas.

C in clean gases = ...   0.3222 lb-atom
C in tar = 0.550 × 0.9 = 0.495 lb or . . . .   0.0412 lb-atom
C in soot = 0.252 lb or . . . .   0.0210 lb-atom

Total . . . .   0.3844 lb-atom

Moles of dry gas per 100 lb of coal = 5.449/0.3844 = 14.18 lb-moles.
Total products in gases. *Basis:* 100 lb of coal charged.

CO_2 = 14.18 × 0.0712 = 1.01 lb-moles or × 44 = ...   44.5 lb
O_2 = 14.18 × 0.0090 = 0.12 lb-mole or × 32 = ...   3.8 lb
CO = 14.18 × 0.2185 = 3.10 lb-moles or × 28 = ...   86.8 lb
CH_4 = 14.18 × 0.0325 = 0.46 lb-mole or × 16 = ...   7.4 lb
H_2 = 14.18 × 0.1365 = 1.94 lb-moles or × 2 = ...   3.8 lb
N_2 = 14.18 × 0.5324 = 7.55 lb-moles or × 28.2 = ...   213.0 lb

Total dry, clean gas 14.18 lb-moles or . . . .   359.3 lb

Water
14.18 × 0.0316 = 0.45 lb-mole, or × 18 = ...   8.1 lb

Tar
C = 14.18 × 0.0412 = 0.58 lb-atom, or × 12 = 7.02 lb
H_2 = 14.18 × 0.055/2 = 0.39 lb-mole, or × 2 = 0.78 lb

Total tar = ...   7.8 lb

Soot
C = 14.18 × 0.0210 = 0.30 lb-atom, or × 12 = ...   3.6 lb

Total products in gases = ...   378.8 lb

2. Weight of air supplied.

Nitrogen Balance. Basis: 100 lb of coal charged.

N_2 in gas = ...	7.55 lb-moles
N_2 in coal = 1.52/28 = ...	0.05 lb-mole
N_2 from air = ...	7.50 lb-moles
Dry air supplied = 7.50/0.79 = ...	9.48 lb-moles
or 9.48 × 29.0 = ...	275 lb
Molal humidity of air (Fig. 9) = ...	0.023
Water in air = 9.48 × 0.023 = ...	0.218 lb-mole
or 0.218 × 18 = ...	3.9 lb
Total wet air = 0.218 + 9.48 = ...	9.70 lb-moles
or 275 + 3.9 =.................................	278.9 lb

3. Weight of steam introduced.

Hydrogen Balance. Basis: 100 lb of coal charged.

Output

H_2 in deposited tar $= \dfrac{591}{709} \times 0.07 = 0.0585$ lb =	0.029 lb-mole
Free H_2 in gas = ...	1.94 lb-mole
H_2 in CH_4 in gas = 0.46 × 2 = ...	0.92 lb-mole
H_2 in H_2O in gas = ...	0.45 lb-mole
H_2 in suspended tar = ...	0.39 lb-mole
Total output of H_2 = ...	3.73 lb-moles

Input

Net H_2 in coal = 3.53/2.02 = ...	1.75 lb-moles
H_2 in water in coal = 23.16/18 = ...	1.28 lb-moles
H_2 in water in air = ...	0.22 lb-mole
Total input in addition to steam = ...	3.25 lb-moles
H_2 from steam = 3.73 − 3.25 = ...	0.48 lb-mole
Steam introduced = 0.48 lb-mole or...........................	8.64 lb

4. Overall material balance.

Input			*Output*	
Coal...................	100	lb	Refuse...................	4.8 lb
Dry air (9.48 lb-moles)...	275	lb	Tar deposited.............	0.83 lb
Water in air (0.22 lb-mole)	3.9	lb	Soot suspended............	3.6 lb
Steam (0.48 lb-mole)....	8.6	lb	Tar suspended.............	7.8 lb
			Dry clean gas (14.18 lb-moles)	359.3 lb
Total...............	387.5 lb		Water in gas (0.45 lb-moles).	8.1 lb
			Total..................	384.43 lb

The slight discrepancy in the totals of the material balance results from inaccuracies of the data and neglect of the sulfur content of the coal. The material balance is also summarized in Fig. 73.

5. Gaseous volumes. *Basis:* 100 lb of coal charged.

$$\text{Volume of wet air} = 9.70 \times 359 \times \frac{535}{492} \times \frac{29.92}{29.75} = \ldots\ldots\ldots \quad 3,800 \text{ cu ft}$$

$$\text{Volume of wet gases} = 14.63 \times 359 \times \frac{1535}{492} \times \frac{29.92}{29.75} = \ldots \quad 16,500 \text{ cu ft}$$

ENERGY BALANCE

Special attention is called to calculations in Parts 2, 3 and 5 of output.

Reference temperature: 75°F.

Basis: 100 lb of coal charged.

Input

1. **Heating value of coal** = $100 \times 11,910 = \ldots\ldots\ldots\ldots\ldots$ 1,191,000 Btu
2. **Sensible enthalpy of coal** = $\ldots\ldots\ldots\ldots\ldots\ldots\ldots\ldots$ 0 Btu
3. **Enthalpy of steam:**
   Pressure = $25.3 + 14.7 = 40.0$ lb per sq in.
   From steam tables, the enthalpy of saturated steam at this pressure referred to 32°F is 1171 Btu per lb
   Referred to 75°F this becomes $1171 - 43 = 1128$ Btu per lb
   Enthalpy of steam = $8.6 \times 1128 = \ldots\ldots\ldots\ldots\ldots$ 9,700 Btu
4. **Enthalpy of dry air** = $\ldots\ldots\ldots\ldots\ldots\ldots\ldots\ldots\ldots$ 0 Btu
5. **Total enthalpy of water vapor in air.**
   Dew point (Fig. 9) = 66°F
   Enthalpy (as in Illustration 4, page 355) = $\ldots\ldots\ldots\ldots$ 4,100 Btu

   Total = $\ldots\ldots\ldots\ldots\ldots\ldots\ldots\ldots\ldots\ldots\ldots\ldots\ldots$ 1,204,800 Btu

Output

1. **Heating value of dry clean producer gas.**   (Calculated from composition)$\ldots\ldots\ldots\ldots\ldots\ldots\ldots\ldots\ldots\ldots\ldots\ldots\ldots$ 794,000 Btu
2. **Heating value of suspended tar** = $7.80 \times 16,000 = \ldots\ldots$ 125,000 Btu
3. **Heating value of soot** = $0.30 \times 96,630 \times 1.8 = \ldots\ldots\ldots$ 52,100 Btu
4. **Heating value of carbon in refuse** = $0.012 \times 96,630$
   $\times 1.8 = \ldots\ldots\ldots\ldots\ldots\ldots\ldots\ldots\ldots\ldots\ldots\ldots\ldots$ 2,100 Btu
5. **Heating value of deposited tar** = $0.83 \times 15,500 = \ldots\ldots$ 12,900 Btu

6. **Enthalpy of dry, clean producer gas.**

Mean heat capacities between 75°F and 1075°F taken from Fig. 37, page 217.

$$
\begin{aligned}
CO_2(1.01)(11.05) &= 11.15 \\
O_2(0.12)(7.59) &= 0.91 \\
CO(3.10)(7.26) &= 22.50 \\
CH_4(0.46)(12.03) &= 5.54 \\
H_2(1.94)(7.01) &= 13.60 \\
N_2(7.55)(7.27) &= 54.89 \\
\hline
\text{Total} &= 108.59
\end{aligned}
$$

Enthalpy = $(108.59)(1075 - 75) = 108,590$ Btu

7. **Sensible enthalpy of tar and soot.**
   Suspended tar and soot = $11.4 \times 0.32(1075 - 75) = \ldots$ 3,650 Btu
   Deposited tar and soot = $0.8 \times 0.32(1075 - 75) = \ldots\ldots$ 250 Btu

8. **Total enthalpy of water vapor in gases.**

Total molal enthalpy calculated by method used in Illustration 4 = ... 27,470 Btu

Total enthalpy = 27,470 × 0.45 = ...   12,360 Btu

9. **Sensible enthalpy of refuse = 0.22 × 4.8 (350 − 75) = .**   290 Btu

Total energy accounted for = ...   1,111,240 Btu

10. **Heat losses, radiation, etc. = 1,204,800 − 1,111,240 = ..**   93,560 Btu

Summary of Energy Balance:

Reference temperature: 75°F.

Basis: 100 lb of coal charged.

Input

1.	Heating value of coal........................	1,191,000 Btu	98.9%
2.	Sensible enthalpy of coal.....................	0 Btu	0%
3.	Sensible enthalpy of air......................	0 Btu	0%
4.	Enthalpy of steam...........................	9,700 Btu	0.8%
5.	Total enthalpy of water in air...............	4,100 Btu	0.3%
	Total...................................	1,204,800 Btu	100.0%

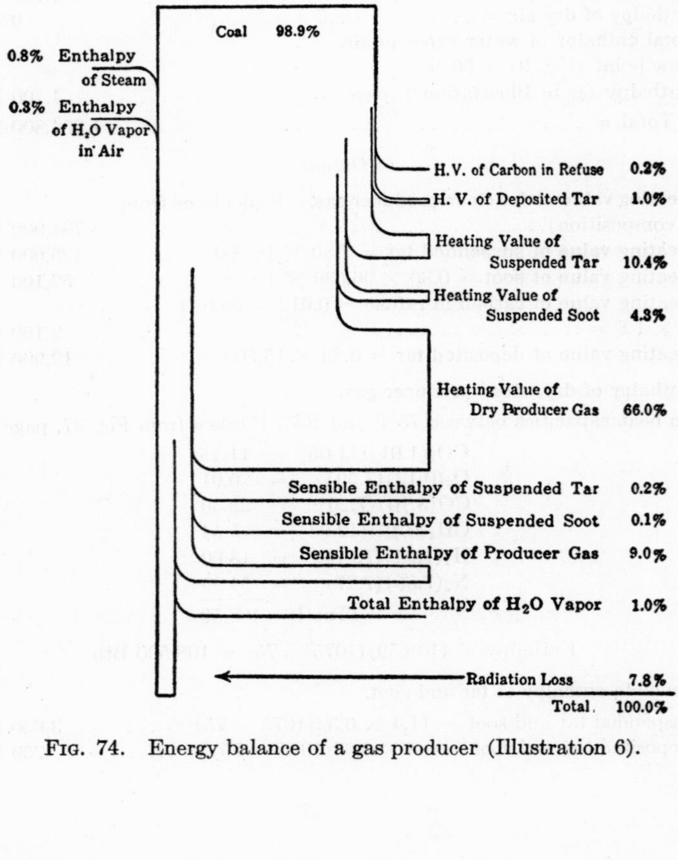

FIG. 74.   Energy balance of a gas producer (Illustration 6).

Output

1. Heating value of clean gas....................	794,000 Btu	66.0%
2. Heating value of suspended tar...............	125,000 Btu	10.4%
3. Heating value of suspended soot..............	52,100 Btu	4.3%
4. Heating value of carbon in refuse.............	2,100 Btu	0.2%
5. Heating value of deposited tar................	12,900 Btu	1.0%
6. Enthalpy of dry, clean gas...................	108,590 Btu	9.0%
7. Enthalpy of tar and soot.....................	3,900 Btu	0.3%
8. Enthalpy of water vapor in gases.............	12,360 Btu	1.0%
9. Enthalpy of refuse..........................	290 Btu	0.0%
10. Heat losses, radiation, etc...................	93,560 Btu	7.8%
Total................................	1,204,800 Btu	100.0%

This energy balance is presented in diagrammatic form in Fig. 74.

THERMAL EFFICIENCY

Cold gas. The effectively utilized energy includes only the heating value of the dry, clean gas.

$$\text{Efficiency} = \frac{794,000}{1,204,800} = 66\%$$

Hot gas. The effectively utilized energy includes the total sensible enthalpy of all materials in the gases and also the heating value of the suspended tar and soot.

Sensible enthalpy of dry, clean gases = ...	108,590 Btu
Sensible enthalpy of water vapor in gases = 12,360 − (0.45 × 18,870) = ...	3,870 Btu
Sensible enthalpy of suspended tar and soot = ...	3,650 Btu
Heating value of dry, clean gas = ...	794,000 Btu
Heating value of tar and soot = ...	177,100 Btu
Effectively utilized energy = ...	1,087,210 Btu

$$\text{Efficiency} = \frac{1,087,210}{1,204,800} = 90.2\%$$

The principal loss in the operation of this particular gas producer is by radiation and conduction of heat from the apparatus.

GRAPHICAL CALCULATION OF COMBUSTION PROBLEMS

Where short-cut methods of combustion calculations are desired a simple graphical solution may be resorted to provided that the following items are negligible: sulfur content of fuel, combustible content of refuse, water vapor in air, suspended tar and soot in gases, and hydrocarbons in flue gases.   The simplified graphical solution of such combustion problems appears in Fig. 75. This method also serves as an approximate solution for other combustion problems where the experimental data or time required do not warrant precise methods. From

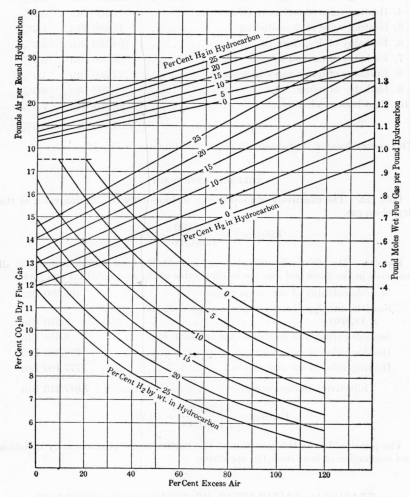

Fig. 75. Combustion chart for fuels.
(Reproduced in " C.P.P. Charts ")

Fig. 75 there may be obtained directly the pounds of air used and the pound-moles of wet flue gas produced and the carbon dioxide content of the dry flue gas on the basis of 1 pound of combustible in oil, coal, gas, or coke.

PROBLEMS

✓ **1.** A Kentucky coal has the following analysis:

		Ultimate analysis of combustible	
Proximate analysis as received		corrected ash-free moisture-free basis	
Moisture.................	2.97%	C.......................	84.39%
Ash (uncorrected)........	2.94%	Net hydrogen............	4.81%
Volatile matter...........	37.75%	N	2.00%
Fixed carbon.............	56.34%	S	1.02%
		Combined H_2O..........	7.78%
	100.00%		100.00%

The combustible referred to above includes those portions of coal as received which are not classified as moisture or corrected ash.

Determine, on the " as received " basis:

(a) The rank of this coal.
(b) The total heating value by Dulong's formula.
(c) The total heating value by Uehling's method.
(d) The estimated available hydrogen content by Uehling's method.
(e) The net heating value by Uehling's method.
(f) Ultimate analysis (as received, corrected ash).

2. In the following table are the analyses of several typical coals, as given in Bureau of Mines Report of Investigations, No. 3296, " Classification Chart of Typical Coals of the United States." Both the ultimate and proximate analyses are on the " as received " basis, and the hydrogen given in the ultimate analysis includes the hydrogen of the free moisture in the coal.

	No. 1	No. 2	No. 3	No. 4	No. 5
Proximate Analysis					
Moisture	4.3%	2.0%	2.2%	10.6%	42.6%
Volatile Matter	3.0	9.6	21.1	36.4	24.7
Fixed Carbon	82.6	77.8	66.2	42.4	27.0
Ash (uncorrected)	10.1	10.6	10.5	10.6	5.7
Ultimate Analysis					
Carbon	82.5%	79.5%	76.1%	62.8%	37.3%
Hydrogen	2.0	3.7	4.8	5.6	7.3
Nitrogen	0.6	0.9	1.2	1.2	0.5
Sulfur	0.5	0.6	2.6	3.3	0.6
Heating Value					
Btu/lb	12,650	13,520	13,530	11,300	6,260

No. 1 Anthracite   No. 4 High volatile bituminous
No. 2 Semi-anthracite   No. 5 Lignite
No. 3 Medium volatile bituminous

For one of the coals, present the following calculations:

(a) Recalculate the ultimate analysis, reporting:

(1) Carbon   (5) Sulfur
(2) Available hydrogen   (6) Nitrogen
(3) Free H_2O (7) Corrected Ash
(4) Combined H_2O

(b) Calculate total heating value by Dulong's formula. Compare the result with the experimental value.

(c) Calculate the total heating value by Uehling's method. Compare the result with the experimental value.

(d) Assuming that the proximate analysis and heating value alone are available, calculate the per cent of carbon and of available hydrogen by Uehling's method.

(e) Calculate the net heating value.

3. A gas has the following composition by volume:

$$
\begin{array}{lr}
\text{Illuminants } (C_2H_4 \text{ and } C_6H_6) \dots \dots & 53.6\% \\
O_2 \dots \dots & 1.6\% \\
CH_4 \dots \dots & 16.9\% \\
C_2H_6 \dots \dots & 24.3\% \\
N_2 \dots \dots & 3.6\%
\end{array}
$$

The heating value of this gas is 1898 Btu per standard cubic foot. Calculate the percentages of C_2H_4 and C_6H_6 in the gas.

4. Calculate the maximum theoretical flame temperature in degree Centigrade when the following gas is burned with the theoretical amount of dry air starting with air and gas at 18°C:

$$
\begin{array}{lr}
CO \dots \dots \dots \dots \dots \dots \dots \dots & 30\% \\
H_2 \dots \dots \dots \dots \dots \dots \dots \dots & 15\% \\
O_2 \dots \dots \dots \dots \dots \dots \dots \dots & 1\% \\
CO_2 \dots \dots \dots \dots \dots \dots \dots \dots & 5\% \\
N_2 \dots \dots \dots \dots \dots \dots \dots \dots & \underline{49\%} \\
& 100\%
\end{array}
$$

5. Calculate the theoretical flame temperature when the above gas is burned with 100% excess air.

6. Calculate the theoretical flame temperature of the above-mentioned gas when burned with the theoretical amount of air and when both gas and air are preheated to 500°C before combustion.

7. Calculate the theoretical flame temperature of the above gas when burned with the theoretical amount of air, the combustion of both CO and H_2 proceeding to only 80% completion. The gas and air are initially at 18°C.

8. A fuel gas has the following analysis:

$$
\begin{array}{lr}
\text{Carbon dioxide} \dots \dots \dots \dots \dots & 3.0\% \\
\text{Ethylene} \dots \dots \dots \dots \dots \dots & 5.8\% \\
\text{Benzene} \dots \dots \dots \dots \dots \dots & 0.8\% \\
\text{Oxygen} \dots \dots \dots \dots \dots \dots & 0.9\% \\
\text{Hydrogen} \dots \dots \dots \dots \dots \dots & 38.8\% \\
\text{Carbon monoxide} \dots \dots \dots \dots \dots & 35.2\% \\
\text{Methane} \dots \dots \dots \dots \dots \dots & 9.5\% \\
\text{Ethane} \dots \dots \dots \dots \dots \dots & 0.7\% \\
\text{Nitrogen} \dots \dots \dots \dots \dots \dots & \underline{5.3\%} \\
& 100.0\%
\end{array}
$$

Calculate the theoretical flame temperature if this dry fuel gas is burned with the theoretical amount of dry air. Gas and air are supplied at 18°C. Assume that combustion is complete. Use 1 g-mole of the dry fuel gas as the basis of calculation.

9. For the fuel gas whose analysis is given in Problem 8, calculate the theoretical flame temperature if the dry gas is burned with the theoretical amount of dry air. The gas is supplied at 18°C, but the air is preheated to 700°C. Assume that combustion is complete. Use 1 g-mole of the dry fuel gas as the basis of calculation.

10. For the fuel gas whose analysis is given in Problem 8, calculate the theoretical flame temperature if the gas is burned with 100% excess air. The gas and air both are dry and are supplied at 18°C. Assume that combustion is complete. Use 1 g-mole of the dry fuel gas as the basis of calculation.

11. The dry fuel gas whose analysis is given in Problem 8 is burned with the theoretical amount of dry air. Both air and gas are supplied at 18°C. If carbon monoxide and hydrogen burn to 80% completion, what is the theoretical flame temperature? Assume that the combustion of the gases other than CO and H_2 is complete. Use 1 g-mole of the dry fuel gas as the basis of calculation.

12. Four thousand (4000) kilograms of coke at an initial temperature of 1400°C has 360 kg of steam blown through it, forming 448 kg CO, 88 kg CO_2, and 40 kg H_2 at an average temperature of 1000°C. The steam is supplied at 120°C, saturated. The radiation loss is 100,000 kcal. Calculate the final temperature of the bed of coke.

13. A fuel gas has the following composition by volume:

CO_2	2.1%
O_2	0.5%
$C_{2.5}H_{4.2}$ (illuminants)	7.0%
CO	33.8%
H_2	40.6%
$C_{1.2}H_{4.4}$ (paraffins)	11.2%
N_2	4.8%
	100.0%

(a) Calculate the analysis of the flue gases formed by burning this gas with 30% excess air, assuming that all combustible components are burned to CO_2 and H_2O.

(b) Calculate the heating value in Btu per standard cubic foot.

14. A fuel oil has a specific gravity of 0.91 and a viscosity of 28 Saybolt Furol seconds at 122°F. Estimate the characterization factor, average molecular weight, average boiling point, hydrogen content, specific heat at 150°F, and heating value of this oil.

15. A petroleum fraction has a specific gravity (60/60°F) of 0.88. When tested in a Saybolt Universal viscometer at 122°F, a reading of 58 seconds was obtained. Using the charts in the text, report the following values:

(a) Characterization factor.
(b) Average molecular weight.
(c) Average boiling point.
(d) Hydrogen content.
(e) Specific heat of the liquid oil at 200°F.
(f) Specific heat of the vapor at 800°F.
(g) Heating value.

16. A Pennsylvania bituminous coal has the following composition:

H..............................	4.71%
C..............................	69.80%
N..............................	1.42%
O_2..........................	7.83%
Ash............................	6.73%
H_2O.........................	9.51%
Total heating value...............	6950 kcal per kg

This coal is gasified in a gas producer using air at 20°C saturated with water vapor. No additional steam or water is admitted into the producer. The barometric pressure is 740 mm. The resulting gas has the following composition:

H_2..........................	0.5%
CO.............................	21.2%
N_2..........................	64.5%
CH_4.........................	5.8%
CO_2.........................	6.2%
O_2..........................	1.8%

It may be assumed that no tar or soot is present in the producer gas. During a test period the total coal charged is 10,500 kg.
The dry refuse formed weighs 825 kg and contains 13.3%C.

Temperature of refuse = 220°C.
Mean specific heat of refuse = 0.25.
Temperature of outgoing gases = 450°C.

(a) Calculate complete material and energy balances of this gas producer on the basis of 100 kg of coal as charged.
(b) Calculate the thermal efficiency on both the hot and cold bases.
(c) Calculate the total volume of gases leaving the producer.
(d) Calculate the heating value of the producer gas in Btu per cubic foot measured at 60°F, 30 in. of Hg, saturated with water vapor.

17. A gas producer is charged with bituminous coal having the following composition:

Proximate analysis:

Moisture.......................	2.70%
Volatile matter..................	25.77%
Fixed carbon....................	62.87%
Ash...........................	8.66%
	100.00%

Ultimate analysis:

Moisture.......................	2.70%
Carbon........................	78.55%
Net hydrogen...................	4.13%
Nitrogen.......................	1.58%
Sulfur.........................	0.69%
Corrected ash..................	8.40%
Combined H_2O................	3.95%
	100.00%

Total heating value = 13,944 Btu per lb.

Air is supplied at 75°F with a percentage humidity of 90%. The barometric pressure is 29.65 in. of Hg. Dry, saturated steam is supplied at a gauge pressure of 50 lb per sq in. The producer gas leaves at a temperature of 1220°F and has the following composition by volume:

CO.............................	25.0%
H₂.............................	22.0%
CH₄.............................	3.6%
C₂H₄.............................	2.8%
CO₂.............................	9.2%
N₂.............................	37.4%
	100.0%

A sample of gas is withdrawn and cooled to 100°F for determination of suspended tar and soot. The tar and soot content is 10 grains per cu ft of gas measured at barometric pressure, 100°F, and saturated with water vapor. The tar and soot contain 95% carbon and 5% hydrogen. Its heating value is 17,100 Btu per lb, and its mean specific heat is 0.34. The dew point of the producer gas is 100°F.

The refuse is discharged at 400°F, moisture-free, and containing 4.52% carbon. The specific heat of the refuse is assumed to be 0.23. The mean molal heat capacity of C₂H₄ (75° to 1220°F) is 20 Btu per lb-mole per °F.

Neglecting deposition of tar in the flues and presence of sulfur, calculate:

(a) Complete energy and material balances of the producer, based on 100 lb of coal charged.

(b) The thermal efficiencies on both the hot and cold bases.

(c) The volume of producer gas, measured at 60°F, 30 in. of Hg, saturated with water, formed per 100 lb of coal charged.

(d) The heating value of the producer gas, per standard cubic foot.

The solution of this problem results in a negative radiation loss. What errors in experimental data are most likely responsible for this condition?

18. An Illinois bituminous coal is burned in a boiler furnace with air at 78°F, 92% percentage humidity. The barometric pressure is 29.40 in. of Hg. The furnace gases leave at 553°F. The refuse is discharged moisture-free, at 440°F. The refuse as analyzed contains 12.2% carbon as coke and 16.1% moisture. The specific heat of the dry refuse is 0.25. The proximate analysis of coal is:

Fixed carbon.....................	50.34%
Volatile matter...................	30.68%
Moisture.......................	9.61%
Ash...........................	9.37%

Heating value of coal = 11,900 Btu per lb.
The ultimate analysis on the moisture-free basis is:

Carbon.........................	73.70%
Hydrogen.......................	4.75%
Oxygen.........................	9.23%
Nitrogen.......................	1.58%
Sulfur.........................	0.55%
Corrected ash..................	10.19%
	100.00%

The flue gas analysis is as follows:

CO_2 .... 12.2%
CO 0.2%
O_2 .... 7.0%
N_2 .... 80.6%
100.0%

Calculate:

(a) Total material and energy balances for this process based on 100 lb of coal charged, neglecting the combustion of sulfur. The heat losses and the heat effectively utilized may be considered together as a single item of heat output.

(b) Percentage excess air used in combustion, based on that required for complete combustion of all coal charged.

(c) The dew point of the flue gases.

(d) The actual volumes in cubic feet at the given condition of temperature, humidity and pressure of the flue gases and air supply, per 100 lb of coal charged.

19. A bituminous coal is burned in a boiler furnace with air at 85°F, 90% percentage humidity. The barometric pressure is 29.20 in. of Hg. The furnace gases leave at 572°F. The refuse leaves the furnace moisture-free at 520°F and when analyzed contains 22.3% moisture, 12.3% volatile matter and 41.4% fixed carbon. The mean specific heat of the refuse is 0.23. The proximate analysis of the coal is:

Fixed carbon 56.34%
Volatile matter 37.75%
Moisture 2.97%
Ash 2.94%

A partial ultimate analysis on the corrected ash- and moisture-free basis is:

Carbon 84.43%
Nitrogen 2.00%
Sulfur 0.82%

The total heating value of the coal is 14,139 Btu per lb. Dry, saturated steam at a gauge pressure of 150 lb per sq in. is produced at the rate of 780 lb per 100 lb of coal charged. Water is fed into the boiler at 72°F. The Orsat analysis of the flue gas is:

CO_2 .... 12.0%
CO 1.2%
O_2 .... 6.2%
N_2 .... 80.6%
100.0%

Calculate:

(a) The net hydrogen content of the coal from an oxygen balance, neglecting combustion of sulfur.

(b) The complete ultimate analysis of the coal.

(c) The complete material balance of the process, based on 100 lb of coal charged.

(d) The complete energy balance of the furnace, based on 100 lb of coal charged.

(e) The thermal efficiencies of the furnace and boiler, based on the total and on the net heating values.

(f) The percentage excess air used, based on the total combustible charged.

20. A heat at interchanger, used for heating the oil in a circulating hot oil heating system, is fired with coal having the following proximate analysis:

Moisture.........................	12.38%
Volatile matter...................	36.88%
Fixed carbon.....................	37.50%
Ash.............................	13.24%
	100.00%

The heating value of the coal is 10,361 Btu per lb, and its sulfur content is 5.1%.

The coal is burned with air at a temperature of 70°F and a percentage humidity of 60%. The barometric pressure is 29.3 in. of Hg.

The refuse from the furnace is discharged at a temperature of 600°F and contains 16% fixed carbon and 84% ash. The sulfur content of the refuse is 7.8%. Its specific heat may be taken as 0.23.

The flue gases leave the furnace at a temperature of 850°F and have the following composition by volume, on the sulfur- and moisture-free basis.

CO_2.............................	11.50%
CO.............................	0.17%
O_2..............................	7.51%
N_2.............................	80.82%
	100.00%

The oil is circulated at a rate of 3800 lb per 100 lb of coal charged and is heated from 155°F to 464°F. The mean specific heat of the oil in this temperature range is 0.55. Calculate:

(a) The complete ultimate analysis of the coal as estimated from the rank and heating value.

(b) The complete material and energy balances of the interchanger, based on 100 lb of coal charged.

(c) The complete analysis, by volume, of the wet flue gases leaving the interchanger.

(d) The percentage excess air, based on the combustible actually burned.

(e) The volume of wet flue gases leaving the interchanger.

(f) The thermal efficiencies of the interchanger, based on both the total and net heating values.

21. A brick kiln, of an intermittent type, is fired with 10,420 lb-moles of dry producer gas. The weight of green ware is 410,000 lb containing 0.52% mechanical water and 3.02% chemically combined water. The gas enters the kiln at 1220°F and a pressure of 29.65 in. of Hg, and contains 10 grains of tar (90% C, 10% H) per cu ft measured at a pressure of 29.65 in. of Hg, 100°F, and saturated with water vapor. The dew point of the producer gas is 100°F.

During the water-smoking period mechanical water is vaporized and leaves the kiln at 300°F, and the chemically combined water leaves at 400°F. During water-smoking the saturation temperature of the gases is 150°F. The flue gas leaves at an average temperature of 720°F. The average temperature of the ware at the end of the burn is 2100°F, and its specific heat is 0.23. The producer gas is burned with air at 75°F, 90% percentage humidity, and a pressure of 29.65 in. of Hg. The average analysis of gases by volume on the moisture-free basis is:

Producer gas		Flue gas	
CO...............	25.00%	CO$_2$...............	12.20%
H$_2$...............	22.00%	CO...............	0.16%
CH$_4$...............	3.60%	H$_2$...............	0.14%
C$_2$H$_4$...............	2.80%	O$_2$...............	7.10%
CO$_2$...............	9.20%	N$_2$...............	80.40%
N$_2$...............	37.40%		100.00%
	100.00%		

Calculate the material balance of the combustion process, the energy balance of the entire unit, and the thermal efficiency. The heat absorbed by the kiln structure may be included with the other undetermined heat losses as a single item.

22. Limestone is burned in a continuous vertical kiln which is heated by coal burned on an external grate located beside the bottom of the kiln shaft. The limestone is charged at the top of the shaft at atmospheric temperature and gradually descends in contact with a rising stream of the flue gases from the grate. The burned lime is discharged from the bottom of the shaft at a temperature of 950°F. The flue gases, mixed with all gases and vapors evolved by the charge, leave the top of the shaft at 560°F. For each 100 lb of coal burned, 161 lb of burned lime are produced. The limestone charged has the following composition:

CaO...........................	51.0%
MgO...........................	2.0%
CO$_2$...........................	42.2%
Al$_2$O$_3$...........................	1.5%
SiO$_2$...........................	1.2%
H$_2$O...........................	2.1%
	100.0%

The ultimate analysis of the coal is as follows:

Moisture.......................	10.69%
C.............................	66.62%
Net H.........................	3.18%
N.............................	1.57%
S.............................	1.91%
Corrected ash.................	6.41%
Combined H$_2$O.................	9.62%
	100.00%

The total heating value of the coal is 11,805 Btu per lb. The flue gases have the following composition by volume:

CO$_2$...........................	16.4%
N$_2$...........................	76.8%
O$_2$...........................	6.8%
	100.0%

The coal is burned with air at a temperature of 70°F having a humidity of 80 per cent. The barometric pressure is 29.4 in. of Hg.

The refuse from the grate contains 4.2% fixed carbon and 95.8% ash. Its sulfur content is 3.1%.

It may be assumed that in the burning process all CO_2 and water are driven from the limestone. The heat of wetting of granular limestone is negligible. It may be assumed that the sulfur burned forms SO_2 which is further oxidized and absorbed by the lime to form $CaSO_4$. The mean specific heat of the burned lime is 0.21.

Calculate the complete energy and material balances of the grate and kiln on the basis of 100 lb of coal fired.

Calculate the thermal efficiency of the process, considering the effectively utilized heat to be consumed in the decomposition of the limestone.

23. A 12-hour test was conducted on a steam-generating plant with four of the boilers in operation. The data for the 12 hour test are as follows:

Proximate analysis of coal

Moisture............	4.38%	Fixed carbon........	48.98%
Volatile matter.......	29.93%	Ash (uncorrected).....	16.71%
			100.00%

Half of the sulfur of the coal appears in the volatile matter.

Ultimate analysis of coal

Carbon.............	65.93%	Combined water......	6.31%
Available hydrogen....	3.50%	Free moisture.........	4.38%
Nitrogen.............	1.30%	Ash (corrected,	
Sulfur...............	2.99%	sulfur free)........	15.59%
			100.00%

Heating value of coal as fired (total)..............	11,670 Btu per lb
Weight of coal fired............................	119,000 lb
Temperature of coal...........................	65°F

Proximate analysis of refuse

Moisture.............	4.77%	Fixed carbon........	12.51%
Volatile matter.......	2.08%	Ash (uncorrected).....	80.64%
			100.00%

Flue gas

CO_2..................	11.66%	CO.................	0.04%
O_2..................	6.52%	N_2.................	81.78%
			100.00%

Temperature...	488°F

Air

Dry-bulb temperature...................	73°F
Wet-bulb temperature...................	59.4°F
Barometer...........................	29.08 in. Hg

Steam

Feed water..........................	193°F
Water evaporated......................	1,038,400 lb
Steam pressure.......................	137.4 lb per sq in. gauge
Quality	98.3%

Calculate the complete material and energy balances for this steam-generating plant.

24. Calculate the complete energy balances for Illustration 5.

25. A steam superheater is to be designed to heat 10,000 pounds per hour of low pressure process steam from 250°F to 700°F. The fuel is a residual cracked oil having an API gravity of 8.5 and a viscosity of 200 Saybolt Furol Seconds at 122°F. The design bases are as follows:

Percentage excess air.................. 40
Air temperature °F..................... 70
Relative humidity of air............... 60
Stack temperature °F.................. 600

Assuming a radiation loss of 5% of the heating value of the fuel burned, calculate complete material and energy balances for 1.0 hour of the combustion operation and the thermal efficiency based on the total heating value of the fuel. Fig. 75 may be used for establishing the material balance of the combustion.

Hydrochloric Acid

$H_2 + Cl_2 \rightarrow HCl$ (controlled combustion)

Absorbed HCl in H_2O

Sulfur Dioxide

Sulfur from Frasch Process, mines,

$C + SO_2 \rightarrow CO_2 + S$ in smelter gas.

SO_2 from S and FeS_2, sulfide ores, gypsum ($CaSO_4 + C \rightarrow CaO + SO_2 + CO$

Rotary, spray burners.

Cotrell precipitators remove dust.

Chamber H_2SO_4

$2 HO\cdot SO_2\cdot ONO + H_2O \rightleftharpoons H_2SO_4 + NO + NO_2$ (Glover tower, Chamber

$SO_2 + H_2O \rightleftharpoons H_2SO_3$

$H_2SO_3 + NO_2 \rightarrow H_2SO_4 \cdot NO$ } (Chambers)

$H_2SO_4 \cdot NO + \frac{1}{2}O_2 (NO_2) \rightarrow 2 HO\cdot SO_2\cdot ONO + H_2O (+NO)$

$H_2SO_4 \cdot NO \rightleftharpoons H_2SO_4 + NO$ (Chamber walls)

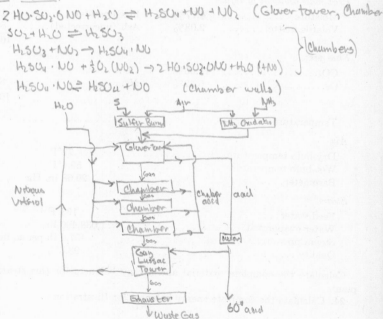

CHAPTER X

CHEMICAL, METALLURGICAL, AND PETROLEUM PROCESSES

The methods applied in calculating material and energy balances are alike in principle for all industrial processes, differing only in detail. Three illustrative material and energy balances are presented in this chapter, representing the procedure applicable to chemical processes in general. Typical illustrations in the chemical, metallurgical, and petroleum industries have been selected. The chemical and metallurgical illustrations summarize the principles involved in dealing with complex chemical reactions where the intermediate courses of the reactions are unknown. The petroleum process illustrates the use of an energy balance to establish a material balance and the principles of recycling to increase yield and effect temperature control. The chemical and metallurgical processes illustrate the analysis of experimental data from existing plants while the petroleum process is presented as a problem in the design of a new plant.

CHAMBER SULFURIC ACID PLANT

The material and energy balances of a chamber process sulfuric acid plant are selected for examination as representative of the chemical industries. The operating conditions have been taken from data published by Kaltenbach.[1] In this particular problem iron pyrites is burned with air in a shelf burner. The burner gases consisting of sulfur dioxide, oxygen, and nitrogen pass through a dust chamber where suspended matter is removed, and then into the Glover tower. In the Glover tower the hot gases meet a descending stream of acids from the Gay-Lussac tower and chambers. Oxides of nitrogen lost in the system are replaced by nitric acid introduced at the top of the tower. In passing through the Glover tower the hot gases are cooled and some conversion of SO_2 to sulfuric acid takes place, the chamber acid is concentrated, and the oxides of nitrogen are evolved from the acids for recirculation. The mixed gases leaving the Glover tower pass into a series of large lead chambers where conversion to H_2SO_4 is completed. Finally, the oxides of nitrogen are recovered in the Gay-Lussac tower

---

[1] *Chemical Age*, **28**, 295 (1920).

by passing the spent gases from the chamber countercurrent to a stream of cold sulfuric acid from the Glover tower. These different steps in the manufacture and the material and energy balances of the burner, Glover tower, chambers and Gay-Lussac tower are each discussed separately. The balances for each unit are based upon 100 kilograms of dry pyrites charged into the burner. The reference temperature for the energy balances is taken at 18°C.

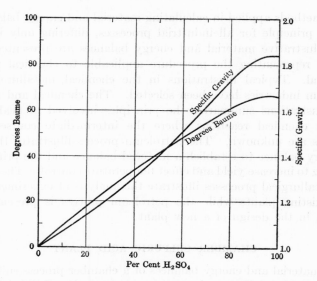

Fig. 76. Density of aqueous solutions of sulfuric acid at 15°/15°C.

The concentration of a strong aqueous solution of sulfuric acid is usually determined by measurement of its specific gravity. Concentrations are usually expressed in terms of specific gravities or of degrees Baumé rather than in percentages. In Fig. 76 are curves relating percentages of H_2SO_4 by weight to the degrees Baumé and specific gravities, 15°/15°C, of the aqueous solutions.

Illustration 1. Material and Energy Balances of a Chamber Sulfuric Acid Plant. Pyrites, containing 85.3% FeS_2, 2% H_2O, and 12.7% inert gangue, is burned in a shelf furnace yielding a gas containing 8.5% SO_2, 10.0% O_2, and 81.5% N_2 by volume. The pyrites is charged at 18°C and the air is supplied at the same temperature with a percentage humidity of 49% and at a pressure of 750 mm of Hg. The cinder leaves the burner at 400°C containing 0.42% sulfur as unburned pyrites. The pyrites burned forms Fe_2O_3 and SO_2 and the gangue passes into the cinder unchanged. The mean specific heat of the cinder is 0.18. The gases from the burner, after passing through the dust chamber, enter the Glover tower at 450°C and leave at 91°C. In the Glover tower 16% of the SO_2 in the gas is converted to H_2SO_4.   There are sprayed into the top of the Glover tower, per 100 kg of moisture-free pyrites charged:

    182 kg of aqueous sulfuric acid at 25°C from the chambers, containing
        64.0% H_2SO_4 and 36% H_2O.
    580 kg of mixed acid at 25°C from the Gay-Lussac tower, containing 77%
        H_2SO_4, 22.1% H_2O, and 0.885% N_2O_3.
    1.31 kg of aqueous nitric acid at 25°C, containing 36% HNO_3 and 64% H_2O.

Acid leaves the bottom of the Glover tower, free from oxides of nitrogen, at a temperature of 125°C, and containing 78.0% H_2SO_4 and 22% H_2O.   This acid is cooled to 25°C, part of it is returned to the top of the Gay-Lussac tower, and the remainder is withdrawn as the final product of the plant.   The gases leaving the Glover tower are passed through a series of four chambers and finally enter the Gay-Lussac tower, at 40°C.   Spray water is introduced at the tops of the various chambers at 18°C. The acid formed in the chambers is withdrawn from the first chamber at 68°C, containing 64.0% H_2SO_4.   This acid is cooled to 25°C and all fed into the top of the Glover tower.   Part of the Glover acid after cooling to 25°C is returned to the top of the Gay-Lussac tower.   The Gay-Lussac acid leaves the bottom of the tower at 27°C and is all fed to the top of the Glover tower at 25°C.   The spent gases leave the top of the Gay-Lussac tower at 30°C.   The flow chart of the entire process is shown diagrammatically in Fig. 77.

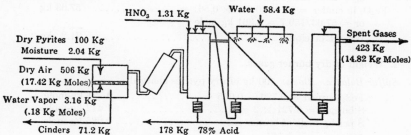

FIG. 77.   Material balance of an entire sulfuric acid plant.

Calculate material and energy balances of the entire plant and of each of the following units, all based on 100 kg of moisture-free pyrites charged:

    (a) The burner.
    (b) The Glover tower.
    (c) The four chambers as a single unit.
    (d) The Gay-Lussac tower.

MATERIAL BALANCE OF ENTIRE PLANT

Before discussing the material and energy balances of the individual units in the sulfuric acid plant it is desirable to calculate the material balance of the entire plant in order to have in mind a perspective of the whole process.   This balance is represented by the following entries:

	Input	Output
Dry ore	Moisture in air	Acid produced
Moisture in ore	Nitric acid	Dry spent gases
Dry air	Spray water	Cinder

1. Weight of cinder formed.

Basis: 100 kg of moisture-free pyrites charged.

Weight of pyrites as charged $= \dfrac{100}{0.98} =$ .... 102 kg

Weight of FeS_2 charged $= 102 \times 0.853 =$ ...... 87 kg

or $87/120 =$ .... 0.726 kg-mole

Weight of gangue charged $= 102 \times 0.127 =$ .... 13 kg

Percentage S in cinder $=$ 0.42%

Percentage FeS_2 in cinder $=$

$\dfrac{0.42}{32} \times \dfrac{1}{2} \times 120 =$ 0.78%

Let $x =$ kilograms of FeS_2 in cinder.

Weight of FeS_2 oxidized $= (87 - x)$ kg or $(87 - x)/120$ kg-moles

Fe_2O_3 formed $= \dfrac{87 - x}{120} \times \dfrac{1}{2} \times 159.7 = (58 - 0.667x)$ kg

Weight of cinder $= 13.0 + x + (58 - 0.667x) = (71 + 0.333x)$ kg

Weight of FeS_2 in cinder $=$

$x = 0.0078(71 + 0.333x)$ or, $x =$ .... 0.56 kg

or $= 0.56/120 = 0.0048$ kg-mole

Fe_2O_3 in cinder $= 58 - (0.667 \times 0.56) =$ .... 57.63 kg

or $= 57.63/159.7 = 0.361$ kg-mole

Weight of cinder $=$ ...., 71.2 kg

2. Weight of dry burner gases.

Sulfur Balance. Basis: 100 kg of dry pyrites charged.

FeS_2 burned $= 0.726 - 0.005 =$ .... 0.721 kg-mole

S burned $= 0.721 \times 2 =$ .... 1.442 kg-moles

S per kg-mole of burner gas $=$ .... 0.085 kg-mole

Dry burner gas $= \dfrac{1.442}{0.085} =$ .... 16.95 kg-moles

$SO_2 = 16.95 \times 0.085 = 1.441$ kg-moles or $\times 64 = 92.2$ kg

$O_2 = 16.95 \times 0.100 = 1.70$ kg-moles or $\times 32 = 54.4$ kg

$N_2 = 16.95 \times 0.815 = 13.81$ kg-moles or $\times 28.2 = 389.8$ kg

Total dry gases $=$ 16.95 kg-moles or 536.4 kg

3. Weight of dry air used.

Nitrogen Balance:

Nitrogen in burner gas $=$ .... 13.81 kg-moles

Air introduced $= \dfrac{13.81}{0.79} =$ .... 17.5 kg-moles

or $17.42 \times 29 =$ .... 506 kg

4. Weight of water vapor in dry air. The air supply enters at 18°C, 49% percentage humidity, and 750 mm pressure. From Fig. 9 the molal humidity is 0.0101.

Water vapor in air $= 0.0101 \times 17.5 =$ .... 0.176 kg-mole

or $0.176 \times 18 =$ .... 3.16 kg

5. H_2SO_4 produced in system.   Sulfuric acid is formed only in the Glover tower and in the chambers.

SO_2 entering Glover tower = ...   1.44 kg-moles

SO_2 converted to acid in Glover tower = 1.44

    $\times$ 0.16 = ...   0.230 kg-mole

H_2SO_4 formed in Glover tower = 0.230 $\times$ 98.1 =   22.6 kg

H_2SO_4 in acid from chambers = 182 $\times$ 0.64 = .   116.5 kg

Total H_2SO_4 formed = ...   139.1 kg

   or = 139.1/98 = ...   1.42 kg-moles

Total product of aqueous acid, 78% H_2SO_4 =

   139.1/0.78 = ...   178 kg

6. Weight of spray water in chambers.

Water Balance:

Output

The small amount of water vapor in the gases from the Gay-Lussac tower may be neglected.   On this basis all water leaves the process in the 78% acid product.

H_2O used in forming H_2SO_4 = 1.42 kg-moles

   or 1.42 $\times$ 18 = ...   25.5 kg

H_2O in aqueous acid = 178 $\times$ 0.22 = ...   39.0 kg

     Total H_2O output = ...   64.5 kg

Input

H_2O in ore = 102 $\times$ 0.02 = ...   2.04 kg

H_2O in air = ...   3.16 kg

H_2O from aqueous nitric acid = 1.31 $\times$ 0.64 = ...   0.84 kg

It is assumed that the HNO_3 introduced is completely decomposed into H_2O, NO, and O_2.

HNO_3 introduced = 1.31 $\times$ 0.36 = 0.47 kg or

   0.47/63 = 0.0075 kg-mole

H_2O formed from HNO_3 = 0.0075/2 = 0.0037 kg-

   mole or = ...   0.067 kg

     Total H_2O input accounted for = ...   6.11 kg

Water supplied by sprays = 64.5 − 6.1 = ...   58.4 kg

7. Gases leaving Gay-Lussac tower.   The gases leaving the acid plant consist of SO_2, O_2, and N_2 from the burner and the oxides of nitrogen which are supplied by the nitric acid and lost from the system.   Most of the SO_2 and a corresponding amount of oxygen are removed from the burner gases to form H_2SO_4.   It is assumed that no water vapor leaves the Gay-Lussac tower because of the great affinity of strong, cold sulfuric acid for water.

SO_2 from burner = ...   1.441 kg moles

SO_2 forming H_2SO_4 = ...   1.42 kg-moles

SO_2 in gases leaving = ...   0.021 kg-mole =   1.34 kg

O_2 in burner gases = ...   1.70 kg-moles

O_2 used in oxidizing SO_2 = 1.42/2 = ...   0.71 kg-mole

O_2 from burner in gases leaving = ...   0.99 kg-mole =   31.9 kg

N_2 in gases from burner = ...   13.81 kg-moles = 389.8 kg

HNO_3 decomposed = ...   0.0075 kg-mole

According to the reaction,

$$2HNO_3 = H_2O + 2NO + 3/2 O_2$$

O_2 from HNO_3 = 0.0075(3/4) = ...	0.0056 kg-mole	= 0.18 kg
NO from HNO_3 = ...	0.0075 kg-mole	= 0.22 kg

Total gases leaving:

SO_2........................	0.021	kg-mole	1.3 kg
O_2...........................	0.995	kg-mole	31.9 kg
NO..........................	0.0075	kg-mole	0.22 kg
N_2...........................	13.81	kg-moles	389.8 kg
Total..................	14.82	kg-moles	423.2 kg

Material Balance of Entire Plant.

Input		Output	
Dry pyrites............	100.0 kg	Cinder..............	**71.2 kg**
H_2O in pyrites.........	2.04 kg	Spent gases (14.82 kg-	
Dry air (17.42 kg-moles)	506.00 kg	moles)............	423.2 kg
H_2O in air (0.176 kg-mole)	3.16 kg	Acid (78% H_2SO_4)....	178.0 kg
Nitric acid.............	1.31 kg	Total...........	672.4 kg
Spray water............	58.4 kg		
Total.............	670.9 kg		

This material balance is summarized in Fig. 77.

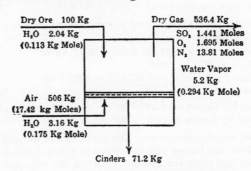

FIG. 78. Material balance of pyrites burner.

MATERIAL BALANCE OF BURNER

Input		Output	
Dry ore..............	100 kg	Cinder............	71.2 kg
Moisture in ore.......	2.04 kg	Dry gases..........	536.4 kg
Dry air (17.42 kg-moles)	506. kg	Water vapor (3.16 +	
Moisture in air.......	3.16 kg	2.04) =.........	5.2 kg
Total...........	611.2 kg	Total.......	612.8 kg

This material balance is summarized in Fig. 78.

ENERGY BALANCE OF BURNER

The energy balance of the burner includes the heat of combustion of FeS_2 to Fe_2O_3 and SO_2 and the enthalpy of the water vapor in the air as the only important sources of heat. The energy output is distributed as sensible enthalpy of the outgoing cinders and dry gases and as enthalpy of outgoing water vapor. The heat losses include radiation from the dust chamber and flues up to the entrance of the Glover tower where the temperature of the burner gases, 450°C, was measured.

1. Heat evolved in the combustion of pyrites.

The reaction involved in the combustion of pyrites and the corresponding standard heat of reaction are as follows:

$$4FeS_2 + 11O_2 = 2Fe_2O_3 + 8SO_2$$
$$\Delta H_{18} = 2(-198,500) + 8(-70,920) - 4(-35,500)$$
$$\Delta H_{18} = -822,360$$

FeS_2 actually burned $= 0.721$ kg-mole

Heat evolved in combustion of FeS_2 burned $= 0.721 \times \dfrac{822,360}{4} = 148,200$ kcal

2. Enthalpy of water vapor in air.

From Fig. 9, dew point $= 7$°C.
Heat of vaporization at 7°C $= 10,630$ kcal per kg-mole
Total enthalpy $= 0.176\ [10,630 - 18(18 - 7) + 8.0(18 - 7)] = 1,850$ kcal

3. Enthalpy of cinder $= 71.2 \times .16(450 - 18) = 4,900$ kcal

4. Enthalpy of dry burner gas.

Mean heat capacities between 18°C and 450°C taken from Table VI, page 216.

$$SO_2 = (1.44)\ (11.02) = 15.8$$
$$O_2\ \ = (1.70)\ (7.45)\ = 12.6$$
$$N_2\ \ = (13.81)\ (7.12) = \underline{98.2}$$
$$\text{Total} \qquad = 126.6$$

Enthalpy $= (126.6)\ (450 - 18) = 54,700$ kcal

5. Enthalpy of water vapor in burner gases.

Water vapor present $= 5.2/18 = 0.29$ kg-mole.
Molal humidity $= 0.29/16.95 = 0.017$
From Fig. 9, dew point $= 15$°C
Heat of vaporization at 15°C $= 10,580$ kcals per kg-mole
Mean molal heat capacity of water vapor (15°C–450°C) $= 8.42$
Enthalpy $= 0.29\ [10,580 - 18(18 - 15) + 8.42(450 - 15)] = 4110$ kcal

Summary of Energy Balance of Burner.

Input

1. Heat evolved in combustion of pyrites $= \ldots\ldots$	148,200 kcal	98.8%
2. Enthalpy of water vapor in air $= \ldots\ldots\ldots$	1,850 kcal	1.2%
Total input $= \ldots\ldots\ldots\ldots\ldots\ldots\ldots$	150,050 kcal	100.0%

Output

1. Enthalpy of cinder = ...   4,900 kcal   3.3%
2. Enthalpy of dry burner gas = ...   54,700 kcal   36.5%
3. Enthalpy of water vapor in gases = ...   4,110 kcal   2.7%
4. Heat losses (by difference) = ...   86,340 kcal   57.5%

   Total output = ...   150,050 kcal   100.0%

This energy balance is summarized in Fig. 79.

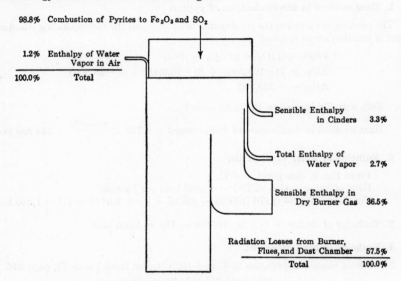

FIG. 79.   Energy balance of pyrites burner.

MATERIAL BALANCE OF GLOVER TOWER

The useful functions of the Glover tower are as follows:

1. Cooling of burner gases before being blown into the chambers.
2. Conversion of about 16% of SO_2 in burner gas to H_2SO_4.
3. Mixing of Gay-Lussac and chamber acids and nitric acid.
4. Evolution of oxides of nitrogen from Gay-Lussac acid.
5. Concentration of chamber acid.

At the top of the tower, mixing of the Gay-Lussac and chamber acids takes place with the release of the oxides of nitrogen from the Gay-Lussac acid when this acid is heated and diluted.   Nitric acid is also added to make up for the losses of the oxides of nitrogen through leakage in the system and incomplete absorption in the Gay-Lussac tower. The oxides of nitrogen are present as a mixture of NO and NO_2, but since the equilibrium mixture gradually changes owing to variation in temperature and in oxygen content it is customary to assume that the oxides of nitrogen leave the Glover tower as NO.   The released oxides of nitrogen react in the vapor state with SO_2, oxygen and water vapor, forming liquid nitrosyl-sulfuric acid

$$2NO + \tfrac{3}{2}O_2 + 2SO_2 + H_2O \rightarrow 2NO_2SO_2OH$$

Because of the high concentration of SO_2 in the Glover tower, the decomposition of this nitrosyl-sulfuric acid is complete, forming sulfuric acid according to the equation

$$2H_2O + 2NO_2SO_2OH + SO_2 = 3H_2SO_4 + 2NO$$

The second reaction proceeds much more rapidly than the first in moles per unit volume of space, chiefly because it is a reaction which proceeds in the liquid phase. The high concentration in the liquid state permits a more rapid rate of reaction, other conditions being the same. The final concentration of the Glover acid takes place at the hottest zone near the bottom of the tower and the acid finally leaves as 78% H_2SO_4.

The input of the material balance includes the gases from the burner, the chamber acid, the make-up nitric acid introduced, and the Gay-Lussac acid. The output includes the Glover acid and the gases leaving to enter the chambers.

Input to Glover Tower

Basis: 100 kg of moisture-free pyrites charged.

1. Dry burner gases = 16.95 kg-moles = ...   536.4 kg
2. Water vapor in burner gases = 0.29 kg-mole = ...   5.2 kg
3. Chamber acid (64% H_2SO_4; 36% H_2O) = ...   182 kg
4. Nitric acid (36% HNO_3) = ...   1.31 kg
5. Gay-Lussac acid (77% H_2SO_4; 22.1% H_2O; and 0.885% N_2O_3) = ...   580 kg

Total input. . . . .   1304.9 kg

Output

1. Weight of Glover acid. The H_2SO_4 leaving the Glover tower in 78% acid includes that from the chamber and Gay-Lussac acids and that formed by conversion of SO_2 to H_2SO_4 in the tower already calculated to be 0.230 kg-mole.

H_2SO_4 from chambers = 182 × 0.64 = ...   116.4 kg
H_2SO_4 from Gay-Lussac acid = 580 × 0.77 = ...   446 kg
H_2SO_4 formed in tower = 0.230 × 98 = ...   22.6 kg

Total H_2SO_4. . . . .   585.0 kg

Weight of 78% acid leaving Glover tower = $\dfrac{585}{0.78}$ = ...   750 kg

2. Weight of dry gases leaving the Glover tower. The weight and composition of the dry burner gases in passing through the Glover tower is changed owing to the disappearance of some SO_2 and the corresponding amount of oxygen in the formation of SO_3 and to the evolution of NO and O_2 due to the decomposition of nitric acid and the release of the oxides of nitrogen from the Gay-Lussac acid.

Dry burner gases entering tower:

SO_2 = ...   1.44 kg-moles
O_2 = ...   1.70 kg-moles
N_2 = ...   13.81 kg-moles

Total = ...   16.95 kg-moles

SO_2 converted to H_2SO_4 in tower = ...   0.230 kg-mole
SO_2 leaving tower = 1.44 − 0.230 = ...   1.21 kg-moles
O_2 used in forming SO_3 = 0.230/2 = ...   0.115 kg-mole
O_2 remaining from burner gases = 1.70 − 0.115 = ...   1.58 kg-moles

Nitric acid decomposed =

1.31 × 0.36 = 0.472 kg or............................   0.0075 kg-mole

$2HNO_3 = H_2O + 2NO + 1\frac{1}{2}O_2$

O_2 from HNO_3 = 0.0075 × $\frac{3}{4}$ = ...   0.0056 kg-mole

NO from HNO_3 = ...   0.0075 kg-mole

H_2O from HNO_3 = 0.0075/2 = ...   0.0037 kg-mole

N_2O_3 from Gay-Lussac acid =

580 × 0.00885 = 5.13 kg or ....   0.0675 kg-mole

$N_2O_3 = 2NO + \frac{1}{2}O_2$

O_2 from N_2O_3 = 0.0675/2 = ...   0.0338 kg-mole

NO from N_2O_3 = ...   0.1350 kg-mole

Total O_2 leaving = 1.58 + 0.0056 + 0.0338 = ...   1.62   kg-moles

Total NO leaving = 0.0075 + 0.1350 = ...   0.143   kg-mole

Total dry gases leaving

SO_2 =   1.21 kg-moles or × 64 = ...   77.5 kg

O_2 =   1.62 kg-moles or × 32 = ...   51.8 kg

NO =   0.143 kg-mole or × 30 = ...   4.29 kg

N_2 = 13.81 kg-moles or × 28.2 = ...   390.0 kg

Total 16.78 kg-moles or..........................   523.6 kg

Total weight of dry gases leaving Glover tower = ...   523.6 kg

3. Water vapor in the gases leaving the Glover tower. The weight of water vapor in the gases leaving the Glover tower is calculated on the basis of a water balance. Water enters as vapor in the gases, as water in the Gay-Lussac and chamber acids and associated with the nitric acid as HNO_3 and as water. The 0.230 kg-mole of H_2SO_4 formed in the tower requires 0.230 kg-mole of H_2O or 4.15 kg. The 78% acid leaving requires (750) (0.22) = 165 kg water.

The 1.31 kg of 36% nitric acid charged yields upon dehydration and decomposition 0.84 + 0.07 = 0.91 kg H_2O.

Water Balance:

Input

In gas (0.289 kg-mole) = ...   5.2 kg

In chamber acid (182) (0.36) = ...   65.5 kg

In Gay-Lussac acid 580 × 0.221 = ...   129.0 kg

From nitric acid = ...   0.9 kg

Total.......................................   200.6 kg

Output

In 78% acid....................................   165.0 kg

In formation of H_2SO_4...........................   4.1 kg

In water vapor (by difference)....................   31.5 kg

Total.......................................   200.6 kg

Total water vapor in gases leaving = ...   31.5 kg

or 31.5/18 = ...   1.75 kg-moles

Molal humidity of gases leaving = $\dfrac{1.75}{16.78}$ = 0.103

Dew point of gases leaving (from Fig. 8) = 113°F or 45°C

Summary of Material Balance of Glover Tower.

Input

Dry gases (16.95 kg-moles) = ...	536.4	kg
Water vapor (0.29 kg-mole) = ...	5.2	kg
Chamber acid (64% H_2SO_4) = ...	182.0	kg
Nitric acid (36% HNO_3) = ...	1.31	kg
Gay-Lussac acid = ...	580.0	kg
Total = ...	1304.9	kg

Output

Dry gases = ...	523.6	kg
Water vapor = ...	31.5	kg
Glover acid (78% H_2SO_4) = ...	750.0	kg
Total = ...	1305.1	kg

This material balance is summarized in Fig. 80.

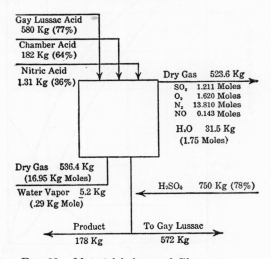

Fig. 80.   Material balance of Glover tower.

ENERGY BALANCE OF GLOVER TOWER

In addition to the enthalpy of all materials entering and leaving, the input side of the energy balance of the Glover tower includes heat evolved in the formation of H_2SO_4 within the tower. This acid goes into solution but at the same time some water leaves the solution, the net effect being one of concentration and hence requiring the input of heat. In addition, heat is required to remove the oxides of nitrogen from solution and to decompose them into NO and O_2. It will be assumed that all oxides of nitrogen, both from the Gay-Lussac and the make-up nitric acids, are decomposed and leave the Glover tower as NO.

1. Heat evolved in formation of H_2SO_4. Sulfuric acid is formed from SO_2 gas, liquid H_2O, and oxygen. Actually the conversion takes place in two steps with intermediate formation of nitrosyl-sulfuric acid. However, the net effect is the same as though the reaction proceeded as follows:

$$SO_2(g) + \tfrac{1}{2}O_2(g) + H_2O(l) = H_2SO_4$$
$$\Delta H_{18} + (-70{,}920) + 0 + (-68{,}320) = -193{,}750$$
$$\Delta H_{18} = -54{,}510 \text{ kcal}$$

Heat evolved in formation of H_2SO_4 = 0.230(54,510) = 12,530 kcal

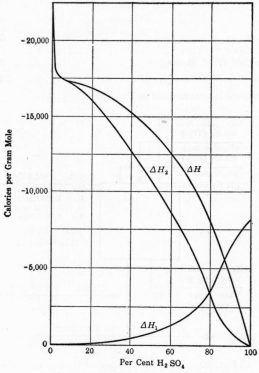

ΔH = Integral heat of solution of H_2SO_4, cal per g-mole
ΔH_1 = Differential heat of solution of H_2O, cal per g-mole
ΔH_2 = Differential heat of solution of H_2So_4, cal per g-mole

Fig. 81.   Differential and integral heats of solution of aqueous solutions of sulfuric acid at 20°C.

2. Heat absorbed in concentrating acid. The sulfuric acid formed in the tower is diluted by the acid already present to form 78% acid leaving. The chamber acid entering at 64% is concentrated to 78%. The Gay-Lussac acid yields aqueous sulfuric acid containing 77/0.991 or 77.7% H_2SO_4. This is concentrated to 78% acid. The net result of these changes is a concentrating effect.

The net heat effect of the concentration changes in the Glover tower may be calculated from integral heat of solution data by the method discussed on page 274. The heat evolved is the difference between the total heat evolved in forming each of the entering acid solutions from H_2SO_4 and H_2O and that evolved in forming the solution leaving from H_2SO_4 and H_2O. These thermal effects may be calculated from the integral heat of solution data plotted in Fig. 81, page 394. This curve was plotted from data of the International Critical Tables, Vol. V, with permission.

Molal integral heats of solution of H_2SO_4:

$$64\% \ H_2SO_4 = \ldots \quad -11,800 \text{ kcal per kg-mole}$$
$$77.7\% \ H_2SO_4 = \ldots \quad - \ 8,700 \text{ kcal per kg-mole}$$
$$78\% \ H_2SO_4 = \ldots \quad - \ 8,600 \text{ kcal per kg-mole}$$

$$\text{Heat of solution of chamber acid} = \frac{182 \times 0.64}{98} \times -11,800 \quad = -14,000 \text{ kcal}$$

$$\text{Heat of solution of Gay-Lussac acid} = \frac{580 \times 0.77}{98} \times -8700 \quad = -39,700 \text{ kcal}$$

$$\text{Heat of solution of entering acids} \quad = -53,700 \text{ kcal}$$

$$\text{Heat of solution of Glover acid leaving} = \frac{750 \times 0.78}{98} \times -8600 = -51,300 \text{ kcal}$$

$$\text{Heat absorbed in concentration of acid} = 53,700 - 51,300 \quad = +2400 \text{ kcal}$$

The net heat absorbed in concentration is 2400 calories, which should be placed on the output side of the energy balance. The large amount of heat required for the concentration of chamber acid is offset by the heat evolved when the H_2SO_4 formed in the tower is dissolved.

3. Heat absorbed in release and decomposition of N_2O_3 from Gay-Lussac acid. The oxides of nitrogen enter the Glover tower as recovered nitrosyl-sulfuric acid from the Gay-Lussac tower and as make-up nitric acid. The Gay-Lussac acid may be considered to consist of 0.0677 kg-mole of N_2O_3 dissolved in 129 kg (7.15 kg-moles) of water. This concentration corresponds to 106 moles of water per mole of N_2O_3. The thermal effects in the evolution of the N_2O_3 from solution and its decomposition may be calculated from the following thermochemical data:

Formula	State	Heat of Formation kcal per kg-mole	Moles of Water	Heat of Solution, kcal
NO	g	+21,600		
N_2O_3	g	+20,000	100	−28,900
NO_2	g	+8,030		
N_2O_4	g	+3,060		
N_2O_5	g	+3,600		
HNO_3	l	−41,660	6.22	−7,000

Heat *absorbed* in evolution of N_2O_3 from solution in the Gay-Lussac acid = 0.0677 $\times$ 28,900 = 1960 kcal

This entry is not exact since it has been assumed that the heat of solution of N_2O_3 is the same in 77.7% H_2SO_4 acid as it is in water, which is a poor approximation. The N_2O_3 is assumed to break down entirely to NO and O_2; this is also an approximation. However, in view of the relatively small heat effects involved these approximations seem justified.

$$N_2O_3 = 2NO + \tfrac{1}{2}O_2$$
$$\Delta H_{18} + 20,000 = 2(21,600)$$
$$\Delta H_{18} \qquad = 23,200 \text{ kcal}$$

Heat absorbed in decomposition of $N_2O_3 = 0.0677 \times 23,200 = 1570$ kcal
Total heat absorbed in release and decomposition of $N_2O_3 = 1960 + 1570$
$= 3530$ kcal

4. Heat absorbed in decomposition of nitric acid. The nitric acid consists of 0.0075 kg-mole of HNO_3 dissolved in 0.0466 kg-mole of water, corresponding to 6.22 moles of water per mole of HNO_3. The acid may be considered to be separated into its liquid components, HNO_3 and water, and then decomposed according to the following reaction:

$$HNO_3(l) = \tfrac{1}{2}H_2O(l) + NO(g) + \tfrac{3}{4}O_2(g)$$
$$\Delta H_{18} + (-41,660) = -34,160 + 21,600$$
$$\Delta H_{18} \qquad = 29,100 \text{ kcal}$$

Heat absorbed in separating HNO_3 from solution $= 0.0075 \times 7000 = \quad$ 52 kcal
Heat absorbed in decomposition of $HNO_3 = 0.0075\,(29,100) = \ldots$ 218 kcal
Total heat absorbed $= \ldots\ldots\ldots\ldots\ldots\ldots\ldots\ldots\ldots\ldots\ldots\ldots\ldots\ldots\ldots$ 270 kcal

The decomposition of the oxides of nitrogen is of particular interest because of their endothermic heats of formation.

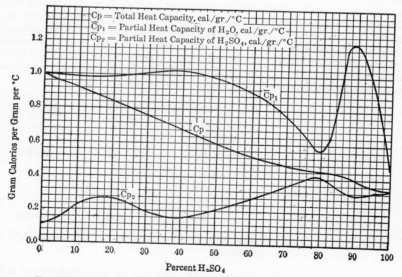

Fig. 82. Partial and total heat capacities of aqueous solutions of sulfuric acid at 20°C.

5. Heat input in burner gases. This has already been calculated as the heat output of the burner, page 390, = ...58,810 kcal

6. Enthalpy of chamber acid.

Specific heat (from Fig. 82) = ... **0.50**
Enthalpy = 182 × 0.50 × (25 − 18) = ... **640 kcal**

7. Enthalpy of Gay-Lussac acid.

Specific heat (Fig. 82) = ... 0.45
Enthalpy = 580 × 0.45 (25 − 18) = ... 1830 kcal

This calculation neglects the effect of the oxides of nitrogen on the heat capacity of the Gay-Lussac acid.

8. Enthalpy of nitric acid. From Fig. 41, page 224, the specific heat of 36% nitric acid is 0.70.

Enthalpy = 1.31 × 0.70 (25 − 18) = ... 7 kcal

9. Enthalpy of dry gases leaving.

Mean heat capacities between 18°C and 91°C taken from Table VI, page 216.

$$
\begin{array}{llr}
SO_2 & (1.21)(9.76) = & 11.8 \\
N_2 & (13.81)(6.97) = & 96.4 \\
NO & (0.14)(7.16) = & 1.0 \\
O_2 & (1.62)(7.05) = & \underline{11.4} \\
& & 120.6
\end{array}
$$

Enthalpy = 120.6 (91 − 18) = ... 8810 kcal

10. Total enthalpy of water vapor leaving Glover tower.

The water vapor leaves at 91°C having a dew point of 45°C.
Heat of vaporization at 45°C (Fig. 8) = 10,280 kcal per kg
Mean molal heat capacity of vapor between 45°C and 91°C = 8.03
Enthalpy = 1.75[10,280 + 18(45 − 18) + 8.03(91 − 45)] = ...19,500 kcal

11. Heat absorbed in cooling outgoing acid.

Heat capacity (Fig. 82) = 0.44 kcal per kg per °C, neglecting the temperature coefficient of the heat capacity.
Heat absorbed = 750 × 0.44 (125 − 25) = ...33,000 kcal

12. Enthalpy of outgoing acid = 750 × 0.44 (25 − 18) =   2310 kcal

Summary of Energy Balance of Glover Tower.

Input

Enthalpy of dry burner gases..................	54,700 kcal	74.0%
Enthalpy of water vapor from burner..........	4,110 kcal	5.6%
Enthalpy of chamber acid..................	640 kcal	0.9%
Enthalpy of Gay-Lussac acid..................	1,830 kcal	2.5%
Enthalpy of nitric acid.....................	7 kcal	0.0%
Heat evolved in formation of H_2SO_4	12,530 kcal	17.0%
Total.....................................	73,817 kcal	100.0%

Output

Enthalpy of dry gases......................	8,810 kcal	11.9%
Enthalpy of water vapor in gases..............	19,500 kcal	26.4%
Enthalpy of acid leaving cooler................	2,310 kcal	3.1%
Heat absorbed in concentration of acid.........	2,400 kcal	3.3%
Heat absorbed in decomposition of nitric acid...	270 kcal	0.4%
Heat absorbed in release and decomposition of N_2O_3 from Gay-Lussac acid.................	3,530 kcal	4.8%
Heat absorbed by outgoing acid cooler..........	33,000 kcal	44.7%
Heat losses (by difference)....................	3,997 kcal	5.4%
Total.....................................	73,817 kcal	100.0%

This energy balance is summarized in Fig. 83.

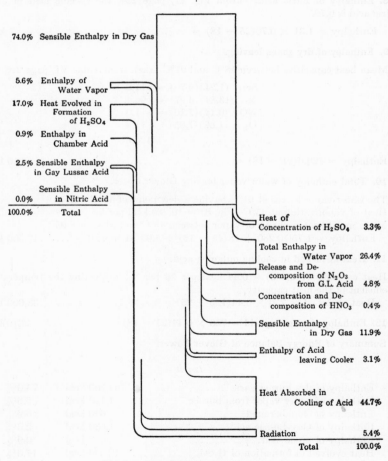

FIG. 83.   Energy balance of Glover tower.

It will be seen that over one-half of the available energy in the Glover tower is absorbed by the concentrated acid. This acid must be cooled before it can be used for absorption of gases in the Gay-Lussac tower and before storage. The cooling of this acid represents one of the difficult problems in acid manufacture, in the development of a heat interchanger which will withstand hot concentrated sulfuric acid and at the same time permit a rapid transfer of heat.

MATERIAL BALANCE OF CHAMBERS

It would be proper to consider the material and energy balances of each chamber separately, but to avoid needless repetition all four chambers will be considered as a unit. In the chambers, H_2O, SO_2, and O_2 are removed from the gases to form H_2SO_4.

Basis: 100 kg of dry pyrites charged.

H_2SO_4 formed in chambers = 116.5 kg or 1.19 kg-moles.

	Gas Entering		Gas Removed	Dry Gas Leaving
SO_2	1.21	kg-moles	1.19 kg-moles	0.021 kg-mole
O_2	1.62	kg-moles	0.595 kg-mole	1.025 kg-moles
NO	0.143	kg-mole		0.143 kg-mole
N_2	13.81	kg-moles		13.81 kg-moles
H_2O	1.75	kg-moles	1.19 kg-moles	
Total =				15.00 kg-moles

Water vapor leaving. The water entering with the gases and from the sprays is used in the formation of H_2SO_4 and dilution of the H_2SO_4 to form the chamber acid.

Water input

H_2O from gases = ...	31.5 kg
H_2O from sprays = ...	58.4 kg
Total = ...	89.9 kg

Water output

H_2O to form $H_2SO_4 = 1.19 \times 18 = $	21.4 kg
H_2O in chamber acid $= 182 \times 0.36 = $	65.5 kg
Output accounted for = ...	86.9 kg
H_2O in gases leaving $= 89.9 - 86.9 = $	3.0 kg
or 3.0/18 = ...	0.167 kg-mole
Moles of dry gases leaving = ...	15.0 kg-moles
Molal humidity = 0.167/15.0 = ...	0.011
Dew point (Fig. 9) = ...	46°F or 8°C

Summary of Material Balance of Chambers.

Entering			Leaving		
SO_2	(1.21 kg-moles)	77.5 kg	SO_2	(0.021 kg-mole)	1.3 kg
O_2	(1.62 kg-moles)	51.8 kg	O_2	(1.025 kg-moles)	32.8 kg
N_2	(13.81 kg-moles)	390.0 kg	N_2	(13.81 kg-moles)	390.0 kg
H_2O	(1.75 kg-moles)	31.5 kg	H_2O	(0.168 kg-mole)	3.0 kg
NO	(0.143 kg-mole)	4.3 kg	NO	(0.143 kg-mole)	4.3 kg
Spray water		58.4 kg	Acid		182.0 kg
Total...........		613.5 kg	Total............		613.4 kg

This material balance is summarized in Fig. 84.

ENERGY BALANCE OF CHAMBERS

The first reaction proceeding in the chambers is a gaseous reaction between the SO_2, H_2O, O_2, and NO gases in contact with the water spray forming nitrosyl-sulfuric acid.

$$2SO_2 + 2NO + 1\tfrac{1}{2}O_2 + H_2O \rightarrow 2NO_2SO_2OH$$

This reaction is favored by high concentrations of SO_2 and NO. The acid spray is swept against the side walls of the chamber where the spray is condensed owing to cooling, and by dilution with water H_2SO_4 is formed with the release of the oxides of nitrogen.

$$2NO_2SO_2OH + H_2O = 2H_2SO_4 + N_2O_3$$
$$N_2O_3 = 2NO + \tfrac{1}{2}O_2$$

The first of these reactions proceeds in the liquid phase and is favored by a high concentration of water brought in by the spray and condensed upon the side walls.

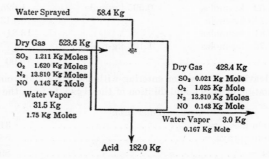

FIG. 84. Material balance of chambers.

In calculating the heat evolved in the chamber reactions only the final, net effects need be considered. It will be assumed that the oxides of nitrogen ultimately leave the chambers in the same form in which they entered, NO. This assumption is not exact because some oxidation of NO to N_2O_3 probably takes place at the relatively low temperatures of the last chamber. The ultimate effects of the reactions in the chambers are then the production of H_2SO_4 from SO_2, O_2, and H_2O and the dissolution of this acid to form an aqueous solution containing 64% H_2SO_4.

Reference temperature: 18°C.

Basis: 100 kg of dry pyrites charged.

1. Heat evolved in formation of H_2SO_4. From item 1 of the energy balance of the Glover tower, the heat evolved is 54,510 kcal per kg-mole of H_2SO_4 formed.

H_2SO_4 formed in chambers = 1.19 kg-moles
Heat evolved = 1.19 × 54,510 = ... **64,870 kcal**

2. Heat evolved in dissolving H_2SO_4. Integral heat of solution (Fig. 81) at a concentration of 64% H_2SO_4 = 11,800 kcal per kg-mole.

Heat evolved in dissolution = 11,800 × 1.19 = ... **14,050 kcal**

3. Enthalpy of dry gases and water vapor entering. This has already been calculated as part of the heat output of the Glover tower = ... **28,310 kcal**

4. Enthalpy of spray water. Since the spray water enters at the reference temperature, 18°C, its enthalpy is equal to. . . . . **0 kcal**

5. Enthalpy of dry gases leaving.

Mean heat capacities between 18°C and 40°C taken from Table VI, page 216.

$$
\begin{aligned}
SO_2 \ (0.021)(9.49) &= 0.2 \\
O_2 \ (1.025)(7.02) &= 7.2 \\
N_2 \ (13.81)(6.97) &= 96.4 \\
NO \ (0.14)(7.16) &= \underline{1.0} \\
\text{Total} &= \overline{104.8}
\end{aligned}
$$

Enthalpy of dry gases = $(104.8)(40 - 18)$ = 2310 kcal

6. Enthalpy of water vapor leaving.

Dew point = 8°C.
Heat of vaporization at 8°C (Fig. 8) = 10,640 kcal per kg-mole.
Total enthalpy =
$0.167[10,640 - 18(18 - 8) + 8.01(40 - 8)]$ = ... **1790 kcal**

7. Heat absorbed in cooling the acid leaving. Heat capacity (Fig. 82) of acid containing 64% H_2SO_4 = 0.50 kcal per kg per °C.

Enthalpy = $0.50 \times 182(68 - 25)$ = ... **3920 kcal**

8. Enthalpy of acid leaving the cooler =

$0.50 \times 182(25 - 18)$ = ... **640 kcal**

Summary of Energy Balance of Chambers.

Input

Enthalpy of dry gases from Glover tower.	8,810 kcal	8.2%
Enthalpy of water vapor in gases entering.	19,500 kcal	18.2%
Enthalpy of spray water.	0 kcal	0%
Heat evolved in forming H_2SO_4.	64,870 kcal	60.5%
Heat evolved in dissolving H_2SO_4.	14,050 kcal	13.1%
Total.	107,230 kcal	100.0%

Output

Enthalpy of dry gases leaving.	2,310 kcal	2.2%
Enthalpy of water vapor in gases.	1,790 kcal	1.7%
Enthalpy of acid leaving the cooler.	640 kcal	0.6%
Heat absorbed by cooler.	3,920 kcal	3.7%
Heat loss from chambers (by difference).	98,570 kcal	91.8%
Total.	107,230 kcal	100.0%

This energy balance is presented diagrammatically in Fig. 85.

It will be seen that nearly the entire source of energy (72.3%) comes from the formation of H_2SO_4 and its dilution whereas nearly the entire energy input is lost by radiation from the extensive lead surfaces of the chambers. More recent developments in the chamber process have provided for more rapid means of heat removal with much less floor space and size of equipment by rapid circulation of both gases and acid in packed towers.

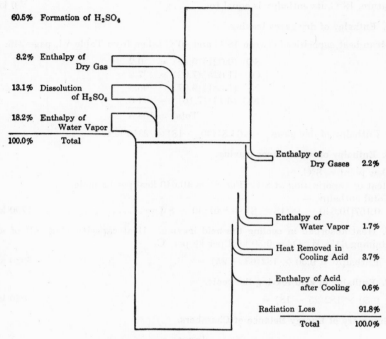

60.5%   Formation of H₂SO₄

8.2%   Enthalpy of
        Dry Gas

13.1%   Dissolution
        of H₂SO₄

18.2%   Enthalpy of
        Water Vapor

100.0%   Total

Enthalpy of
    Dry Gases   2.2%

Enthalpy of
    Water Vapor   1.7%

Heat Removed in
    Cooling Acid   3.7%

Enthalpy of Acid
    after Cooling   0.6%

Radiation Loss   91.8%

Total   100.0%

Fig. 85.   Energy balance of chambers.

MATERIAL BALANCE OF GAY-LUSSAC TOWER

The purpose of the Gay-Lussac tower is to recover the oxides of nitrogen from the chamber gases. These oxides are then returned to the system in the Glover tower. The water remaining in the chamber gases is also absorbed. The conditions favorable for absorption of the oxides of nitrogen are a high concentration of acid, a low temperature, and a low concentration of SO_2 in the residual gas. A high water content in the gas from the chambers causes objectionable dilution of the acid. Small amounts of SO_2 will cause decomposition of nitrosyl-sulfuric acid with release and loss of the oxides of nitrogen. The presence of oxygen is essential to effect the oxidation of NO to N_2O_3 and its absorption in the acid.

The loss of oxides of nitrogen in the gases from the Gay-Lussac tower may be assumed to be equivalent to the make-up nitric acid introduced. It will be assumed that these oxides leave in the form of NO.

1. Input. The input to the Gay-Lussac tower consists of the wet gases from the chambers and the Glover acid which is introduced. All the gases pass through the tower unchanged with the exception of the NO and O_2.

2. NO in gas leaving = 0.008 kg-mole or = ...   0.2 kg

3. O_2 in gases leaving:

NO oxidized to N_2O_3 = 0.143 − 0.008 = ... 0.135 kg-mole

O_2 consumed = 0.135/4 = ... 0.034 kg-mole

O_2 leaving = 1.025 − 0.034 = 0.991 kg-mole = .   31.7 kg

Summary of Material Balance of Gay-Lussac Tower.

Input			*Output*	
SO_2 (0.021 kg-mole)	1.3 kg		SO_2 (0.021 kg-mole)	1.3 kg
O_2 (1.025 kg-moles)	32.8 kg		O_2 (0.991 kg-mole)	31.7 kg
N_2 (13.81 kg-moles)	390.0 kg		N_2 (13.81 kg-moles)	390.0 kg
NO (0.143 kg-mole)	4.3 kg		Acid leaving	580.0 kg
H_2O (0.167 kg-mole)	3.0 kg		NO (0.008 kg-mole)	0.2 kg
Glover acid	572.0 kg		Total	1003.2 kg
Total	1003.4 kg			

This material balance is summarized diagrammatically in Fig. 86.

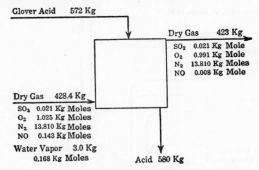

FIG. 86.   Material balance of Gay-Lussac tower.

ENERGY BALANCE OF GAY-LUSSAC TOWER

1. Heat evolved in forming and dissolving N_2O_3.   The N_2O_3 released from the Gay-Lussac acid and decomposed in the Glover tower is reformed and recovered in the Gay-Lussac tower evolving 3530 kcal calculated as part of the energy balance of the Glover tower, page 398.

2. Heat evolved in the dissolution of the water vapor absorbed.   The water vapor leaving the last chamber is completely absorbed in the Gay-Lussac tower.   Since the resulting concentration change in this absorption is negligible, it is necessary to determine the heat evolved in the dissolution of the water from data on the differential heat of solution of water in sulfuric acid solutions.   From Fig. 81, the differential molal heat of solution of water in a sulfuric acid solution containing 77.7% H_2SO_4 is 3100 kcal per kg-mole.   This value neglects the effect of the dissolved oxides of nitrogen.

Heat evolved = 3100 × 0.167 = ...   520 kcal

3. Enthalpy of Glover acid introduced.

Heat capacity (Fig. 82) = 0.44 kcal per kg per °C.

Enthalpy = 572 × 0.44(25 − 18) = ...   1760 kcal

4. Enthalpy of entering gases. Already calculated as output items in the energy balance of the chambers = ... 4100 kcal

5. Enthalpy of gases leaving.

Mean heat capacities between 18°C and 30°C taken from Table VI, page 216.

$$
\begin{array}{lll}
SO_2 & (0.021)(9.44) = & 0.2 \\
O_2 & (0.99)(7.01) = & 6.9 \\
N_2 & (13.81)(6.96) = & 96.1 \\
NO & (0.01)(7.16) = & \underline{0.0} \\
& & 103.2
\end{array}
$$

Enthalpy = $(103.2)(30 - 18)$ = ... 1240 kcal

6. Heat absorbed in cooling acid.

$580 \times 0.45(27 - 25)$ = ... 520 kcal

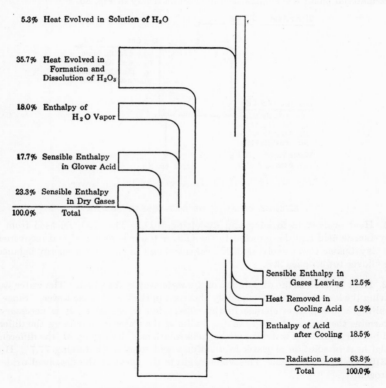

5.3% Heat Evolved in Solution of H_2O

35.7% Heat Evolved in Formation and Dissolution of H_2O_3

18.0% Enthalpy of H_2O Vapor

17.7% Sensible Enthalpy in Glover Acid

23.3% Sensible Enthalpy in Dry Gases

100.0% Total

Sensible Enthalpy in Gases Leaving 12.5%

Heat Removed in Cooling Acid 5.2%

Enthalpy of Acid after Cooling 18.5%

Radiation Loss 63.8%

Total 100.0%

Fig. 87.   Energy balance of Gay-Lussac tower.

7. Enthalpy of acid leaving cooler.

$580 \times 0.45(25 - 18)$ = ... 1830 kcal

Summary of Energy Balance of Gay-Lussac Tower.

Input

Enthalpy of dry gases entering = ...	2310 kcal	23.3%
Enthalpy of water vapor entering = ...	1790 kcal	18.1%
Enthalpy of Glover acid entering = ...	1760 kcal	17.7%
Heat evolved in forming N_2O_3 = ...	1570 kcal	15.9%
Heat evolved in dissolving N_2O_3 = ...	1960 kcal	19.8%
Heat evolved in dissolving H_2O = ...	520 kcal	5.2%
Total = ...	9910 kcal	100.0%

Output

Enthalpy of gases leaving = ...	1240 kcal	12.5%
Heat absorbed in cooling acid = ...	520 kcal	5.2%
Enthalpy of acid leaving cooler = ...	1830 kcal	18.5%
Heat loss (by difference) = ...	6320 kcal	63.8%
Total = ...	9910 kcal	100.0%

This energy balance is shown diagrammatically in Fig. 87.

SUMMARIZED ENERGY BALANCE FOR ENTIRE PLANT

Reference temperature: 18°C.

Basis: 100 kg of dry pyrites charged.

Input

1. Heat evolved in combustion of pyrites = ...	148,200 kcal	61.2%
2. Enthalpy of water vapor in air = ...	1,850 kcal	0.8%
3. Heat evolved in formation of H_2SO_4 in Glover tower	12,530 kcal	5.2%
4. Sensible enthalpy of nitric acid..................	7 kcal	0.0%
5. Heat evolved in formation of H_2SO_4 in chamber...	64,870 kcal	26.8%
6. Heat evolved in solution of H_2SO_4 in chamber.....	14,050 kcal	5.8%
7. Heat evolved in solution of H_2O in Gay-Lussac tower	520 kcal	0.2%
Total.................................	242,027 kcal	100.0%

Output

1. Enthalpy of cinder............................	4,900 kcal	2.0%
2. Net heat absorbed in concentrating acid in Glover tower.....................................	2,400 kcal	1.0%
3. Concentration and decomposition of nitric acid....	270 kcal	0.1%
4. Radiation losses from burners....................	86,340 kcal	35.7%
Radiation losses from Glover tower..............	3,997 kcal	1.7%
Radiation losses from chamber..................	98,570 kcal	40.8%
Radiation losses from Gay-Lussac tower..........	6,320 kcal	2.6%
5. Cooling of Glover acid........................	33,000 kcal	13.6%
Cooling of chamber acid......................	3,920 kcal	1.6%
Cooling of Gay-Lussac acid.....................	520 kcal	0.2%
6. Enthalpy of acid product 178(0.44)(25 − 18)......	550 kcal	0.2%
7. Enthalpy of spent gases leaving Gay-Lussac tower	1,240 kcal	0.5%
Total.....................................	242,027 kcal	100.0%

The energy balance for the entire sulfuric acid plant is shown diagrammatically in Fig. 88.

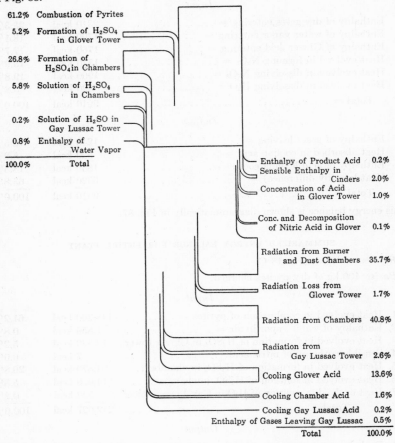

61.2% Combustion of Pyrites

5.2% Formation of H_2SO_4 in Glover Tower

26.8% Formation of H_2SO_4 in Chambers

5.8% Solution of H_2SO_4 in Chambers

0.2% Solution of H_2SO in Gay Lussac Tower

0.8% Enthalpy of Water Vapor

100.0% Total

Enthalpy of Product Acid 0.2%
Sensible Enthalpy in Cinders 2.0%
Concentration of Acid in Glover Tower 1.0%
Conc. and Decomposition of Nitric Acid in Glover 0.1%
Radiation from Burner and Dust Chambers 35.7%
Radiation Loss from Glover Tower 1.7%
Radiation from Chambers 40.8%
Radiation from Gay Lussac Tower 2.6%
Cooling Glover Acid 13.6%
Cooling Chamber Acid 1.6%
Cooling Gay Lussac Acid 0.2%
Enthalpy of Gases Leaving Gay Lussac 0.5%
Total 100.0%

FIG. 88. Energy balance of an entire sulfuric acid plant.

THE MATERIAL AND ENERGY BALANCES OF A BLAST FURNACE

A blast furnace is essentially a huge gas producer where, in conjunction with the partial combustion and distillation of a carbonaceous fuel, the reduction of ore and the formation of slag occur simultaneously. The charge, consisting of iron ore, coke, and limestone in proper proportions, is fed into the top of the blast furnace. Preheated air, preferably free from water vapor, is blown through the tuyères near the bottom of the furnace into the descending stream of solids. This results in combustion of the coke to carbon dioxide. The carbon dioxide gas in the presence of excess coke is reduced at the high prevailing temperature to carbon monoxide. A great many chemical reactions occur within the furnace. As the charge descends the shaft and as its temperature is gradually increased, dehydration of the ore, coke, and limestone takes place, followed by distillation of the remaining volatile matter of the coke, calcination of magnesium and calcium carbonates present in the limestone or ore, and reduction of the higher oxides of manganese and iron to manganous and ferrous oxides by the rising stream of reducing gases. The carbon dioxide formed by reduction of the ore with carbon monoxide is reduced in the presence of excess coke. As the temperature of the descending charge becomes still higher the lower oxides of iron and manganese are reduced to the metallic state.

At the highest temperature of the tuyères, part of the silica present is reduced to the metallic state and is dissolved in the molten iron. The excess silica and alumina of the charge are fluxed by reaction with the metallic bases present resulting in the formation of a fusible slag consisting of complex silicates and aluminates of calcium, magnesium, and iron. The high temperature at the tuyères produces a fluid slag and molten metal which readily flow through the solid reacting charge, separate into two layers at the bottom of the furnace, and are periodically run out in two separate streams as molten pig iron and as molten slag. A high temperature at the tuyères favors a ready separation of the slag and removal, as CaS in the slag, of much sulfur which was originally present in the coke. The temperature of the blast furnace is too low and insufficient coke is present to reduce the oxides of calcium, magnesium, and aluminum and the silicates. Hence, these compounds pass into the slag. Silica is used as a flux in a few exceptional cases where the alkaline earths and alumina predominate in the gangue present in the ore.

The purpose of preheating the air used in combustion of the coke is to permit the attainment of the high temperatures necessary for the final reduction of the ore and the fusion of the pig iron and slag. Any water vapor present in the incoming air will lower the temperature in the fusion zone on account of the heat absorbed in its reduction to hydrogen and carbon monoxide. For this reason it is desirable to use a blast of dried air.

The products of the blast furnace consist of molten pig iron, slag, and blast-furnace gas. The outgoing gas consists essentially of nitrogen, carbon monoxide, carbon dioxide, and water vapor, with small amounts of hydrogen and methane, and also carries in suspension a considerable amount of dust. The heating value of this gas is very low because of its high content of nitrogen and the small amount of volatile matter in the coke which is used for reduction. The free moisture in the incoming charge is distilled off near the top of the furnace and escapes into the blast furnace gas without reduction.

The slag contains all the lime, magnesia, alumina, and alkalies originally present in the ore and flux, together with most of the silica and some ferrous and manganous oxides. The exact mineralogical compositions of ore, flux, and slag are usually not known completely, so that some uncertainty exists concerning the exact thermal energy involved in reduction and chemical transformations. The molten pig iron contains, in addition to iron, some carbon present as cementite and lesser amounts of silicon and manganese. On cooling, the cementite partly decomposes into graphite and iron.

In order to establish the energy balance of a blast furnace it is necessary to know the masses and chemical compositions of the ore, flux, dust, and pig iron, and the analysis of the dry blast-furnace gas. The masses of slag, air, and water vapor can then be calculated.

The material balance includes:

Input	Output
Iron ore	Dry gases
Flux	Water vapor in gases
Coke	Pig iron
Air	Dust
Water vapor	Slag

As an illustration of the calculations involved in the material and energy balance of a blast furnace, the data for the reduction of a basic iron ore with charcoal and an acid flux will be given. An example of the more usual operation with a limestone flux is given in the problems at the end of this chapter. The balances are worked out on a basis of 100 kilograms of pig iron produced.

Illustration 2. A blast furnace is charged with 212.7 kg of ore, 110.0 kg of charcoal, and 13.9 kg of flux per 100 kg of pig iron produced. The compositions of these materials are as follows:

Ore (212.7 kg)		Charcoal (110.0 kg)	
Fe_2O_3	54.93%	C	86.89%
FeO	8.48%	O	3.15%
CaO	9.58%	H	0.45%
Mn_3O_4	4.97%	N	0.51%
Al_2O_3	3.00%	H_2O	7.00%
MgO	1.83%	Ash	2.00%
SiO_2	4.92%		100.00%
H_2O	4.48%		
CO_2	7.81%		
	100.00%		

Flux (13.9 kg)

SiO_2...............	78.38%
Al_2O_3..............	13.99%
CaO...............	0.53%
Fe_2O_3.............	3.90%
H_2O..............	3.20%
	100.00%

Pig Iron (100.0 kg)

C.................	3.12%
Si.................	1.52%
Mn...............	2.22%
Fe................	93.14%
	100.00%

The total heating value of the charcoal is 7035 kcal per kg.

The clean gas produced has the following composition by volume on the moisture free basis:

CO_2............................	12.62%
CO............................	25.56%
CH_4............................	0.69%
H_2............................	1.34%
N_2............................	59.79%
	100.00%

The ore, flux, and charcoal are charged to the furnace at an average temperature of 18°C. The air blast is dried and enters the tuyères at a temperature of 300°C and moisture-free.

The gases leave the furnace at a temperature of 173°C and contain only negligible quantities of dust.

The slag and pig iron are poured at an average temperature of 1360°C.

In order to cool the outside of the bosh of the furnace and thereby protect the refractories from excessive heating, water is circulated in a pipe passing around the circumference of the bosh. On the basis of 100 kg of pig iron produced 576 kg of water are circulated and heated through a temperature rise of 13°C.

Calculate the complete material and energy balances of this furnace.

MATERIAL BALANCE

Distribution of Ore Materials. The mineralogical composition of the ore is not given, although it is customary to assume that the carbon dioxide present is combined with the lime and the magnesia present as limestone. This limestone is calcined at about 900°C to CaO, MgO, and CO_2. The silica is present chiefly as silicates of aluminum and magnesium. Any free moisture present in the ore is driven off at the top of the furnace without reduction. However, the chemically combined water, as in the minerals kaolinite and limonite, will be retained until the ore reaches a hotter zone, where it will be partly reduced by coke to hydrogen and carbon monoxide. It will be assumed that all the elements of the carbon dioxide and water of the ore leave the blast furnace in the gases.

The alumina, lime, and magnesia pass into the slag without reduction but combine with the silica of the flux to form complex silicates of calcium, magnesium, and aluminum. The exact thermal energy of this latter change is unknown since the mineralogical composition of the slag as well as the heats of formation of complex silicates are unknown.

The higher oxides of iron are reduced at a relatively low temperature to the lower oxides. In the hot zone of the furnace the oxides are further reduced to the metallic state to supply the iron requirements of the pig iron. It is assumed that the remaining iron, as ferrous oxide, passes into the slag forming ferrous silicate. This assumption is somewhat inexact because it is known that a part of the iron exists as metallic particles included in the slag.

Sufficient silica is reduced to metallic silicon to supply the silicon content of the pig iron. The remainder combines with the basic oxides to form silicates in the slag.

The Mn_3O_4 of the ore is in part reduced to metallic manganese, supplying that present in the pig iron. The remainder of the manganese is assumed to enter the slag as MnO, forming silicates.

The oxygen given up in the reduction of the oxides of iron, silicon, and manganese will be present in the gases as CO, CO_2 or H_2O. The gases also contain the CO_2 and H_2O of the ore.

Basis: 100 kg of pig iron produced.

Distribution of Fe_2O_3 and FeO:

Fe_2O_3 in ore $= 212.7 \times 0.5493 = 116.8$ kg or $\dfrac{116.8}{159.7} = ..$ 0.732 kg-mole

FeO in ore $= 212.7 \times 0.0848 = 18.03$ kg or $\dfrac{18.03}{71.8} = ..$ 0.2515 kg-mole

Total Fe in ore $= 2 \times 0.732 + 0.2515 = $ .... 1.7155 kg-atoms

Fe in pig iron $= 93.14$ kg or $\dfrac{93.14}{55.8} = $ .... 1.669 kg-atoms

Fe into slag as FeO $= $ .... 0.0465 kg-atom
FeO into slag $= 0.0465$ kg-mole or $0.0465 \times 71.8 = ..$ 3.33 kg
Oxygen in iron oxides of ore $= (3/2 \times 0.732)$
$+ (\frac{1}{2} \times 0.2515) = $ .... 1.223 kg-moles
Oxygen in FeO of slag $= \frac{1}{2} \times 0.0465 = $ .... 0.023 kg-mole
Oxygen into gases $= $ .... 1.20 kg-moles
or $1.20 \times 32 = $ .... 38.4 kg

Distribution of SiO_2:

SiO_2 in ore $= 212.7 \times 0.0492 = 10.47$ kg or $\dfrac{10.47}{60.1} = ...$ 0.174 kg-mole

Si in pig iron $= 1.52$ kg or $\dfrac{1.52}{28.1} = $ .... 0.054 kg-atom

Si into slag as $SiO_2 = $ .... 0.120 kg-atom
SiO_2 into slag $= 0.120$ kg-mole or $0.120 \times 60.1 = $ ..... 7.2 kg
O_2 into gases $= 0.054$ kg-mole or $0.054 \times 32 = $ .... 1.7 kg

Distribution of Mn_3O_4:

Mn_3O_4 in ore $= 212.7 \times 0.0497 = 10.57$ kg or $\dfrac{10.57}{228.8} = $ 0.0462 kg-mole

Mn in ore $= 0.0462 \times 3 = $ .... 0.1386 kg-atom

Mn into pig iron $= 2.22$ kg or $\dfrac{2.22}{54.9} = $ .... 0.0405 kg-atom

Mn into slag $= $ .... 0.0981 kg-atom
MnO into slag $= 0.0981$ kg-mole or $0.0981 \times 70.9 = ..$ 6.95 kg
Oxygen in Mn_3O_4 of ore $= 2 \times 0.0462 = $ .... 0.0924 kg-mole
Oxygen in MnO of slag $= \frac{1}{2} \times 0.0981 = $ .... 0.0491 kg-mole
Oxygen into gases $= $ .... 0.0433 kg-mole
or $0.0433 \times 32 = $ .... 1.38 kg

Distribution of H_2O and CO_2:

Both compounds enter the gases in partly reduced forms.

H_2O in ore = $212.7 \times 0.0448 = 9.52$ kg or $\dfrac{9.52}{18.02} = \ldots$ .... 0.529 kg-mole

CO_2 in ore = $212.7 \times 0.0781 = 16.62$ kg or $\dfrac{16.62}{44} = \ldots$ ... 0.378 kg-mole

H_2 into gases = 0.529 kg-mole or $0.529 \times 2.02 = \ldots\ldots$ 1.06 kg

C into gases = 0.378 kg-atom or $0.378 \times 12 = \ldots\ldots$ 4.54 kg

Total O_2 into gases = $0.378 + (\frac{1}{2} \times 0.529) = 0.642$ kg-mole or. . . . . 20.54 kg

Summary of distribution of ore materials:

Into pig iron:

Fe = ...   93.14 kg
Mn = ...   2.22 kg
Si = ...   1.52 kg

Into slag:

FeO = ...   3.33 kg
SiO_2 = ...   7.2 kg
MnO = ...   6.95 kg
CaO = 212.7×0.0958 = ...   20.4 kg
Al_2O_3 = 212.7×0.0300 = ...   6.4 kg
MgO = 212.7×0.0183 = ...   3.9 kg

Into gases:

O = $38.4 + 1.7 + 1.38 + 20.54 = \ldots$ 62.0 kg
C = ...   4.54 kg
H = ...   1.06 kg
    Total = ...   212.6 kg

Distribution of Flux Materials. Since it has been assumed in the preceding calculations that all the iron, manganese, and silicon of the pig iron were derived from the ore materials, all the materials of the flux must pass into the slag or the gases. The silica, lime, and alumina enter the slag unchanged. It will be assumed that the ferric oxide is reduced to FeO and enters the slag in this form. The oxygen evolved in the reduction of the iron oxide and the water present in the flux enter the gases.

Basis: 100 kg of pig iron produced.

Distribution of iron oxide:

Fe_2O_3 in flux = $13.9 \times 0.0390 = 0.542$ kg or. . . . .   0.0034 kg-mole
FeO into slag = $2 \times 0.0034 = 0.0068$ kg-mole or. . . . . .   0.49 kg
O_2 into gases = $\frac{1}{2} \times 0.0034 = 0.0017$ kg-mole or. . . . . . .   0.05 kg

Distribution of water:

H_2O in flux = $13.9 \times 0.0320 = 0.445$ kg or. . . . .   0.0247 kg-mole
H_2 into gases = 0.0247 kg-mole or. . . . .   0.05 kg
O_2 into gases = $\frac{1}{2} \times 0.0247 = 0.0123$ kg-mole or. . . . . . .   0.39 kg

Summary of distribution of flux materials:

Into slag:

$$SiO_2 = 13.9 \times 0.7838 = \dots\dots\dots\dots \quad 10.90 \text{ kg}$$
$$Al_2O_3 = 13.9 \times 0.1399 = \dots\dots\dots\dots \quad 1.95 \text{ kg}$$
$$CaO = 13.9 \times 0.0053 = \dots\dots\dots\dots \quad 0.07 \text{ kg}$$
$$FeO = \dots\dots\dots\dots\dots\dots\dots\dots\dots\dots\dots \quad 0.49 \text{ kg}$$

Into gases:

$$O = 0.05 + 0.39 = \dots\dots\dots\dots\dots \quad 0.44 \text{ kg}$$
$$H = \dots\dots\dots\dots\dots\dots\dots\dots\dots\dots\dots \quad \underline{0.05 \text{ kg}}$$
$$\text{Total} = \dots\dots\dots\dots\dots\dots\dots \quad 13.90 \text{ kg}$$

Distribution of Charcoal. The carbon in the pig iron will be assumed to be derived from the charcoal. The ash enters the slag while the remainder of the charcoal constituents passes into the gases.

Distribution of water.

$$H_2O \text{ in charcoal} = 110.0 \times 0.0700 = 7.70 \text{ kg or} \dots\dots\dots \quad 0.428 \text{ kg-mole}$$
$$H_2 \text{ into gases} = 0.428 \text{ kg-mole or} \dots\dots\dots\dots\dots \quad 0.86 \text{ kg}$$
$$O_2 \text{ into gases} = 0.214 \text{ kg-mole or} \dots\dots\dots\dots\dots \quad 6.85 \text{ kg}$$

Summary of distributions of charcoal materials.

Into pig iron:

$$C = \dots\dots\dots\dots\dots\dots\dots\dots\dots\dots\dots \quad 3.12 \text{ kg}$$

Into gases:

$$C = (110.0 \times 0.8689) - 3.12 = \dots\dots \quad 92.44 \text{ kg}$$
$$H = (110.0 \times 0.0045) + 0.86 = \dots\dots \quad 1.36 \text{ kg}$$
$$O = (110.0 \times 0.0315) + 6.85 = \dots\dots \quad 10.32 \text{ kg}$$
$$N = 110.0 \times 0.0051 = \dots\dots\dots\dots\dots \quad 0.56 \text{ kg}$$

Into slag:

$$\text{Ash} = 110.0 \times 0.02 = \dots\dots\dots\dots\dots \quad \underline{2.20 \text{ kg}}$$
$$\text{Total} = \dots\dots\dots\dots\dots\dots\dots \quad 110.00 \text{ kg}$$

Weight and Composition of Slag. Since it may be assumed that the slag contains only materials derived from the ore, flux, and charcoal, its total weight and composition may be obtained by adding together the weights of materials entering the slag from these three sources. Actually these materials are in the slag in the form of complex compounds.

Component	Weight, kg	Percentage	Kg equivalents
FeO = 3.33 + 0.49 = ...	3.82	6.0	0.106
SiO₂ = 7.2 + 10.90 = ...	18.10	28.4	0.602
MnO = ...	6.95	10.9	0.196
CaO = 20.4 + 0.07 = ...	20.47	32.1	0.730
Al₂O₃ = 6.4 + 1.95 = ...	8.35	13.1	0.246
MgO = ...	3.90	6.1	0.194
Ash = ...	2.20	3.4	
Total = ...	63.79 kg	100.0%	

In calculating the kilogram equivalents, alumina is assumed to behave as an acid. Neglecting the charcoal ash:

Total kilogram equivalents of metallic bases = 1.226
Total kilogram equivalents of acids (SiO_2 and Al_2O_3) = 0.848

It is apparent that the slag is distinctly basic in character, despite the fact that an acid flux was used.

In order to verify the above calculations it is possible to determine the weight and composition of the slag experimentally.

Weight of Dry, Clean Blast Furnace Gas. The direct measurement of the volume of blast furnace gas is difficult and rarely undertaken. The volume or weight of gases can be calculated from a carbon balance as in obtaining the material balance of a combustion process. The carbon in the blast furnace gas comes from the carbon dioxide of the limestone and the ore and from the charcoal used for reduction.

Carbon Balance. Basis: 100 kg of pig iron produced.

Total C into gases = 4.54 + 92.44 = 96.98 kg or 8.08 kg-atoms
Carbon per kg-mole of dry, clean gas (from gas analysis) =
  0.1262 + 0.2556 + 0.0069 =.................... 0.3887 kg-atom
Total dry gas = 8.08/0.3887 =.................... 20.78 kg-moles

CO_2 = 20.78 × 0.1262 =	2.62 kg-moles or ×44 =...	115.2	kg
CO = 20.78 × 0.2556 =	5.30 kg-moles or ×28 =...	148.3	kg
CH_4 = 20.78 × 0.0069 =	0.14 kg-mole or ×16 =....	2.2	kg
H_2 = 20.78 × 0.0134 =	0.28 kg-mole or ×2.02 =..	0.57	kg
N_2 = 20.78 × 0.5979 =	12.44 kg-moles or ×28.2 =.	351.5	kg
Total =........	20.78 kg-moles or........	617.8	kg

Average molecular weight = 617.8/20.78 = 29.7

Weight of Air Introduced. The weight of dry air introduced is calculated from a nitrogen balance as in a combustion process. The nitrogen in the gas is derived from only two sources, the charcoal and the air.

N_2 in gases = ... 351.5 kg
N_2 from charcoal = ... 0.56 kg
N_2 from air =.................... 350.9 kg or 12.42 kg-moles
Air introduced = 12.42/0.79 =.............. 15.72 kg-moles
  or 15.72 × 29.0 =....................... 456 kg

Since the air is dried, containing no water vapor, this is the total weight of the air introduced.

Weight of Water Vapor in Blast Furnace Gas. The weight of water vapor in the blast furnace gas is calculated from a hydrogen balance. All hydrogen of the entering materials is present in the gas either as H_2, CH_4, or H_2O.

Basis: 100 kg of pig iron produced.

Total H introduced = 1.06 + 0.05 + 1.36 = 2.47 kg

Output of H:
As H_2 = ... 0.57 kg
As CH_4 = 2 × 0.14 × 2.02 =.................... 0.57 kg
  Total =.................................... 1.14 kg
As H_2O = 2.47 − 1.14 = 1.33 kg or.............. 0.66 kg-mole

H_2O in gases = 0.66 kg-mole or...................... 11.9 kg
H_2O introduced in charge = $9.52 + 0.44 + 7.70 =$ ...... 17.66 kg
H_2O decomposed in blast furnace = $17.66 - 11.9 =$ ...... 5.76 kg
   or 5.76/18 = ... 0.320 kg-mole

Overall Material Balance.

Input		*Output*	
Ore..............	212.7 kg	Slag.............	63.8 kg
Flux.............	13.9 kg	Dry gas (20.78 kg-	
Charcoal........	110.0 kg	moles)........	617.8 kg
Air (15.72 kg-moles)	456.0 kg	Water vapor in gas	
Total......	792.6 kg	(0.66 kg-mole)..	11.9 kg
		Pig iron.........	100.0 kg
		Total........	793.5 kg

This material balance is summarized in Fig. 89.

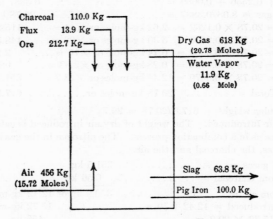

Fig. 89.   Material balance of a blast furnace.

ENERGY BALANCE

An energy balance might be established by considering the enthalpies and heats of formation of all components of the charge and all components of the slag, pig iron, and furnace gas together with the heat loss by radiation.   Such an energy balance would be disproportionate since the chemical energies of formation of the oxides and silicates which pass through the process unchanged, such as the oxides of aluminum, silicon, calcium, and magnesium, are of no interest.   More valuable information is obtained by including in the energy balance only the *net* chemical and thermal changes which take place during the process.

During the course of reduction many intermediate chemical reactions take place each accompanied by a certain thermal change.   Examples are the progressive reduction of the higher oxides of iron and manganese, the oxidation of carbon at the tuyères and its subsequent reduction by coke to carbon monoxide, the reduction of metallic oxides by carbon and by carbon monoxide, and the reduction of water to carbon monoxide, hydrogen, and carbon dioxide. However, in any chemical process the total change in energy is dependent only upon the initial and final states of chemical constitution, temperature, pressure, and state of aggregation and is independent of any intermediate state. Hence, in calculating the energy balance of a blast furnace, the numerous intermediate reactions involved need not be considered. It is sufficient to know the temperature, state of aggregation, and composition of each material charged and each product formed, without knowing how the various components of the products are actually produced.

The oxides of calcium, magnesium, and aluminum pass through the furnace apparently unchanged so that the heats of formation of these oxides need not be considered, but the state of the oxide is much different in the slag from that in the ore or flux. For example, in the ore, the oxides of iron and manganese exist as oxides but in the slag as silicates, so that the net heat effect accompanying the formation of silicates from the oxides should be considered. However, accurate calculation of this quantity requires data which are not ordinarily available.

The energy balance is calculated with a reference temperature of 18°C, based on 100 kg of pig iron produced.

1. Heat absorbed in reduction of iron oxides at 18°C. The heat absorbed is obtained by subtracting the total heat of formation of the reactants from that of the products. Iron oxides enter the process in both the ore and the flux. The necessary heat of formation data are obtained from Table XIV, page 253.

Reactants	Kg-moles	Molal Heat of Formation	Total Heat of Formation, kcal
Fe_2O_3	0.735	$-198,500$	$-145,900$
FeO	0.2515	$-64,300$	$-16,170$
Total =			$-162,070$ kcal

Products			
Fe	1.669	0	
FeO	0.0533	$-64,300$	$-3,430$
Oxygen	1.25	0	

Heat absorbed in reduction of iron oxides =

$$-3,430 - (-162,070) = \ldots\ldots\ldots\ldots\ldots\ldots\ldots\ldots +158,640 \text{ kcal}$$

2. Heat absorbed in reduction of Mn_3O_4 at 18°C.

Reactant	Kg-moles	Molal Heat of Formation	Total Heat of Formation, kcal
Mn_3O_4	0.0462	345,000	$-15,940$ kcal

Products			
Mn	0.0405	0	
MnO	0.0981	$-96,500$	-9470 kcal
Oxygen	0.0433	0	

Heat absorbed in reduction of $Mn_3O_4 = -9470 + 15,940 = \qquad +6,470 \text{ kcal}$

3. Heat absorbed in reduction of SiO_2 at 18°C.

Reactant	Kg-moles	Molal Heat of Formation	Total Heat of Formation, kcal
SiO_2 reduced	0.054	−203,340	−10,980 kcal
Products	Kg-atom		
Si	0.054	0	
O	0.054	0	

Heat absorbed in reduction of SiO_2 = ...   +10,980 kcal

4. Heat absorbed in calcination of carbonates at 18°C. Carbonates are present only in the ore. It is assumed that all CaO of the ore is combined with CO_2 as the carbonate and that the surplus CO_2 is combined with MgO.

CO_2 in ore = ...	0.378 kg-mole
CaO in ore = 20.4/56 = ...	0.364 kg-mole
MgO in ore = 3.9/40.3 = ...	0.097 kg-mole
$CaCO_3$ in ore = ...	0.364 kg-mole
$MgCO_3$ in ore = 0.378 − 0.364 = ...	0.014 kg-mole
MgO in ore = 0.097 − 0.014 = ...	0.083 kg-mole

Reactant	Kg-moles	Molal Heat of Formation	Total Heat of Formation, kcal
$CaCO_3$	0.364	−289,500	−105,400
$MgCO_3$	0.014	−268,000	−3,750
Total = ...			−109,150 kcal
Products			
CO_2	0.378	−94,030	−35,540 kcal
CaO	0.364	−151,700	−55,220 kcal
MgO	0.014	−146,100	−2,040 kcal
Total = ...			−92,800 kcal

Heat evolved in calcination = −92,800 + 109,150 = ...   +16,350 kcal

5. Heat evolved in the partial combustion of charcoal. The purpose of introducing charcoal into the charge is to furnish heat for all endothermic reactions involved in reduction, to supply heat for producing the slag and pig iron in a molten state, and to supply carbon for the reduction of the various metallic oxides of the ore. The charcoal is not burned completely to carbon dioxide and water vapor, and its total heating value is not rendered available in the blast furnace. The actual products resulting from the partial combustion of the charcoal include carbon dioxide, water vapor, carbon monoxide, methane, and hydrogen in the outgoing gases and graphite in the solidified pig iron.

Since the heats of formation of the materials making up the charcoal are not known it is necessary to calculate the heat evolved in its partial combustion from standard heat of combustion data. As pointed out in Chapter VIII, page 267, the heat evolved in a reaction is the difference between the sum of the heats of combustion of the products and that of the reactants. The actual reactants entering into the combustion of the charcoal include carbon dioxide, water vapor, and oxygen from both the air and the ore. However, the heats of combustion of these materials are zero, and they need not be considered in calculating the heat of reaction.

Heat of combustion of charcoal $= 110.0 \times -7035 = \ldots\ldots\ldots -773,850$ kcal

Product	Kg-moles	Molal Heat of Combustion	Total Heat of Combustion, kcal
CO	5.30	−67,410	−357,300
CH₄	0.14	−212,805	−29,790
H₂	0.28	−68,320	−19,130
Graphite =	0.26	−94,030	−24,450
Total = ...			−430,670 kcal

Heat evolved in partial combustion of charcoal =

$-430,670 - (-773,850) = \ldots\ldots\ldots\ldots\ldots\ldots\ldots$ 343,180 kcal

6. Heat evolved in formation of slag. Blast-furnace slags consist of complex mixtures of silicates and aluminates of calcium, magnesium, iron, and manganese. In order to calculate the heat of reaction accompanying the formation of a slag it would be necessary to determine the complete mineralogical composition of the ore, flux, and slag, together with the heats of formation of all the compounds present. Such data are rarely available, and it is ordinarily necessary to neglect the heat evolved in forming the slag from the oxides. It would not be expected that this thermal effect is large. The formation of the monosilicate of calcium, manganese, iron, or magnesium from SiO_2 and the respective oxide is in every case accompanied by an evolution of heat. On the other hand, the formation of dicalcium silicate is accompanied by an absorption of heat. These effects will frequently tend to compensate each other.

7. Heat absorbed in the formation of cementite. In the blast furnace some carbon combines with iron to form cementite, Fe_3C, with simultaneous absorption of heat. The formation of cementite is an endothermic reaction absorbing 4000 kilo calories per kg-mole of Fe_3C. Upon slow cooling of pig iron this cementite is partly decomposed with liberation of heat. In this instance the carbon present in the pig iron is assumed to remain present as cementite because of rapid cooling during solidification.

Heat absorbed in formation of cementite $= 4000 \times \dfrac{3.12}{12} = \ldots\ldots$ 1,040 kcal

This item of heat absorbed is neglected in the output of the present energy balance.

8. Enthalpy of air.

Air supplied = 15.72 kg-moles

Mean molal heat capacity between 18°C and 300°C (from Table VI, page 216) = 7.06.

Enthalpy of air $= 15.72 \times 7.06(300 - 18) = \ldots\ldots\ldots\ldots\ldots$ 31,300 kcal

9. Enthalpy of dry blast-furnace gas.

Mean heat capacities between 18°C and 173°C taken from Table VI, page 216.

$$CO_2 \quad (2.62) \ (9.60) \ = \ 25.2 \text{ kcal}$$
$$CH_4 \quad (0.14) \ (9.27) \ = \ 1.3 \text{ kcal}$$
$$H_2 \quad (0.28) \ (6.94) \ = \ 1.9 \text{ kcal}$$
$$CO \quad (5.30) \ (6.99) \ = \ 37.1 \text{ kcal}$$
$$N_2 \quad (12.44) \ (6.99) \ = \ \underline{87.0} \text{ kcal}$$
$$152.5 \text{ kcal}$$

Total enthalpy = 152.5 (173 − 18) = 23,640 kcal

10. Enthalpy of water vapor in gas.

Moles of H_2O in gas = ...   0.66 kg-mole
Molal humidity = 0.66/20.78 = ...   0.0318
Dew point (Fig. 9) = ...   76°F (24°C)
Heat of vaporization at 76°F = 18,880/1.8 =   10,485 kcal per kg-mole
Total enthalpy of water vapor =
0.66[10,485 + 18(24 − 18) + 8.11(173 − 24)] = ...   7790 kcal

11. Enthalpy of slag. S. Umino[2] determined the relative enthalpies of various blast-furnace and open-hearth-furnace slags and has found that the enthalpies of these two types of slags are nearly the same when measured at the same temperature. These values are shown graphically in Fig. 90. It will be seen that there is no sudden break in the enthalpy-temperature curve, thus indicating that no sudden transformations take place in cooling and that the slag exists essentially in the form of a glass   From Fig. 90 it will be seen that at a temperature of 1360°C the enthalpy of slag is 365 kcal per kg.

Total enthalpy of slag = 365 × 63.8 = ...   23,290 kcal

12. Enthalpy of pig iron. The average enthalpy of molten pig iron from blast furnaces has also been determined by S. Umino.[3] Although the composition of the pig iron used in his experiments is not identical with the one under discussion, nevertheless the compositions are sufficiently alike to justify use of the same values of enthalpy. At a pouring temperature of 1360°C this enthalpy is 300 kcal per kg and includes the sensible enthalpy of the solid and liquid states, the latent heat of fusion, and the heat evolved in the separation and decomposition of cementite, Fe_3C, from solution in the iron.

Total enthalpy of pig iron = 100 × 300 = ...   30,000 kcal

13. Heat absorbed by cooling-water = 576 × 13 = ...   7,490 kcal

Summary of Energy Balance.

Input

Partial combustion of charcoal.	343,180 kcal	91.6%
Enthalpy of air.	31,300 kcal	8.4%
Total.	374,480 kcal	100.0%

[2] S. Umino, *Science Repts. Tôhoku Imp. Univ.*, **17,** 985 (1928).
[3] *Idem*, **16,** 575 (1927).

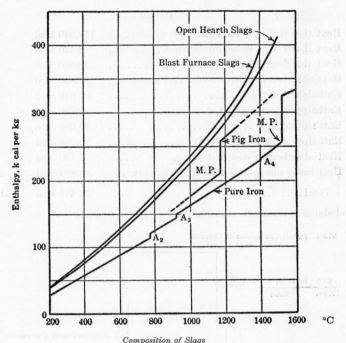

Composition of Slags

	Blast-Furnace Slags			Open Hearth Slags		
	1	2	3	4	5	6
SiO₂	34.50	37.26	34.22	18.20	18.28	20.28
FeO	1.58	0.82	0.74	13.45	10.27	10.47
Fe₂O₃	0.29	0.84	0.20	2.40	3.63	2.50
CaO	40.92	43.15	41.80	42.63	43.55	44.20
MgO	3.90	1.80	4.56	9.14	11.84	10.47
P	trace	trace	trace	0.33	0.25	0.25
S	0.98	0.11	0.90	0.54	0.45	0.40
MnO	2.24	1.82	1.88	6.97	6.60	6.60
Al₂O₃	15.48	13.15	15.60	5.00	4.68	4.79
Apparent sp. gr. 20°C	2.80	2.94	2.97	3.46	3.56	3.12

Composition of Pig Iron

C = 4.31%
Si = 1.11%
Mn = 0.53%
P = 0.12%
S = 0.022%
Cu = 0.21%

Heat of Fusion = 46.63 kcal per kg

Thermal Data on Pure Iron

Heat of transition A₂ (α to β)	6.5 kcal per kg
Heat of transition A₃ (β to γ)	5.6 kcal per kg
Heat of transition A₄ (γ to δ)	1.9 kcal per kg
Heat of fusion	65.6 kcal per kg

FIG. 90. Enthalpies of iron and slags, referred to 18°C. Data taken from S. Umino, *Science Repts. Tôhoku Imp. Univ.*, **17**, 985 (1928) for slag; **16, 575** (1927) for pig iron; **18**, 91 (1929) for pure iron.

Output

Heat absorbed in reduction of iron oxides.....	158,640 kcal	42.4%
Heat absorbed in reduction of Mn_3O_4.........	6,470 kcal	1.7%
Heat absorbed in reduction of SiO_2..........	10,980 kcal	2.9%
Heat absorbed in calcination of carbonates....	16,350 kcal	4.4%
Enthalpy of dry gas.......................	23,640 kcal	6.3%
Enthalpy of water vapor..................	7,790 kcal	2.1%
Enthalpy of slag...........................	23,290 kcal	6.2%
Enthalpy of pig iron......................	30,000 kcal	8.0%
Heat absorbed by cooling-water............	7,490 kcal	2.0%
Heat losses unaccounted for (by difference)...	89,830 kcal	24.0%
Total................................	374,480 kcal	100.0%

This balance is summarized in Fig 91.

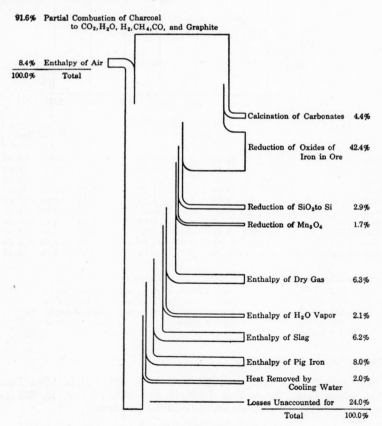

FIG. 91. Energy balance of a blast furnace.

PETROLEUM CRACKING PROCESS

In a so-called " vapor phase cracking " process a clean, well fractionated gas oil cut from petroleum, containing no material of gasoline boiling range, may be decomposed to form gasoline and gas by heating in the tubes of a furnace designed to provide the necessary reaction time at elevated temperatures. In order to arrest the reaction and minimize coke formation, the hot vapor mixture of gas, gasoline, oil, and tar from the heater is " quenched " by a relatively cool stream of oil. The resulting mixture passes to an evaporator where further cooling and rough fractionation is accomplished by means of a reflux stream of cool oil sprayed in at the top of the vessel. The quantity of reflux is regulated to obtain such a temperature in the evaporator as to produce the desired quality of " tar " which is withdrawn from the bottom of the evaporator and generally sold as heavy fuel oil.

The vapors from the evaporator pass to a fractionating tower the lower section of which is utilized for preheating the fresh charge by direct heat exchange. This tower is operated to produce well fractionated gasoline and gas as the overhead product. All partially decomposed feed in the boiling range between gasoline and the tar is condensed as a bottom product. This " recycle stock " mixes in the bottom section of the tower with the fresh feed to form the " combined feed," part of which is charged to the furnace, part used for quenching the hot vapors leaving the furnace, and part used as a reflux in the evaporator. It may be assumed that the mixture leaving the furnace is completely vaporized and that negligible condensation is produced by the quench. It may also be assumed that all the oil used for quench and reflux is vaporized under the conditions of the evaporator forming no tar.

Illustration 3. The flow diagram of such a process is shown in Fig. 92, on which are indicated significant temperatures and characteristics of the streams. The furnace is of the gas-fired type, provided with both radiant and convection heating sections. An air preheater transfers heat from the stack gases to the air used for combustion.

The gas used for heating the furnace is supplied at 65°F and has the following composition by volume:

$$CH_4 \dots\dots\dots\dots\dots\dots\dots\dots\dots\dots \quad 14.6\%$$
$$C_2H_6 \dots\dots\dots\dots\dots\dots\dots\dots\dots\dots \quad 77.3$$
$$CO \dots\dots\dots\dots\dots\dots\dots\dots\dots\dots \quad 1.2$$
$$H_2 \dots\dots\dots\dots\dots\dots\dots\dots\dots\dots \quad 6.1$$
$$N_2 \dots\dots\dots\dots\dots\dots\dots\dots\dots\dots \quad 0.8$$

The temperatures shown on the flow diagram are either arbitrarily set as bases for design or are derived from previous pilot plant or commercial experience indicating the temperatures necessary for the desired reaction rates and separations.

Pilot plant tests indicate that when cracking a gas oil having a gravity of 29°API and a characterization factor of 11.8 the following yields are obtained, expressed as percentage by liquid volume of the fresh feed:

Gasoline (58°API; K = 11.9) 61.0%

Tar (2°API; K = 10.2) 24.5

It may be assumed that the operation is conducted with 100% material recovery and that the yield of gas is determined by difference.

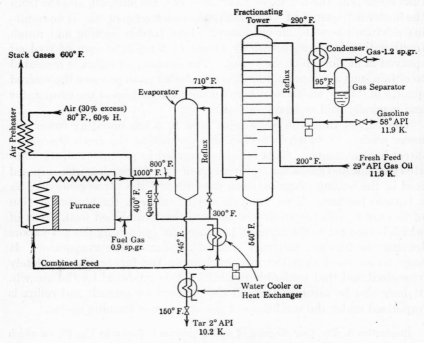

Fig. 92.   Vapor phase cracking unit.

Previous commercial experience has indicated that such an operation may be carried out satisfactorily under reaction conditions which result in the conversion per pass into gasoline and gas of 22.0 weight per cent of the combined feed entering the furnace.   The recycle stock from such an operation is found to have a gravity of 19° API and a characterization factor of 10.7.   The standard heat of the endothermic reaction from liquid charge to liquid gasoline, recycle stock and tar, and gas at 60°F and atmospheric pressure may be taken as 600 Btu per pound of gasoline plus gas formed.   Thus $\Delta H = 600$.

On the basis of the above information it is desired to develop material and energy balances of the separate parts and of the entire plant for the design of a unit to process 5000 barrels (42 gallons) per day of fresh charge.   Radiation losses may be neglected except from the furnace where a loss corresponding to 5% of the fuel is assumed. Although the pressures throughout will be somewhat above atmospheric it will be assumed that they are sufficiently low that all enthalpies may be taken as at atmospheric pressure.

In calculating the enthalpies of mixtures, heats of mixing in both gas and liquid phases may be neglected and also the effect of pressure upon enthalpies and upon the heat of cracking. It should be noted that the characterization factors and degrees API are additive on a weight basis.

The following information is required: All flow rates should be expressed in pounds per hour and in barrels per day (at 60°F) for the liquids and in cubic feet (at 60°F, 30 inches Hg, saturated) per hour for the gas. Heat rates should be expressed in Btu per hour.

(a) Production rates of gasoline, gas, and tar.
(b) Flow rate of the combined feed to the furnace.
(c) Properties of the combined feed.
(d) Heat absorbed in heating and cracking oil in the furnace.
(e) Flow rate of the combined feed used for quenching vapors from the furnace.
(f) Flow rate of the combined feed used for the reflux in the evaporator.
(g) Reflux rate of gasoline in the fractionating column.
(h) Temperature of flue gases leaving the convection section of the furnace.
(i) Fuel gas burned in the furnace.
(j) Thermal efficiency of the combined furnace and preheater.
(k) Heat transfer duties of the condenser and coolers.
(l) Cooling water required by the condenser and coolers in gallons per minute. (Assume a 20°F temperature rise of the water.)

SOLUTION

For ready reference the physical and thermal properties of the various petroleum fractions are tabulated in Tables A and B. The characterization factors and degrees API of a mixture are additive properties on a weight basis. The average molecular weights and boiling points are obtained from values of API and K by use of Fig. 63. Latent heats of vaporization are calculated from Equation (32), Chapter VII, page 233. The mean specific heats of liquids and vapors are obtained from Figs. 66 and 67. Enthalpies at various temperatures are then calculated from the above data using 65°F as the reference temperature.

All calculations are based upon 1 hour of operation.

TABLE A

PHYSICAL PROPERTIES OF OIL FRACTIONS

	°API	K	G 60/60°	Av Molecular Weight	Av Boiling Point — °F
Gasoline	58.0	11.9	0.747	109	240
Tar	2.0	10.2	1.06	320	800
Fresh Feed	29.0	11.8	0.88	300	675
Recycle	19.0	10.7	0.940	220	560
Gas			1.22 (air)		
Combined Feed (calculated)	22.1	11.04	0.922	240	585

TABLE B

Thermal Properties of Oil Fractions

	Boiling Point °F	Heat of Vaporization Btu/lb	Mean Specific Heat of Liquids (65° to t°F)		Mean Specific Heat of Vapors (240° to t°F)		Enthalpies 65°F reference	
			t°F	C_p	t°F	C_p	t°F	Btu/lb
Gasoline	240	134.8	95	0.475	290	0.464	95(l)	14.3
			240	0.518	710	0.574	240(l)	90.7
					800	0.596	240(v)	225.5
					1000	0.640	290(v)	248.7
							710	495.2
							800	559.0
							1000	711.7
Tar	800	100			(800°F to t°F)		150(l)	43.4
			150	0.377	800	0.620	745(l)	353.6
			745	0.520	1000	0.660	800(l)	388.1
			800	0.528			800(v)	488.1
							1000(v)	620.3
Gas					(65°F to t°F)			
					95	0.444	95	13.3
					290	0.498	290	112.1
					710	0.608	710	392.3
					800	0.630	800	463.0
					1000	0.675	1000	631.3
Fresh Feed	675	88	200	0.464	200		200(l)	62.6
Combined Feed	585	104.4			(585°F to t°F)			
			300	0.462	710	0.610	300(l)	108.6
			540	0.523	800	0.630	540(l)	248.4
			585	0.534			585(l)	277.7
							585(v)	382.1
							710(v)	458.5
							800(v)	517.4
Recycle	560	110.6	560	0.519	710	0.586	560(l)	256.9
					800	0.605	560(v)	367.5
					1000	0.644	710(v)	455.3
							800(v)	512.7
							1000(v)	650.7

1. Rates of Production. Rates of production are calculated from the yield statement, as follows:

	Charge	Gasoline	Tar	Gas
Bbl/day	5000	3050	1225	....
Gal/hr	8750	5340	2145	....
Sp. gravity	0.882	0.747	1.06	....
Lb/hr	64,290	33,210	18,930	12,150
M cu ft/day	....	....	....	3230

2. Flow rate of combined feed to furnace.

Of the combined feed entering the furnace 22% by weight is converted into gas plus gasoline.

Gas plus gasoline, lb per hour = $(33,210 + 12,150) = 45,360$

Combined feed to furnace, lb per hour $= \dfrac{45,360}{0.22} = 206,170$

Recycle stock produced in furnace, lb per hour = 206,170 − $(33,210 + 18,930 + 12,150) = 141,880$

3. Properties of combined feed.

	Recycle Stock	Fresh Feed
Gravity, °API	19°	29°
Characterization Factor K	10.7	11.8
Lb per hour	141,880	64,290
% by wgt	68.8	31.2

K of combined feed = $(10.7)(0.688) + (11.8)(0.312) = 11.04$
°API of combined feed = $(19)(0.688) + (29)(0.312) = 22.1$
From Fig. 63 average molecular weight = 240

Average boiling point = 585°F

The thermal properties are tabulated in Table B.

4. Heat absorbed in heating and cracking oil in furnace.

Input

Enthalpy of combined feed at 540°F = (206,170) (248.4) = ...	51,212,600 Btu
Heat supplied to oil from combustion of fuel (by difference) = ...	111,318,600
	162,531,200

Output

Enthalpy of gas = (12,150)(631.3) = ...	7,670,000 Btu
Enthalpy of gasoline = (33,210)(711.7) = ...	23,635,600
Enthalpy of tar = (18,930)(620.3) = ...	11,742,300
Enthalpy of recycle = (141,800)(650.7) = ...	92,269,300
Energy absorbed in cracking oil (600)(45,360)	27,214,000
Total......................................	162,531,200

5. Weight of combined feed used for quenching hot vapors from furnace.

The energy balance about the quench point is shown diagrammatically in Fig. 93. The required quantity of quench is fixed by this balance.

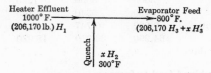

FIG. 93.   Flow chart about quench point.

Let x = lb quench supplied to vapor from furnace.

Input

Enthalpy of vapors at 1000°F = ... 135,317,200 Btu

Enthalpy of quench at 300°F = ... 108.6x

Total = ... 135,317,200 + 108.6x

Output

Enthalpy of gas = (12,150)(462.97) = ... 5,625,100

Enthalpy of gasoline = (33,210)(559) = ... 18,564,400

Enthalpy of tar vapor = (18,930)(488.1) = ... 9,239,300

Enthalpy of recycle vapors = (141,880)(512.7) = ... 72,741,900

Enthalpy of combined feed quench = (x)(517.42) = . 517.42x

Total = ... 106,170,700 + 517.42x

From the above balance:

$$106,170,700 + 517.42x = 135,317,200 + 108.6x$$
$$408.82x = 29,146,500$$
$$x = 71,294 \text{ lb of combined feed for quench}$$

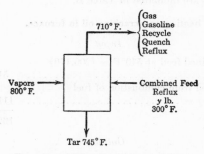

FIG. 94.   Flow chart of evaporator.

6. Flow rate of combined feed used for reflux in the evaporator.

Let y = lb of combined feed used for reflux in evaporator.

The reflux rate required in the evaporator is determined by an energy balance following the flow chart shown in Fig. 94.

Energy Balance of Evaporator

Input

Enthalpy of vapors at 800°F = 106,170,700 +

(71,294)(517.42) = ... 143,059,600

Enthalpy of combined feed for reflux = ... 108.6y

Total = ... $\overline{143,059,600 + 108.6y}$

Output

Enthalpy of gas = (12,150)(392.29) = ... 4,766,300

Enthalpy of gasoline = (33,210)(495.23) = ... 16,468,800

Enthalpy of recycle = (141,880)(455.34) = ... 64,636,400

Enthalpy of combined feed quench = (71,294)(458.45) = 32,684,700

Enthalpy of combined feed reflux = (y)(458.45) = .. + 458.45y

Enthalpy of tar at 745°F = (18,930)(353.6) = ... 6,693,600

Total = ... $\overline{125,249,800 + 458.45y}$

From the energy balance:

$$125,249,800 + 458.45y = 143,059,600 + 108.6y$$
$$349.85y = 17,809,800$$
$$y = 50,906 \text{ lb.}$$

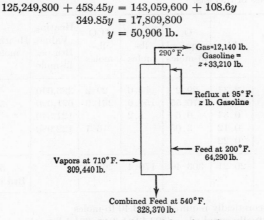

FIG. 95.   Flow chart of fractionating tower.

7. Rate of gasoline reflux in fractioning tower.

The reflux rate required in the fractionating tower is determined from an energy balance following the flow chart shown in Fig. 95.

Energy Balance

Let z = reflux, lb per hour.

Input

Enthalpy of gas = ... 4,766,300 Btu

Enthalpy of gasoline = ... 16,468,800

Enthalpy of recycle = ... 64,636,400

Enthalpy of combined feed recycle and quench

(122,200)(458.45) = ... 56,022,600

Enthalpy of fresh feed = (64,290)(62.57) = ... 4,022,600

Enthalpy of gasoline reflux = ... 14.25z

$\overline{145,916,700 + 14.25z}$

Output

Enthalpy of gas $= (12,150)(112.05) =$ ....   1,361,400 Btu
Enthalpy of gasoline $= (33,210 + z)(248.65) =$ ..... $8,257,700 + 248.65z$
Enthalpy of combined feed $= (328,370)(248.42) =$ ...   81,573,700

     Total $=$ .... $91,192,800 + 248.65z$

From the energy balance:

$$91,192,800 + 248.65z = 145,916,700 + 14.25z$$
$$234.4z = 54,723,900$$
$$z = 233,463 \text{ lb of gasoline reflux per hour}$$

8. Temperature of stack gases leaving furnace.

The temperature of the stack gases leaving the furnace is calculated from an energy balance of the air preheater based upon the combustion of 100 lb moles of fuel gas with 30% excess air.

Basis: 100 lb-moles of fuel gas.

	Moles	Mol Wt	Lb	O_2 required lb-moles	CO_2 lb-moles	H_2O lb-moles	Heating Value Btu per lb-mole	Heating value × mole fraction
CH_4	14.6	16	2.34	29.2	14.6	29.2	383,040	55,930
C_2H_6	77.3	30	23.19	270.55	154.6	231.9	671,090	518,160
CO	1.2	28	0.34	0.6	1.2		121,340	1,460
H_2	6.1	2	0.12	3.05		6.1	122,980	7,510
N_2	0.8	28	0.22					
			26.21	303.40	170.4	267.2		583,060 Btu per lb-mole

Oxygen theoretically required $= 303.40$ lb-moles
Oxygen actually supplied $= (303.40)(1.3) = 394.4$ lb-moles

Nitrogen actually supplied $= 394.4 \times \dfrac{0.79}{0.21} = 1483$ lb-moles

Total moles air supplied per 100 moles fuel $= 1877$ lb-moles
Molal humidity, lb moles water per lb mole dry gas (Fig. 9) $= 0.022$
Water from air $= (0.022)(1877)$ lb-moles $= 41.3$ lb-moles

Products of combustion:

Carbon dioxide...........................   170.4 lb-moles
Water $= 267.2 + 41.3$.....................   308.5
Oxygen $= 394.4 - 303.40$..................   91.0
Nitrogen $= 0.8 + 1483$....................   1483.8

                           2053.7 lb-moles

Partial pressure of water vapor $= \dfrac{(308.5)}{(2053.7)} (760) = 114$ mm

Dew point $= 130°F$

Enthalpy of water vapor in gas streams.   *Enthalpy*
                                                              Btu/lb-mole

In entering air, 80°F, dew point 66°F.....................19,190
In leaving air, 400°F, dew point 66°F.....................21,820
In waste gases, t°F, dew point 130°F.....................$18,500 + 8.30t$
In waste gases, 600°F, dew point 130°F.................23,490

The temperature of the gases from the furnace is determined by an energy balance
of the air preheater, illustrated in Fig. 96. As a first approximation it may be
assumed that the temperature drop of the gases in the preheater approximately equals
the temperature rise of the air. Mean heat capacities are based on this assumed
temperature which may be corrected by a second approximation if the energy balance
shows it to be seriously in error.

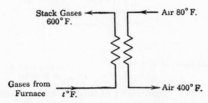

FIG. 96.   Flow chart of air preheater.

Energy Balance of Preheater

Input

Basis: 1.0 lb-mole of fuel gas

Enthalpy of air $(18.75)(80 - 65)(7.0) = 1,971$
Enthalpy of water vapor in air $(0.413)(19,190) = 7,925$
Enthalpy of hot waste gases:

$$\begin{aligned}
CO_2 &= 1.704(t - 65)(10.9) &=& \quad\quad +\ 18.57(t - 65) \\
H_2O &= (3.085)(18,500 + 8.30t) &=& \ 57,085 + \ 25.6t \\
O_2 &= (0.910)(t - 65)(7.5) &=& \quad\quad 6.82(t - 65) \\
N_2 &= 14.838(t - 65)(7.1) &=& \quad\quad \underline{105.35(t - 65)} \\
& & & 66,981 + 130.74(t - 65) + 25.6t
\end{aligned}$$

or $58,483 + 156.34t$

Output

Enthalpy of waste gases leaving (at 600°F)

$$\begin{aligned}
CO_2 &= (1.704)(600 - 65)(10.1) &=& \ 9,207 \\
H_2O &= (3.085)(23,490) &=& 72,466 \\
O_2 &= (0.910)(600 - 65)(7.3) &=& \ 3,554 \\
N_2 &= (14.838)(600 - 65)(7.1) &=& 56,362
\end{aligned}$$

Enthalpy of air at 400°F

$$\begin{aligned}
\text{Air} &= (18.77)(400 - 65)(7.0) &=& \ \ 44,015 \\
\text{Water vapor} &= (0.413)(21,820) &=& \quad \underline{9,011} \\
&\quad \text{Total} & =& \ 194,615 \ \text{Btu/lb-mole of flue gas}
\end{aligned}$$

From the energy balance:

$$156.34t + 58,483 = 194,615$$
$$t = 871°F \text{ temperature of gas leaving furnace}$$

9. Fuel gas burned in furnace per hour.

The amount of fuel gas burned in the furnace may be calculated from an energy balance either of the furnace or the furnace plus preheater based upon one hour of operation. The latter balance is illustrated in Fig. 97.

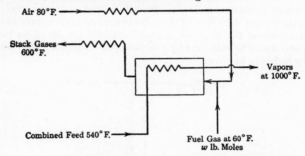

Fig. 97.   Flow chart of furnace and preheater.

Let w = lb-moles of fuel gas burned per hour.

Energy Balance of Furnace plus Preheater

Input

Enthalpy of dry air at 80°F = ...	1,971w Btu
Enthalpy of H_2O in air = ...	7,925w Btu
Enthalpy combined feed = ...	+ 51,212,600
Heating value of fuel = ...	583,060w Btu
Total = ...	592,956w + 51,212,600 Btu/hr

Output

Enthalpy of vapors at 1000°F..........	135,317,200	
Enthalpy of waste gases.............		141,509w
Heat absorbed in cracking...........	27,214,000	
Heat losses = (0.05)(583,060)w.......		29,153w
Total =.....................	162,531,200 +	170,662w

From the energy balance,

$$592,956w + 51,212,600 = 162,531,200 + 170,662w$$
$$422,294w = 111,318,600$$
$$w = 263.6 \text{ lb-moles dry fuel gas per hour}$$

Cubic feet per hour of gas, at 60°F and 30 in. Hg saturated with H_2O = (263.6) (385.5) = 101,618 cu ft per hour

10. Thermal efficiency of furnace.

From part (4) the heat supplied to oil = 111,318,600 Btu

Total heating value of fuel gas (583,060)(263.6) = 153,694,600 Btu

Thermal efficiency based on heating value of gas $\dfrac{111,318,600}{153,694,600} \times 100 = 72.4\%$

11. Heat transfer duties of condenser and coolers.

1. Condenser on fractionation tower

Gasoline product = (33,210)(248.7 − 14.3) = ...   7,784,420 Btu/hr
Gasoline reflux = (233,463)(248.7 − 14.3) = ...   54,723,730
Gas = (12,150)(112.1 − 13.3) = ...   1,199,690

Total = ...   63,707,840

2. Tar cooler (18,930) (253.6 − 43.4) = ...   5,873,200
3. Combined feed cooler (122,200)(248.4 − 108.6) =   17,086,000

Total = ...   86,667,040 Btu/hr

12. Cooling water required by condenser and cooler.

Heat absorbed per gallon of water = 20 × 8.33 = 166.6 Btu

1. Condenser on fractionating tower = $\dfrac{63,707,840}{(166.6)(60)}$ = ...   6373 gpm

2. Tar cooler = $\dfrac{5,873,200}{(166.6)(60)}$ = ...   587

3. Combined feed cooler = $\dfrac{17,086,000}{(166.6)(60)}$ = ...   1709

Total cooling water required = ...   8669 gpm

PROBLEMS

1. In a plant for the manufacture of sulfuric acid by the chamber process pyrites is burned in a shelf burner. The gases from the burner enter the Glover tower at 480°C and leave this tower at 105°C, entering the first chamber. The gases leave the last chamber at 42°C and finally leave the Gay-Lussac tower at 21°C. On the basis of 100 kg of pyrites, as charged, there are charged into the Glover tower, 175 kg of chamber acid, 65.2% H_2SO_4 (51.8°Bé) at 30°C; 610 kg of Gay-Lussac acid at 23°C, and 1.30 kg of 40% nitric acid at 20°C. The Gay-Lussac acid contains 78.0% H_2SO_4 (60.0°Bé), 0.984% N_2O_3 in solution, and 21.0% H_2O.

The analyses of the pyrites, cinder, and the moisture-free gases leaving the burner are as follows:

Pyrites		Cinder		Gases (by volume)	
FeS_2	90.00%	Fe_2O_3	89.80%	O_2	9.32%
SiO_2	4.80%	FeS_2	1.65%	N_2	82.38%
H_2O	5.20%	SO_3	1.93%	SO_2	8.00%
	100.00%	SiO_2	6.62%	SO_3	0.30%
			100.00%		100.00%

The pyrites is charged to the burner at 20°C. The air enters at 20°C, under a barometric pressure of 722 mm of Hg and with a percentage humidity of 40%. The cinder is withdrawn at 320°C.

For each 100 kg of pyrites as charged, 768 kg of acid containing 79.4% H_2SO_4 leave the Glover tower at 100°C and are cooled to 23°C. The chamber acid leaves the first chamber at 65°C and is cooled to 30°C before entering the Glover tower. The acid leaves the Gay-Lussac tower at 30°C and is cooled to 23°C for recirculation. The spray water enters the chambers at 20°C.

From the flow chart and assumptions of Illustration 1, calculate individual material and energy balances, on the basis of 100 kg of pyrites as fired, of:

    (a) The burner.
    (b) The Glover tower.
    (c) The chambers.
    (d) The Gay-Lussac tower.
    (e) The entire plant.

2. The charge delivered to a blast furnace, on the basis of 1000 lb of pig iron, consists of 1810 lb of ore, 361 lb of limestone, and 892 lb of coke. The analyses of various components of the charge are as follows:

Ore (1810 lb)		Limestone (361 lb)		Coke (892 lb)	
Fe_2O_3	62.10%	CaO	51.12%	Carbon	88.20%
FeO	19.07%	MgO	2.10%	Hydrogen	2.00%
Mn_3O_4	2.12%	SiO_2	2.89%	Fe_2O_3	2.10%
Al_2O_3	2.89%	Al_2O_3	4.12%	SiO_2	1.98%
SiO_2	8.62%	Fe_2O_3	0.52%	CaO	2.32%
H_2O	5.20%	CO_2	35.05%	MgO	1.10%
	100.00%	H_2O	4.20%	S	0.20%
			100.00%	H_2O	2.10%
					100.00%

The total heating value of the coke is 14,200 Btu per lb.

On the basis of 1000 lb of pig iron produced, 51 lb of dust are collected from the gases leaving the furnace. The analyses of the products are as follows:

Pig iron (1000 lb)		Flue dust (51 lb)		Gas analysis (by volume)	
Fe	92.28%	FeO	83.2%	CH_4	0.80%
Si	2.10%	C	10.1%	CO_2	12.10%
Mn	1.38%	CaO	3.1%	CO	29.30%
S	0.03%	SiO_2	3.6%	H_2	2.12%
C	4.21%		100.0%	O_2	0.20%
	100.00%			N_2	55.48%
					100.00%

The surrounding air is at 70°F, 40% percentage humidity and a barometric pressure of 29.2 in. of Hg. This air is heated and supplied to the tuyères at 850°F.

The ore, flux, and coke are charged at an average temperature of 65°F.

The gases leave the furnace at a temperature of 422°F. The molten slag and pig iron are tapped from the furnace at a temperature of 2500°F. The sensible enthalpy of the flue dust is negligible.

Calculate the complete material and energy balances of this furnace, using the assumptions of Illustration 2.

3. In producing 1 ton (2000 pounds) of steel in an open-hearth furnace the following charge was supplied:

Hot metal from blast furnace (2400°F)...........	814 pounds
Cold scrap iron..................................	1250 pounds
Limestone (95.5% $CaCO_3$, 4.5% H_2O)..........	118 pounds
Iron ore (94% Fe_2O_3, 6% H_2O)................	56 pounds
Fuel oil..	28.2 gallons

Air was supplied to the regenerators at 80°F, 60% relative humidity, atmospheric pressure.   The air (85% of total supply) was preheated to 2000°F in the regenerators. The hot gaseous products of combustion left the hearth at 2860°F, entered the regenerators at 2560°F and entered the stack at 1000°F.   The average analysis of the flue gases measured over the nine hour run was as follows:

$$CO_2 = 17.0\%$$
$$O_2 = 0.8\%$$
$$\underline{N_2 = 82.2\%}$$
$$100.0\%$$

The metals analyzed as follows:

	Hot Metal	Cold Scrap	Steel
Carbon................	4.25%	0.15%	0.15%
Silicon................	1.92	0.50	0.25
Manganese............	0.32	0.02	0.02
Phosphorus............	0.65	0.065	0.02
Iron.................	92.86	99.20	99.56

The steel and slag were poured at 2800°F.   From experience it is known that 15% of the air used in the furnace leaks in through the doors and brickwork of the hearth. The fuel oil had the following properties:

Characterization factor...................   11.1
API gravity.............................   12.0

The following information is desired on the basis of 1 ton of steel produced: (Use 80°F as basis of enthalpies.)

1. Material balance of solids charged to the hearth.
2. Weight of flue gas.
3. Weight of dry air used.
4. Weight of water vapor in air supply and in flue gas.
5. Overall material balance of combustion and refining processes.
6. Air theoretically required for combustion.
7. Percentage excess air used based on requirements of fuel oil.
8. Overall energy balance of process.
9. Energy balance of reactions on hearth.

4. It is desired to prepare a preliminary engineering design study of a proposed catalytic unit for the " cracking " of higher boiling petroleum fractions into gasoline and gases rich in recoverable olefins.   A flow diagram is shown in Fig. 97a.[4]

The charge, a well-fractionated gas oil, is pumped through a heater where its temperature is raised to 900°F at a pressure somewhat above atmospheric and passed to a reactor B where it is contacted with a refractory catalyst in the form of small pellets. The reactor is provided with a heat exchange system through which a molten salt is circulated for temperature control and is so designed that heat exchange surface is uniformly distributed throughout the entire catalyst bed.   The decomposition reaction is endothermic, requiring the supply of heat from the salt system in order to maintain isothermal conditions.

[4] The process illustrated in Fig. 97a is not in commercial application as shown but represents a possible combination of features described in the many patents relating to this subject.

As the cracking reaction progresses a carbonaceous deposit accumulates on the catalyst which reduces its activity.   In order to maintain continuous operation an alternate reactor A is provided in which the hydrocarbon stream may be processed while the catalyst deposit is removed from reactor B by oxidation with air.   Thus, one reactor system is at all times in the *process period* of the operation while the other is in the burning or *reactivation period*.   In Fig. 97a the solid lines represent the various streams while reactor B is processing and the broken lines indicate the changes in flow when reactor A is processing and B is undergoing reactivation.

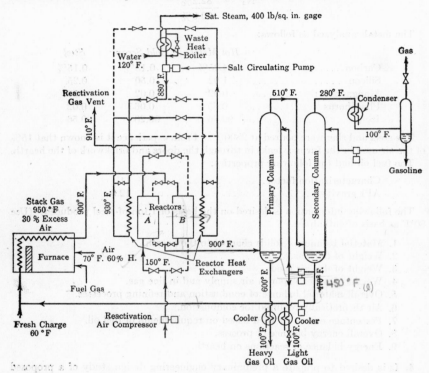

FIG. 97a.   Flow chart of catalytic cracking process.

The burning of the catalyst deposit is highly exothermic and during the reactivation period the circulating salt heat exchange system serves to cool the catalyst to prevent destructive overheating.   Since the heat liberated in reactivation is generally greater than that required to maintain isothermal conditions in the cracking reaction, a waste heat boiler is provided for removing heat from the circulating salt by the generation of steam.   The salt system is provided with switch valves so arranged that the cooled salt from the waste heat boiler is pumped first to the reactor under reactivation and then to the reactor which is processing.   The temperature in the processing reactor is controlled by a by-pass valve permitting regulation of the quantity of salt passed through the heat exchanger.   The entire circulating salt stream passes through the reactor which is undergoing reactivation.

The hydrocarbon products from the processing reactor pass to a primary fractionating column in which a heavy gas oil fraction is removed as bottoms. The overhead from this column passes to a secondary fractionating column in which well fractionated gasoline and gases are the overhead products. The bottoms, a light gas oil, is in part recycled to the cracking process, in part used to reflux the primary column, and in part withdrawn as final product.

Laboratory tests indicate that at the operating temperatures indicated in Fig. 97a the following products and yield are obtained from the indicated charging stock:

	°API	K	% by volume of charge
Charge	30	11.9	100
Gasoline	60	12.0	49
Light Gas Oil	32	11.4	24
Heavy Gas Oil	25	11.5	18

The gas has a specific gravity of 1.6.

The catalyst deposit is found to correspond to 3% by weight of the charge and contains 4% hydrogen and 96% carbon by weight.

The hydrocarbon combined feed, consisting of fresh charge and recycled light gas oil, is passed through the processing reactor at a rate of 1.2 volumes of oil, measured at 60°F, per hour per volume of catalyst. This method of expressing reactor feed rates is commonly used in catalytic processes and is termed the *liquid hourly space velocity*. The catalyst volume includes the actual volume of the catalyst pellets and the voids between the pellets in the catalyst bed. The density of the catalyst bed is 55 pounds per cubic foot.

At the above operating conditions it is found that 42% by weight of the combined feed is decomposed into gasoline and gas in a single pass through the unit.

The average standard heat of reaction at 60°F and one atmosphere from liquid combined feed to liquid gasoline and gas oils and gaseous gas is found to be +220 Btu per pound of gasoline plus gas formed, corresponding to an endothermic reaction.

The fuel gas burned in the furnace has the following composition:

	Mole per cent
Hydrogen	22.4
Methane	26.0
Ethylene	6.8
Ethane	7.2
Propylene	29.5
Propane	8.1
	100.0

In accordance with the process information given above, develop the following design factors and evaluate complete material and energy balances for a plant to charge 10,000 barrels (42 gal.) per day. Base the balances on one hour of operation and express rates in barrels per day and pounds per hour for liquids and pounds per hour and thousands of cubic feet per day for gases. Heat losses may be neglected except from the furnace.

(a) The production rates of all net hydrocarbon products and the catalyst deposit.

(b) The rates and properties of the combined feed to the heater and the light gas oil recycle stream.

(c) The heat absorbed by the oil in the furnace, assuming complete vaporization but no decomposition in the heater.

(d) The rate of fuel consumption in the furnace, assuming complete combustion and a radiation loss of 5% of the heating value of the fuel burned.

(e) The thermal efficiency of the furnace.

(f) The reflux rates to the primary and secondary fractionating columns.

(g) The heat transfer duties of the gasoline condenser and the light and heavy gas oil coolers.

(h) The cooling water requirements of the plant in gallons per minute with a 30°F temperature rise for the water.

(i) The volume and weight of the catalyst required in each reactor.

(j) The length of the process period in minutes if it is desired to limit the deposit on the catalyst to a maximum of 2.5% by weight of the catalyst.

(k) The rate at which air must be supplied in order that reactivation may be completed in the time of the process period. It may be assumed that the oxygen of the air is 100% utilized, going to carbon dioxide and water under the catalytic combustion conditions.

(l) The rate at which heat must be removed by the circulating salt from a reactor under reactivation, neglecting changes in the enthalpy of the catalyst bed.

(m) The rate at which salt must be circulated through a reactivating reactor heat exchanger. The heat capacity of the molten salt may be taken as 0.25 Btu per pound per °F.

(n) The temperature of the salt entering the waste heat boiler.

(o) The quantity of steam generated in the waste heat boiler, in M lb per hr.

(p)

Contact H_2SO_4

Makes stronger acid, requires purer SO_2.

$$SO_2 + \tfrac{1}{2}O_2 \longrightarrow SO_3 + \triangle$$

Catalysts are Pt : lower temp ,

V : Cheaper, not poisoned , longer life,

Oleum in steel
83 - 78 in C.I.
75 - Lead

H_2O air   air   Burned   Cooler   Filter

Drying Tower

Cooler

Converter   Cooler

Mixer   Oleum Storage   Oleum Tower

Cooler

Cooler   98% acid Storage   98% acid Tower

Cooler

Oleum   96% acid   Waste gas

APPENDIX

ATOMIC WEIGHTS OF THE MORE COMMON ELEMENTS

Aluminum	Al	26.97	Manganese	Mn	54.93
Antimony	Sb	121.76	Mercury	Hg	200.61
Argon	A	39.944	Molybdenum	Mo	95.95
Arsenic	As	74.91	Neon	Ne	20.183
Barium	Ba	137.36	Nickel	Ni	58.69
Bismuth	Bi	209.00	Nitrogen	N	14.008
Boron	B	10.82	Oxygen	O	16.000
Bromine	Br	79.916	Phosphorus	P	30.98
Cadmium	Cd	112.41	Platinum	Pt	195.23
Calcium	Ca	40.08	Potassium	K	39.096
Carbon	C	12.010	Selenium	Se	78.96
Chlorine	Cl	35.457	Silicon	Si	28.06
Chromium	Cr	52.01	Silver	Ag	107.880
Cobalt	Co	58.94	Sodium	Na	22.997
Copper	Cu	63.57	Strontium	Sr	87.63
Fluorine	F	19.00	Sulfur	S	32.06
Gold	Au	197.2	Tellurium	Te	127.61
Helium	He	4.003	Tin	Sn	118.70
Hydrogen	H	1.008	Titanium	Ti	47.90
Iodine	I	126.92	Tungsten	W	183.92
Iron	Fe	55.85	Uranium	U	238.07
Lead	Pb	207.21	Vanadium	V	50.95
Lithium	Li	6.940	Zinc	Zn	65.38
Magnesium	Mg	24.32	Zirconium	Zr	91.22

Source: *J. Am. Chem. Soc.*, **63**, 850 (1941).

CONVERSION FACTORS AND CONSTANTS

Analysis of Air

By weight: oxygen, 23.2%; nitrogen, 76.8%
By volume: oxygen, 21.0%; nitrogen, 79.0%
Average molecular weight of air.............................. 29
Average molecular weight of atmospheric nitrogen............... 28.2

Physical Constants

The Gas Law Constant R

$$R = 1.987 \text{ (calories)}/(\text{g-mole})(°K)$$
$$R = 82.06 \text{ (cu cm)}(\text{atm})/(\text{g-mole})(°K)$$
$$R = 10.71 \text{ (lb/sq in.)}(\text{cu ft})/(\text{lb-mole})(°R)$$
$$R = 0.729 \text{ (atm)}(\text{cu ft})/(\text{lb-mole})(°R)$$

1 faraday = 96,500 coulombs
Avogadro constant = 6.023×10^{23} per gram-atom

Density

1 gram-mole of an ideal gas at 0°C, 760 mm of mercury = 22.414 liters
1 pound-mole of an ideal gas at 0°C, 760 mm of mercury = 359.0 cubic feet
Density of dry air at 0°C, 760 mm of
mercury = 1.293 grams per liter or.. 0.0807 pound per cubic foot
1 gram per cc........................ 62.4 pounds per cubic foot
1 gram per cc........................ 8.337 pounds per U. S. gallon

Energy*

	Calories	Btu	Joules	Foot-pounds	Kilogram-meters
Calorie	1	3.968×10^{-3}	4.185	3.087	0.4267
Btu	252	1	1055	777.9	107.5
Joule	0.2389	9.482×10^{-4}	1	0.73756	0.1019
Foot-pound	0.3240	1.286×10^{-3}	1.356	1	0.13826
Kilogram-meter	2.343	9.298×10^{-3}	9.806	7.2327	1
Liter-atmos	24.21	9.607×10^{-2}	101.32	74.733	10.333
1 Chu	453.6	1.8	1899	1400	193.5

	Liter-atm	Cu ft-atm	Foot-poundals	Horsepower hours
Calorie	4.130×10^{-2}	1.459×10^{-3}	99.31	1.5591×10^{-6}
Btu	10.41	0.3676	25030	3.929×10^{-4}
Joule	9.869×10^{-3}	3.485×10^{-4}	23.73	3.725×10^{-7}
Foot-pound	1.3381×10^{-2}	4.7253×10^{-4}	32.174	5.0505×10^{-7}
Kilogram-meter	9.678×10^{-2}	3.4177×10^{-3}	232.7	3.6529×10^{-6}
Liter-atmos	1	3.5319×10^{-2}	2403.8	3.7734×10^{-5}
1 Chu	13.74	0.6617	45054	7.072×10^{-4}

* From Perkins, *Introduction to General Thermodynamics*, John Wiley & Sons, Inc., Publishers, with permission.

Length

1 inch...	2.540 centimeters
1 micron...	10^{-6} meter
1 Ångström unit..................................	10^{-10} meter

Mass

1 pound*..	16 ounces*
1 pound*..	7000 grains
1 pound*..	453.6 grams
1 ton (short)......................................	2000 pounds*
1 gram...	15.43 grains
1 kilogram...	2.205 pounds*

* Avoirdupois.

Mathematical Constants

e...	2.7183
π...	3.1416
$\ln N$..	2.303 log N

Power

1 kilowatt.....................	56.92 British thermal units per minute
1 kilowatt.....................	1.341 horsepower
1 horsepower.................	550 foot-pounds per second
1 watt........................	44.24 foot-pounds per minute
1 watt........................	14.34 calories per minute

Pressure

1 pound per square inch.................	2.04 inches of mercury
1 pound per square inch................	2.31 feet of water
1 atmosphere..........................	14.7 pounds per square inch
1 atmosphere..........................	760 mm of mercury
1 atmosphere..........................	29.92 inches of mercury

Temperature Scales

Degrees F......................................	1.8 (degrees C) + 32
Degrees K......................................	degrees C + 273.15
Degrees R......................................	degrees F + 459.7

Volume

1 cubic inch................................	16.39 cubic centimeters
1 liter......................................	61.02 cubic inches
1 cubic foot................................	28.32 liters
1 cubic meter..............................	1.308 cubic yards
1 cubic meter..............................	1000 liters
1 U. S. gallon..............................	4 quarts
1 U. S. gallon..............................	3.785 liters
1 U. S. gallon..............................	231 cubic inches
1 British gallon............................	1.20094 U. S. gallons
1 cubic foot...............................	7.481 gallons
1 liter.....................................	1.057 quarts
1 U. S. fluid ounce.........................	29.57 cubic centimeters

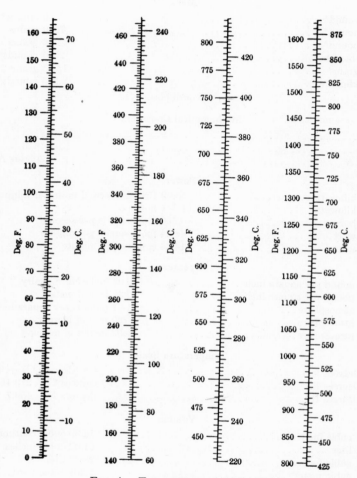

FIG. A.   Temperature conversions.

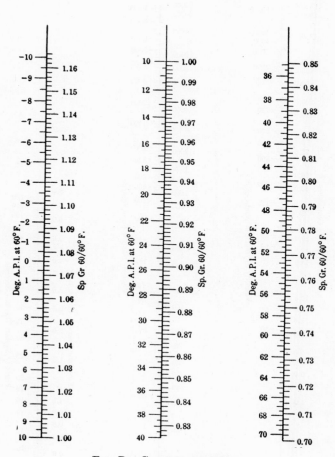

Fig. B.   Gravity conversions.

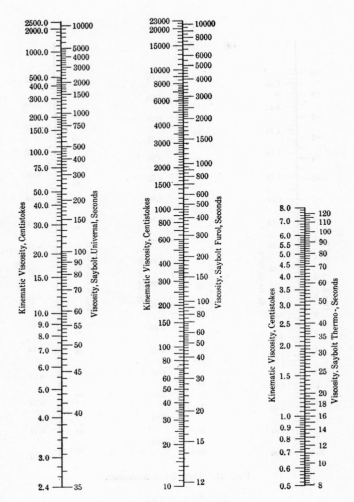

FIG. C. Viscosity conversions.

AUTHOR INDEX

Nitrogen Fixation

Arc process (obsolete)

Cyanamide Process

$$CaCO_3 \rightarrow CaO + CO_2$$
$$CaO + 3C \rightarrow CaC_2 + CO \quad (arc\ furnace)$$
$$CaC_2 + N_2 \rightarrow CaCN_2 + C + \Delta \quad (oven\ furnace,\ own\ ht)$$

Synthetic Ammonia Process.

$$N_2 + 3H_2 \rightleftharpoons 2NH_3 + \Delta$$

H_2 from:

Electrolysis of H_2O

Coal: $C + H_2O \rightarrow CO + H_2$
$CO + H_2O \rightarrow CO_2 + H_2 \quad (FeO\ cat)$
O_2, CO, CO_2 removed - $Cu(NH_3)OC\overset{\cdot}{H}$ sol. regenerated by ht.

Natural Gas: $CH_4 + H_2O \rightleftharpoons CO + 3H_2$, $CH_4 + 2H_2O \rightleftharpoons CO_2 + 4H_2$
endothermic

$3H_2 + N_2$
350 ATM

Catalyst: Fe, Ni, M promoter

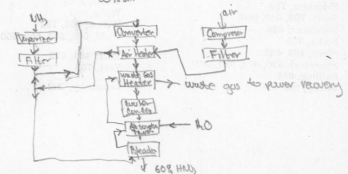

Nitric Acid

Divided into a) [O] of NH_3 (b) [O] and absorption of N oxides.

(a) $4NH_3 + 5O_2 \rightarrow 4NO + 6H_2O + \Delta$
Side reactions with $3O_2$ or $6NO$ to N_2, decomposition.

(b) $2NO + O_2 \rightleftharpoons 2NO_2 + \Delta$; $2NO_2 \rightleftharpoons N_2O_4 + \Delta$
$3NO_2 + H_2O \rightleftharpoons 2HNO_3 + NO + \Delta$
60% sol.

NH_3 → Vaporizer → Filter

Converter → Air Heater ← air

Compressor → Filter

Waste Gas Heater → waste gas to power recovery

Cooler Condenser

Absorption Towers ← H_2O

Bleacher

↓ 60% HNO_3

SUBJECT INDEX